Drug Repurposing

Naveen Chella • Om Prakash Ranjan •
Amit Alexander
Editors

Drug Repurposing

Innovative Approaches to Drug Discovery and Development

Springer

Editors
Naveen Chella
Department of Pharmaceutical Technology (Formulations)
NIPER Guwahati
Guwahati, Assam, India

Om Prakash Ranjan
Department of Pharmaceutical Technology (Formulations)
NIPER Guwahati
Guwahati, Assam, India

Amit Alexander
Department of Pharmaceutics
NIPER Guwahati
Guwahati, Assam, India

ISBN 978-981-97-5015-3 ISBN 978-981-97-5016-0 (eBook)
https://doi.org/10.1007/978-981-97-5016-0

This Springer imprint is published by the registered company Springer Nature Singapore Pte Ltd.
The registered company address is: 152 Beach Road, #21-01/04 Gateway East, Singapore 189721, Singapore

Preface

Drug discovery and development processes have flipped towards a more innovative and efficient direction from the traditional arena of drug research. Among these approaches, drug repurposing stands out as a novel strategy with the potential to revolutionize the way we tackle diseases and discover new treatments.

The concept of drug repurposing is characterized by the exploration of novel therapeutic benefits of currently available drug candidates in a variety of human diseases, including cancer, inflammatory disorders, neurodegenerative disorders, etc., which is becoming a significant part of the drug development process. Drug repurposing offers several benefits over traditional drug discovery approaches, including shorter development time, lower costs, and a higher probability of success by utilizing existing pharmacological data and clinical experience.

In this book, readers will embark on a journey through the diverse landscape of drug repurposing strategies. From in silico methods of predicting drug-disease associations to experimental approaches to the integrations of artificial intelligence (AI) and machine learning (ML) in drug repurposing for validating repurposed candidates. Each chapter offers valuable insights into the methodologies and tools driving this innovative field of drug research forward.

Moreover, this book discussed the case studies of successful drug repurposing initiatives in individual chapters, showcasing how this approach has led to the discovery of new treatments for a wide range of diseases, including cancer, inflammatory disorders, CNS disorders, rare genetic diseases, and emerging infectious diseases. These case studies highlight the transformative potential of drug repurposing in addressing unmet medical needs and improving patient outcomes.

However, despite its potential, medication repurposing has many drawbacks and difficulties. A multidisciplinary team effort is needed to navigate the complexity of repurposing pharmaceuticals, from concerns about intellectual property rights to substantial clinical validation requirements. In order to fully realize the potential of drug repurposing as a potent tool in the drug discovery and development pipeline, this book examines the regulatory obstacles related to drug repurposing and clinical trials, with separate chapters addressing their future perspectives.

The editors are grateful to all the contributors of each of the individual chapters of this book who have shared their expertise and insights to compile it successfully. We hope that the book titled *Drug Repurposing: Innovative Approaches to Drug Discovery and Development* serves as a valuable resource for researchers, clinicians, students, and industry professionals interested in harnessing the benefits of drug repurposing to address global health challenges.

Finally, we wish to conclude by saying, let us together explore the boundless possibilities of drug repurposing and open the door for a future where every drug has the potential to become a lifesaving therapy.

Guwahati, Assam, India Naveen Chella
Guwahati, Assam, India Om Prakash Ranjan
Guwahati, Assam, India Amit Alexander

Contents

1 **Drug Repurposing: History, Significance, Benefits, Approaches, and Challenges** 1
Lasya Priya Katukam, Amruta Prabhakar Padakanti, and Naveen Chella

2 **Methodologies Adopted in Drug Repurposing** 13
Sateesh Kumar Vemula, Sunil Kumar Kadiri, Merugumolu Vijay Kumar, Nagarjuna Narala, Rajendra Kumar Jadi, Mounika Kuchukuntla, Sagar Narala, and Michael A. Repka

3 **In Silico Methods Used in Drug Repurposing** 29
Neelanjan Chowdhury and Satheesh Kumar Nanjappan

4 **Role of Artificial Intelligence (AI) and Machine Learning (ML) in Drug Purposing** 43
Neelanjan Chowdhury and Satheesh Kumar Nanjappan

5 **Drug Repurposing in Cancer Therapy** 57
Srijita Sen, Hemant Kushwah, and Om Prakash Ranjan

6 **Drug Repurposing in Inflammatory Disorders** 93
Keshari Kumar Sriwastawa, Vaishnavi Sawkare, Ayaan Ansari, and Ashutosh Kumar

7 **Drug Repurposing in CNS Disorders** 109
Nazir Hussain, Snigdha Singh, and Amit Alexander

8 **Reshaping the Therapeutic Landscape: Drug Repurposing for Emerging Infectious Diseases and Rare Diseases** 135
Amritha G. Nambiar, Anuj A. Biswas, Ashish K. Agrawal, Brahmeshwar Mishra, and Dinesh Kumar

9 **Clinical Trials on Repurposed Drugs: An Overview** 173
Ruchira Patil, Harshad Takate, Gaurav Shanbhag, Harshada Kiran Sonawane, Amruta Prabhakar Padakanti, and Naveen Chella

10 Intellectual Property Rights and Regulatory Considerations Throughout Drug Repurposing 201
Prajakta Bule, Lal Bahadur Pal, Raghunandhan Painoori, and Naveen Chella

11 Redirection to the Drug Discovery: Antidiabetic Drugs Repurposing in Cancer 217
Prajakta Bule, Tejaswini Kolipaka, Shital Ranvare, and Naveen Chella

Editors and Contributors

About the Editors

Naveen Chella is an Assistant Professor in the Department of Pharmaceutical Technology (Formulations) at the National Institute of Pharmaceutical Education and Research (NIPER), Guwahati, under the Ministry of Chemicals and Fertilizers, GoI. His research interests include developing drug delivery systems to improve the physicochemical and biopharmaceutical properties of BCS class II and IV molecules. He also works on developing various delivery systems for the repurposed drugs to enhance their deliverability and therapeutic efficacy. He is a recipient of many awards like "Young Scientists" award from the Telangana Academy of Sciences, best oral and poster awards, travel grant from ICMR and DST and two extramural grants from ICMR. He is also Associate Fellow of the Telangana Academy of Sciences. He has a total 13 years of academic and research experience. He has published over 35 peer-reviewed research articles and 7 book chapters. He also serves as a reviewer for more than 20 SCI/SCOPUS-indexed high-impact factor journals. He filed about 7 Indian patents and has 1 granted patent to his credit.

Om Prakash Ranjan is currently working as an assistant professor in the Department of Pharmaceutical Technology (Formulations), National Institute of Pharmaceutical Education and Research (NIPER), Guwahati. He obtained his Ph.D. in Pharmaceutical Sciences from Manipal College of Pharmaceutical Sciences, Manipal University (now known as MAHE), Karnataka, India. He has a total of 13 years of experience including teaching, research, and industry. He has worked for 7 years in the pharmaceutical industry in the Formulation Research and Development Departments. He is the recipient of an international travel grant from DST and ICMR, Govt. of India. His major focus is on the development of controlled and modified-release drug delivery systems, vesicular drug delivery systems based on the Quality by Design (QbD) approach, and solubility enhancement techniques for poorly soluble drugs. To date, he has authored over 20 research articles in various journals of international repute, 7 book chapters, and 2 patents.

Amit Alexander is presently working as an Associate Professor in the Department of Pharmaceutics, National Institute of Pharmaceutical Education and Research (NIPER-Guwahati), under the aegis of the Department of Pharmaceuticals (DoP), Ministry of Chemicals and Fertilizers, Government of India (GoI). Dr. Alexander has teaching and research experience of more than 12 years. He is elected member of the Royal Society of Chemistry (MRSC), London, United Kingdom (UK), Royal Society of Biology (MRSB), London, United Kingdom (UK), Controlled Release Society-Indian Local Chapter (CRS-IC), Mumbai, Indian Academy of Neuroscience (IAN), and a life fellow of Indian Chemical Society (FICS), Kolkata. He has more than 175 scientific publications in high impact factor, SCI indexed, peer review journals to his credit. Dr. Alexander has received many National and International Awards the Young Scientist Award-2013, Young Researcher Award 2020, UGC-JRF National Fellowship 2011, and secured best research presentation position in many national seminal/conferences. He is recognized as one of the Top 2% Scientists ranking worldwide by Stanford University for three consequent years (2020, 2021, and 2022).

Contributors

Ashish K. Agrawal Department of Pharmaceutical Engineering and Technology, Indian Institute of Technology (BHU) Varanasi, Varanasi, Uttar Pradesh, India

Amit Alexander Department of Pharmaceutics, National Institute of Pharmaceutical Education and Research (NIPER) Guwahati, Guwahati, Assam, India

Ayaan Ansari Department of Medical Devices, National Institute of Pharmaceutical Education and Research (NIPER) S.A.S Nagar, MohaliPunjab, India

Anuj A. Biswas Department of Pharmaceutical Engineering and Technology, Indian Institute of Technology (BHU) Varanasi, Varanasi, Uttar Pradesh, India

Prajakta Bule Department of Pharmaceutical Technology (Formulations), National Institute of Pharmaceutical Education and Research Guwahati, Guwahati, Assam, India

Naveen Chella Department of Pharmaceutical Technology (Formulations), National Institute of Pharmaceutical Education and Research Guwahati, Guwahati, Assam, India

Neelanjan Chowdhury Department of Natural Products, National Institute of Pharmaceutical Education & Research (NIPER) Kolkata, Kolkata, West Bengal, India

Nazir Hussain Department of Pharmaceutics, National Institute of Pharmaceutical Education and Research (NIPER) Guwahati, Guwahati, Assam, India

Rajendra Kumar Jadi Department of Pharmaceutics, School of Pharmacy, Anurag University, Hyderabad, Telangana, India

Sunil Kumar Kadiri Department of Pharmacology, College of Pharmaceutical Sciences, Dayananda Sagar University, Bengaluru, Karnataka, India

Lasya Priya Katukam Department of Pharmaceutical Technology (Formulations), National Institute of Pharmaceutical Education and Research Guwahati, Guwahati, Assam, India

Tejaswini Kolipaka Department of Pharmaceutical Technology (Formulations), National Institute of Pharmaceutical Education and Research Guwahati, Guwahati, Assam, India

Mounika Kuchukuntla Department of Pharmaceutics, School of Pharmacy, Anurag University, Hyderabad, Telangana, India

Ashutosh Kumar Department of Pharmacology and Toxicology, National Institute of Pharmaceutical Education and Research (NIPER) S.A.S Nagar, Mohali, Punjab, India

Dinesh Kumar Department of Pharmaceutical Engineering and Technology, Indian Institute of Technology (BHU) Varanasi, Varanasi, Uttar Pradesh, India

Merugumolu Vijay Kumar Department of Pharmacology, College of Pharmaceutical Sciences, Dayananda Sagar University, Bengaluru, Karnataka, India

Hemant Kushwah Department of Pharmaceutical Technology (Formulations), National Institute of Pharmaceutical Education and Research (NIPER) Guwahati, Guwahati, Assam, India

Brahmeshwar Mishra Department of Pharmaceutical Engineering and Technology, Indian Institute of Technology (BHU) Varanasi, Varanasi, Uttar Pradesh, India

Amritha G. Nambiar Department of Pharmaceutical Engineering and Technology, Indian Institute of Technology (BHU) Varanasi, Varanasi, Uttar Pradesh, India

Satheesh Kumar Nanjappan Department of Natural Products, National Institute of Pharmaceutical Education & Research (NIPER) Kolkata, Kolkata, West Bengal, India

Nagarjuna Narala Department of Pharmaceutics and Drug Delivery, School of Pharmacy, University of Mississippi, University, MS, USA

Sagar Narala Pharmaceutical Technology, Ashland Specialty Ingredients, Wilmington, DE, USA

Amruta Prabhakar Padakanti Department of Pharmaceutical Technology (Formulations), National Institute of Pharmaceutical Education and Research Guwahati, Guwahati, Assam, India

Raghunandhan Painoori Regulatory Affairs, Odin Pharmaceutical, LLC, Somerset, NJ, USA

Lal Bahadur Pal Department of Pharmaceutical Technology (Formulations), National Institute of Pharmaceutical Education and Research Guwahati, Guwahati, Assam, India

Ruchira Patil Department of Pharmaceutical Technology (Formulations), National Institute of Pharmaceutical Education and Research Guwahati, Guwahati, Assam, India

Om Prakash Ranjan Department of Pharmaceutical Technology (Formulations), National Institute of Pharmaceutical Education and Research (NIPER) Guwahati, Guwahati, Assam, India

Shital Ranvare Department of Pharmaceutical Technology (Formulations), National Institute of Pharmaceutical Education and Research Guwahati, Guwahati, Assam, India

Mic**hael A. Repka** Department of Pharmaceutics and Drug Delivery, School of Pharmacy, University of Mississippi, University, MS, USA

Pii Center for Pharmaceutical Technology, University of Mississippi, University, MS, USA

Vaishnavi Sawkare Department of Medical Devices, National Institute of Pharmaceutical Education and Research (NIPER) S.A.S Nagar, Mohali, Punjab, India

Srijita Sen Department of Pharmaceutical Technology (Formulations), National Institute of Pharmaceutical Education and Research (NIPER) Guwahati, Guwahati, Assam, India

Gaurav Shanbhag Department of Pharmaceutical Technology (Formulations), National Institute of Pharmaceutical Education and Research Guwahati, Guwahati, Assam, India

Snigdha Singh Department of Pharmaceutics, National Institute of Pharmaceutical Education and Research (NIPER), Guwahati, Assam, India

Harshada Kiran Sonawane Department of Pharmaceutical Technology (Formulations), National Institute of Pharmaceutical Education and Research Guwahati, Guwahati, Assam, India

Keshari Kumar Sriwastawa Department of Pharmacology and Toxicology, National Institute of Pharmaceutical Education and Research (NIPER) S.A.S Nagar, Mohali, Punjab, India

Harshad Takate Department of Pharmaceutical Technology (Formulations), National Institute of Pharmaceutical Education and Research Guwahati, Guwahati, Assam, India

Sateesh Kumar Vemula Department of Pharmaceutics and Drug Delivery, School of Pharmacy, University of Mississippi, University, MS, USA

Department of Pharmaceutics, School of Pharmaceutical Sciences, Lovely Professional University, Phagwara, Punjab, India

Drug Repurposing: History, Significance, Benefits, Approaches, and Challenges

1

Lasya Priya Katukam, Amruta Prabhakar Padakanti, and Naveen Chella

Abbreviations

DR	Drug repurposing
NCE	New chemical entity
NME	New molecular entity
NPOs	Nonprofit organizations
PoC	Proof of concept
QSAR	Quantitative structure-activity relationship
TRND	Therapeutics for rare and neglected diseases
US FDA	United States Food and Drug Administration

Learning Objectives

- Understanding the concept of drug repurposing
- Necessity of drug repurposing
- Advantages of drug repurposing in drug discovery and development
- Challenges in drug repurposing

L. P. Katukam · A. P. Padakanti · N. Chella (✉)
Department of Pharmaceutical Technology (Formulations), National Institute of Pharmaceutical Education and Research Guwahati, Guwahati, Assam, India
e-mail: naveen@niperguwahati.in

N. Chella et al. (eds.), *Drug Repurposing*,
https://doi.org/10.1007/978-981-97-5016-0_1

1.1 Introduction

Drug discovery refers to the systematic process of developing novel and effective medications in the realms of medicine, biotechnology, and pharmacology. This phenomenon is accomplished by designing compounds explicitly targeting a particular biochemical target or through a systematic screening of compounds until recognizing those that can effectively act against the target's function. The process of drug discovery involves the use of several techniques, including traditional drug design, quantitative structure-activity relationship (QSAR) studies, docking strategy, pharmacophore-based mapping, target identification, and validation, as well as lead identification and optimization [1–4]. Discovering and developing drugs are crucial for better healthcare outcomes [5, 6]. Consequently, the research provides insight into disease underlying processes, which in turn helps to understand their causes better and, eventually, pave the way for potential treatment options for conditions that were previously incurable or unmet medical requirements [7]. This chapter briefly describes the significance of drug repurposing and its potential in uncovering new therapeutic uses for existing drugs and expanding treatment choices for patients in need and impending hazardous pandemics. Also, it explores the benefits of drug repurposing in comparison to traditional drug discovery.

1.1.1 Significance of Drug Repurposing

According to the US FDA, the traditional drug discovery to enter the market is a complex and time-consuming process that involves multiple stages. It entails the discovery and development of a novel therapeutic entity. The drug discovery pipeline comprises the following key stages: Target identification, which aims to identify a specific protein or pathway involved in a disease. Lead discovery, where combinatorial chemistry and assay development are used to screen chemical entities. Subsequently, the lead optimization phase enhances the identified compound's potency, selectivity, pharmacokinetic properties, and safety profile. Preclinical testing is then conducted to evaluate the compound's therapeutic potential value and to address safety concerns. Subsequently, the clinical trials are carried out to provide evidence of the drug's safety and efficacy. Finally, regulatory approval is obtained at every phase to ensure the safety and efficacy of the newly identified compounds.

However, only about 10% of drugs developed using traditional drug discovery processes are granted marketing approval, as depicted in Fig. 1.1. Those that failed in phases 2 and 3, having poor efficacy issues but no safety-related issues in the intended indication were withdrawn from the market. Additional medical benefits for those discontinued drugs will be forfeited, along with a substantial financial loss for the pharmaceutical industry. Therefore, it is imperative to devise an approach to recoup the incurred loss. Significant healthcare difficulties also affect the whole society, including the advent of new diseases or pandemics, the existence of 7000 uncommon and neglected diseases with no effective therapy or insufficient treatment, and the problem of drug resistance [8, 9]. One challenging goal in continuing

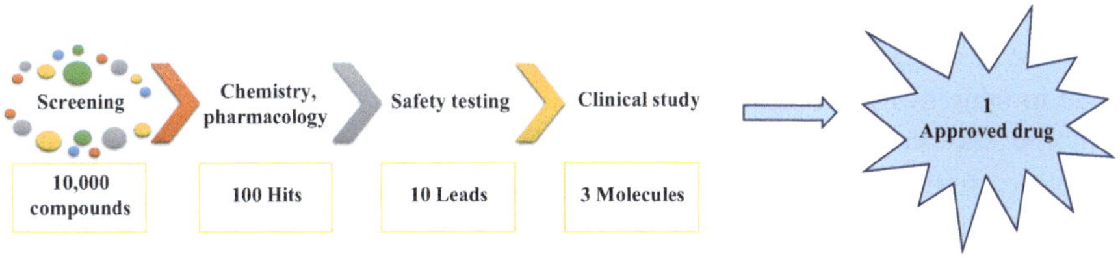

Fig. 1.1 Strategies for drug approval

research and development is to make drugs available for pandemics and uncommon illnesses.

Rare diseases predominantly impact children and pediatrics. Due to the increased complexity of neonatal trials, which are influenced by ethical, financial, and methodological factors, pharmaceutical companies have been hesitant to invest in drug discovery [10]. Furthermore, the conventional approaches to drug discovery encounter various obstacles, such as substantial attrition rates, inadequate resources for disease conditions, exorbitant development costs, and a dearth of commercial interest.

As a result, Therapeutics for Rare and Neglected Diseases (TRND) launched a new program to use drug repurposing to accelerate the discovery of innovative medications for diseases with a high degree of unmet medical needs. "Drug repurposing" is a cutting-edge technique that entails pharmaceuticals already available on the market and has completed essential phases of development. The future of scientific research lies in the concept of "drug repurposing," also known as "drug switching." This innovative approach offers a potential remedy for patients with limited treatment options in addition to saving time and money. It also expedites the search for possible treatments for uncommon illnesses and untreated medical issues. It can potentially revolutionize how we approach drug development and patient care [9].

The rationale of drug repurposing is based on the idea that several diseases might share common molecular pathways [11]. A survey shows that 90% of the medications have relevant indications beyond their initially approved indication. Companies strategically revitalized stagnant or abandoned medication prospects by leveraging a variety of innovative technologies and methodologies, including computational approaches, experimental approaches, and so on. This could significantly strengthen late-stage pipelines, address medical deficiencies, fulfill unmet medical requirements, and convert a sunk cost into a revenue generator. Consequently, these potential benefits substantially affected commercializing the idea of repurposing or repositioning. As a result, many pharmaceutical companies began exploring this strategy to maximize their profits and gain a competitive edge in the market [12].

Drug repurposing refers to identifying novel therapeutic applications for existing drugs or an alternative approach to drug development by leveraging previous research. The concept is appealing because it utilizes existing drugs, often with a well-established safety profile, and accelerates the drug development process. Drug repurposing (DR) is also referred to as drug repositioning, drug re-tasking, drug reprofiling, drug rescue, drug recycling, drug redirection, and therapeutic switching

[13]. The success story of sildenafil, now widely known as Viagra, serves as a prime example of the benefits of drug repurposing through serendipity. Originally developed to address hypertension and angina, the drug's repurposing for erectile dysfunction showcased the potential of this approach in terms of efficiency and facilitated regulatory approval. Instead of looking for entirely new medications, drug repurposing is like recycling. It is the use of well-known medications for purposes other than those for which they were initially intended due to action on multiple targets [14]. FDA-approved medications that may be useful for disorders for which they are not yet licensed and have molecular similarities in disease or target exhibiting pleiotropic effect contribute to successful drug repurposing as depicted in Fig. 1.2). Frequently, it entails finding prospective therapeutic targets, linking already approved medications to those targets, monitoring off-label drug use, researching the drug in clinical trials, and accumulating data to possibly alter FDA approval or treatment recommendations [15]. Researchers claimed that drug repurposing may accelerate drug development by omitting initial stages of safety screening as they have already been completed for the original indication, resulting in faster, more cost-effective, and less risky processes with three times higher success rates compared to traditional drug development cycles [16, 17]. Nevertheless, the success of repurposing drugs is not guaranteed due to the variability in the effectiveness and safety of a drug, depending on the particular condition it is employed for. Additionally, there may be constraints related to patent protection and financial motivations for pharmaceutical companies to allocate resources toward repurposing.

New approaches and technologies are ready to transform shelved compounds into new medicines, thus uncovering hidden treasure in pharma's attic. New technologies that provide a highly systematic, multifaceted approach to finding new development paths for shelved and unsuccessful drug candidates that extend beyond the initial therapeutic area of interest and attractive prospects from a corporate and social perspective are beneficial to drug repositioning. Although not every repositioning treatment will become a blockbuster, the lack of safety-related concerns may give them a significant edge over drug candidates that are already licensed. This has led to productive partnerships and funding opportunities in the past 5 years. As a result, academic institutions, businesses, nonprofit organizations (NPOs), and governmental entities adopted and supported this strategy [11, 18].

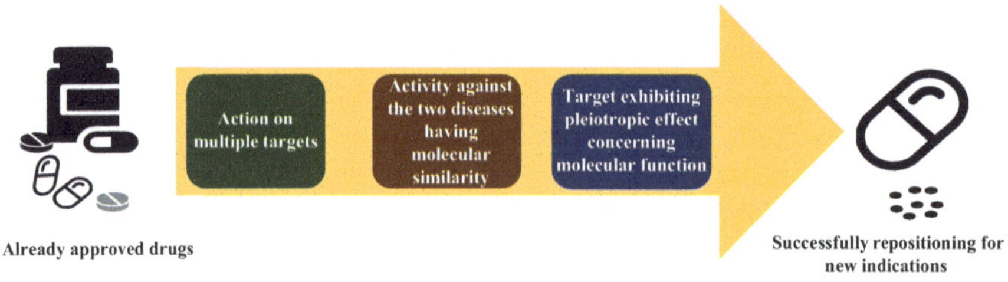

Fig. 1.2 Drugs characteristics that contribute to successful repositioning

1.1.2 Historical Background of Drug Repurposing

Drug repurposing technology first emerged by serendipity or retrospective clinical observations. The observation of the off-target or recently found on-target effects prompted an examination of the drug's potential for commercial exploitation. The development of high-throughput technologies and a broad range of cheminformatics and bioinformatics tools has led to their acknowledgment by academia and the pharmaceutical industry. Drug repurposing has now evolved into a sophisticated, data-driven, cutting-edge futuristic strategy [19]. The most successful drug repurposing cases to date have not applied a systemic approach. The earliest effective repurposed medications include aspirin, thalidomide, and sildenafil. Bayer approved them in 1890 for analgesia, Celgene in 1998 for leprosy, and Pfizer in 1998, respectively. Aspirin was later transformed into an antiplatelet aggregation drug to treat cardiovascular disease in 1980, thalidomide was used to treat multiple myeloma in 2006 based on serendipity, and sildenafil was introduced as Viagra® in 2006 to treat pulmonary arterial hypertension based on retrospective clinical experience. This resulted in widespread sales and significant commercial success. The terms "pleiotropic" allude to drugs like aspirin and thalidomide that display new indications by working on several targets, in contrast to sildenafil, which operates on the same biological process but has distinct indications [20–22]. According to a recent in-depth examination [9, 18], most current repurposing endeavors rely on a systematic approach, utilizing either experimental or computational methodologies or both.

1.1.3 A Repurposing Drug vs. A Conventional Drug

Conventional drug discovery and development provide significant challenges and complexities in biomedical research. This approach involves the de novo identification of new chemical entities (NCE), which can be classified into five stages: preclinical and discovery, safety screening, clinical research, FDA evaluation, and post-market safety surveillance by the FDA [13] as illustrated in Fig. 1.3. Furthermore, the existing methods employed in drug development are not viable due to a high attrition rate above 95%. Additionally, it often requires an average of 10–15 years and costs over US$2 billion before medication becomes available in the market. The pharmaceutical industry strives to uncover innovative treatments by expediting the drug discovery process and discovery approach.

Drug repurposing is one of the most captivating and effective approaches for exploring and advancing pharmaceuticals. It is increasingly employed to surpass the constraints of conventional drug research and development and presents numerous benefits depicted in Fig. 1.4 [15].

Benefits: Drug repurposing leads to faster regulatory approvals due to the availability of existing drugs' preclinical, safety, and tolerability data. Drug repurposing becomes cost-effective since much of the phase I/II and preclinical work has already been completed. Repurposing a medicine is expected to cost an average of $300 million instead of investing $2 to $3 billion in a new chemical entity (NCE). Lower

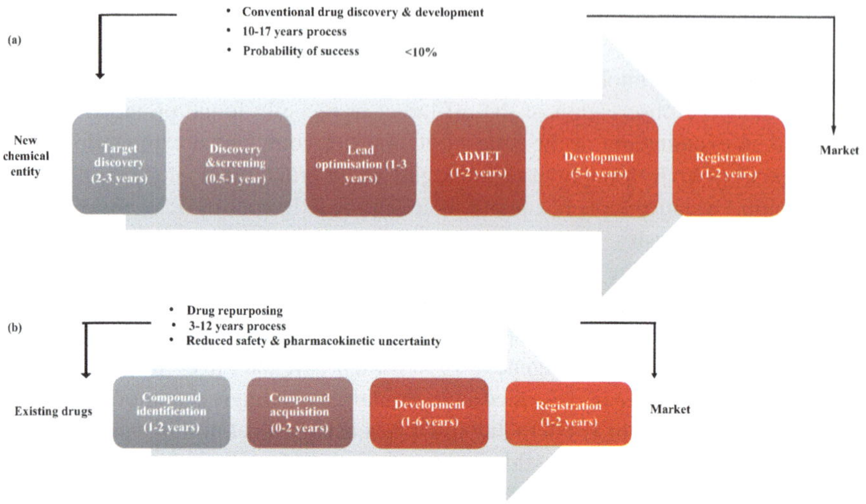

Fig. 1.3 Differences between (**a**) the traditional way of drug discovery and development (**b**) drug repurposing

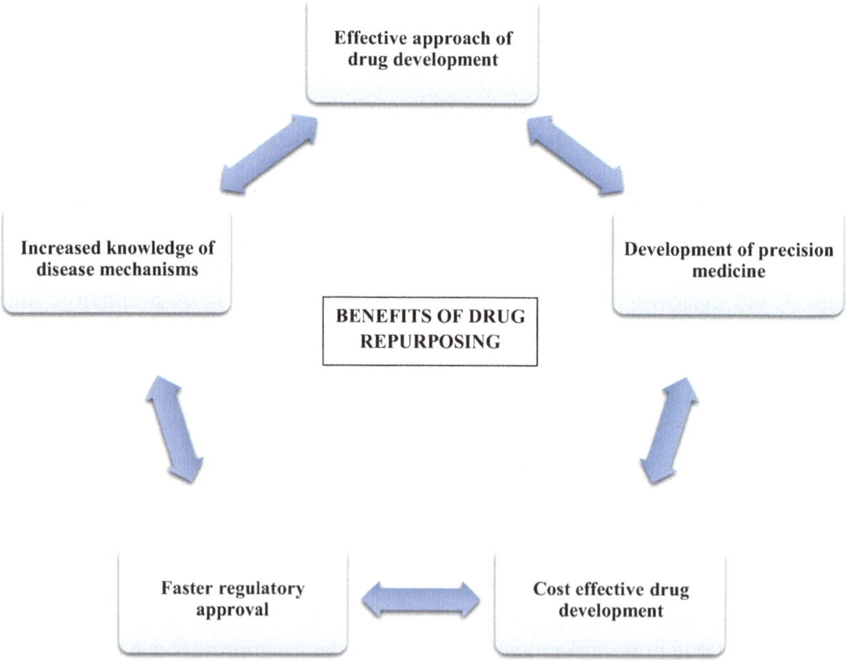

Fig. 1.4 Benefits of drug repurposing

development costs can make quicker returns on investments possible, and faster approval may translate into a more significant return on investment.

Drug repurposing may be advantageous for rare or neglected diseases, where the regular discovery process cannot be cost-effective [23]. Increased knowledge of disease mechanisms results in identifying new targets, pathways, and biomarkers that may help create new molecule structurally related to the repurposed medicine but are more effective [9].

1.2 Approaches of Drug Repurposing

Drug repurposing approaches fall into three categories: drug-centric, disease-centric, and target-centric, depending on different screening methodologies and techniques. Based on the therapeutic action of a drug, the relationship between the drug, disease, and target was explored [24, 25]. Different screening techniques include in silico-based, experimental-based, and mixed strategies. The process of virtual database screening, known as "in-silico-based repurposing," employs several methods, including genetic association, data mining, text mining, molecular docking, and pharmacophore modeling. Because there is less chance of failure, less money spent on research, and a quicker procedure overall, it is more advantageous than experimental methods; yet, the need for structural data on drug targets remained paramount.

In contrast, the repurposing of drugs through an experimental approach includes screening drugs for new therapeutic use based on in vitro assays, in vivo tests, non-invasive imaging, binding assay, and so on that do not require any target structural information. Currently, researchers have combined in silico and experimental approaches for repurposing, known as a mixed approach, as illustrated in Fig. 1.5 [26, 27].

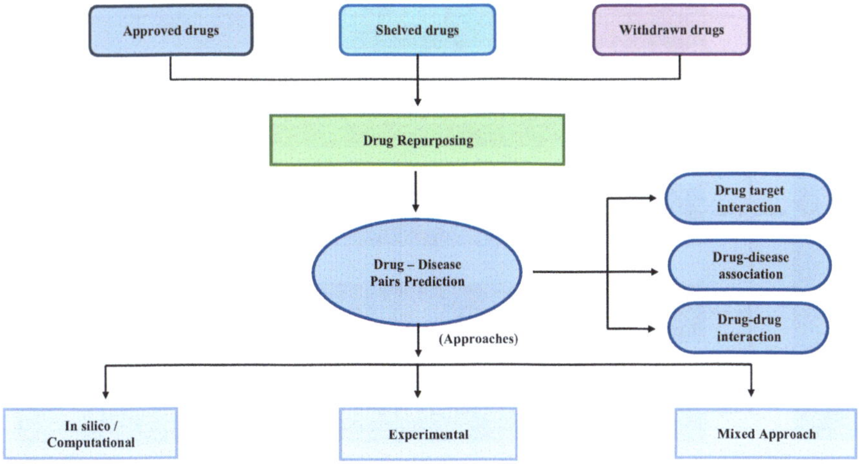

Fig. 1.5 Overview of commonly used techniques for drug repurposing

1.3 Challenges

Repurposing is one innovative approach, but the assumptions that it will be faster, safer, and more effective are not general and require a case-by-case scientific assessment. The following are the main challenges in drug repurposing (Fig. 1.6).

- Selection of appropriate database and compound availability—It might be challenging to identify a robust drug molecule from an acceptable database from a trusted vendor [25, 27].
- Intellectual property and patent considerations—Patent approval necessitates either a different strength or a new formulation, which could impact the effectiveness of the compound [23].
- Regulatory issues—The lack of safety data for the newly introduced indication and the requirement of robust evidence for 505(b) [4] application, which is analogous to NDA filings [28] may create hurdles in the regulatory approvals.
- Affordability—Expensive clinical trials and clinical trial feasibility for rare diseases of repurposing drugs are problematic, and regulatory approval requires expensive Proof of Concept (PoC) experiments that small biotech companies, academics, and nonprofit organizations cannot afford [19].
- Target population selection—Optimization involves complicated inclusion and exclusion criteria that require experimental confirmation.
- Safety challenge—An approved molecule's safety is assumed to have previously been established, which is usually true if the dosage is the same and the route remains same [29].
- The repurposing of drugs aims at different groups of patients with varied physiological conditions, which may result in unforeseen adverse effects and require scrutiny during repurposing [30].

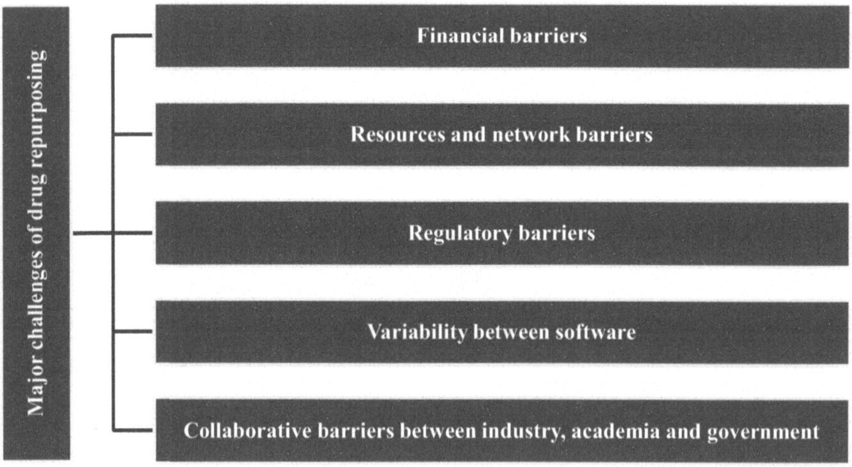

Fig. 1.6 Challenges in drug repurposing

- Approaching approval within timelines and legal framework for drug repurposing is challenging.
- Data requirement—A preliminary information on drug–drug interactions, pharmacodynamics, and pharmacokinetics of both drugs is required before repurposing of drug combinations [23].
- Financial and resource constraints (for example, an organization (academic/small biotech company) lacks financial resources, whereas industries lack personnel with relevant expertise on compound and investigated indications for repurposing).
- Inadequate or lack of collaboration among industry, academics and government may hinder the effective utilization of resources available with different people [17].
- Repurposing only lowers but fails to eliminate drug development risks, necessitating thorough evaluation.
- Biases ("NOT SOLD HERE"- the reluctance of companies to out-license the molecules that may have promising indications).
- Liability risks due to the low revenue generation of the products developed [19].
- Unknown functioning of various diseases and pandemics and patient population heterogeneity make drug target identification difficult in repurposing any compound [31].
- Lack of expertise among rare disease nonprofit leads [32].
- Verification and validation of the tools used—These tools need to be carefully examined and critically reviewed since various software programs employ different methods and scoring functions, which can lead to unpredictability and inconsistency in data related to docking. For example, in vitro data collected in an antiviral assay may not convert to in vivo owing to phospholipidosis [28].
- Limited efficacy of repurposed drugs at an indicated dose (which requires molecular modifications or a higher dose to increase the effectiveness of said drug toward its repurposed target). For example, non-antibiotics repurposing for infectious diseases need high concentration like antibiotics, which may lead to toxicity as shown in (Fig. 1.7) [33].

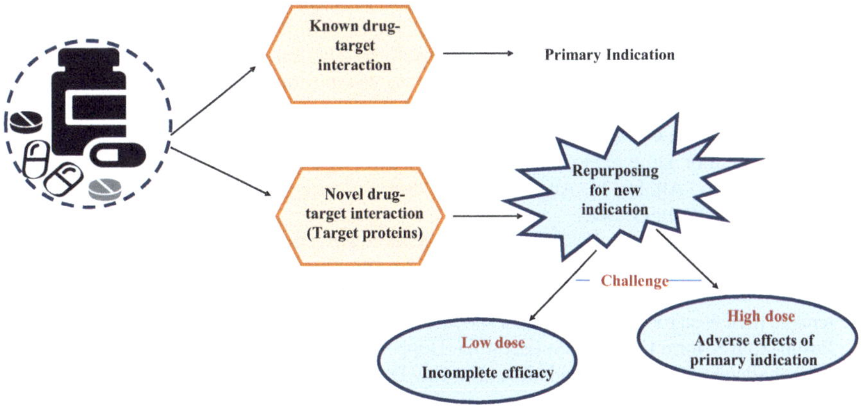

Fig. 1.7 Challenges of dose-related limited efficacy and toxicity of repurposed indication

1.4 Summary

Exploring drug repurposing involves identifying alternative therapeutic applications for drugs that have already received approval from regulatory agencies. This method is becoming increasingly popular in the pharmaceutical sector and among researchers because of its ability to streamline drug development processes and reduce costs. Discovering novel applications for existing drugs can accelerate the development of new therapies and effectively meet healthcare gaps. However, care must be taken to consider the challenges like regulatory issues, knowledge sharing, and financial and resource management. The ongoing exploration of drug repurposing offers a valuable opportunity to optimize healthcare resources and advance precision medicine initiatives with a few limitations that can be resolved.

References

1. Decker S, Sausville EA. Drug discovery. In: Principles of clinical pharmacology. London: Elsevier; 2007. p. 439–47.
2. Guido RVC, Oliva G, Andricopulo AD. Modern drug discovery technologies: opportunities and challenges in Lead discovery. Comb Chem High Throughput Screen. 2011;14(10):830–9.
3. Nagarajan K, Bodla RB. Fundamental principles of drug discovery. London: Lambert Academic Publishing; 2019.
4. Samudrala G. Brief note on drug discovery and development. Res Rev Pharm Pharm Sci. 2022;11:1–4.
5. Akhondzadeh S. The importance of clinical trials in drug development. Avicenna J Med Biotechnol. 2016;8(4):151.
6. Gilbert IH. ChemMedChem. 2011;6(10):1769.
7. Trevelyan O. The importance of discovering new drugs. Manchester: AZoLifeSciences; 2021.
8. Mohs RC, Greig NH. Drug discovery and development: role of basic biological research. Alzheimers Dement. 2017;3(4):651–7.
9. Rudrapal M, Khairnar SJ, Jadhav AG. Drug repurposing (DR): an emerging approach in drug discovery. In: Drug repurposing—hypothesis, molecular aspects and therapeutic applications. London: IntechOpen; 2020.
10. Giannuzzi V, Landi A, Bosone E, Giannuzzi F, Nicotri S, Torrent-Farnell J, et al. Failures to further developing orphan medicinal products after designation granted in Europe: an analysis of marketing authorisation failures and abandoned drugs. BMJ Open. 2017;7(9):e017358.
11. Oprea TI, Bauman JE, Bologa CG, Buranda T, Chigaev A, Edwards BS, et al. Drug repurposing from an academic perspective. Drug Discov Today Ther Strateg. 2011;8(3–4):61–9.
12. Tartaglia LA. Complementary new approaches enable repositioning of failed drug candidates. Expert Opin Investig Drugs. 2006;15(11):1295–8.
13. Dey G. An overview of drug repurposing: review article. J Med Sci Clin Res. 2019;7(2):55–6.
14. Cavalla D, editor. Drug repurposing. Cambridge: The Royal Society of Chemistry; 2022.
15. Berdigaliyev N, Aljofan M. An overview of drug discovery and development. Future Med Chem. 2020;12(10):939–47.
16. Hauser AS, Attwood MM, Rask-Andersen M, Schiöth HB, Gloriam DE. Trends in GPCR drug discovery: new agents, targets and indications. Nat Rev Drug Discov. 2017;16(12):829–42.
17. Krishnamurthy N, Grimshaw AA, Axson SA, Choe SH, Miller JE. Drug repurposing: a systematic review on root causes, barriers and facilitators. BMC Health Serv Res. 2022;22(1):970.
18. Pushpakom S, Iorio F, Eyers PA, Escott KJ, Hopper S, Wells A, et al. Drug repurposing: progress, challenges and recommendations. Nat Rev Drug Discov. 2019;18(1):41–58.

19. Polamreddy P, Gattu N. The drug repurposing landscape from 2012 to 2017: evolution, challenges, and possible solutions. Drug Discov Today. 2019;24(3):789–95.
20. Galiè N, Ghofrani HA, Torbicki A, Barst RJ, Rubin LJ, Badesch D, et al. Sildenafil citrate therapy for pulmonary arterial hypertension. N Engl J Med. 2005;353(20):2148–57.
21. Mohapatra TK, Subudhi BB. Repurposing of aspirin: opportunities and challenges. Res J Pharm Technol. 2019;12(4):2037.
22. Raje N, Anderson K. Thalidomide—a revival story. N Engl J Med. 1999;341(21):1606–9.
23. Talevi A, Bellera CL. Challenges and opportunities with drug repurposing: finding strategies to find alternative uses of therapeutics. Exp Opin Drug Discov. 2020;15(4):397–401.
24. Kulkarni VS, Alagarsamy V, Solomon VR, Jose PA, Murugesan S. Drug repurposing: an effective tool in modern drug discovery. Russ J Bioorg Chem. 2023;49(2):157–66.
25. Xue H, Li J, Xie H, Wang Y. Review of drug repositioning approaches and resources. Int J Biol Sci. 2018;14(10):1232–44.
26. Kumar R, Harilal S, Gupta SV, Jose J, Thomas Parambi DG, Uddin MS, et al. Exploring the new horizons of drug repurposing: a vital tool for turning hard work into smart work. Eur J Med Chem. 2019;182:111602.
27. Ng YL, Salim CK, Chu JJH. Drug repurposing for COVID-19: approaches, challenges and promising candidates. Pharmacol Ther. 2021;228:107930.
28. Sharma PP, Bansal M, Sethi A, Poonam, Pena L, Goel VK, et al. Computational methods directed towards drug repurposing for COVID-19: advantages and limitations. RSC Adv. 2021;11(57):36181–98.
29. Novack GD. Repurposing medications. Ocul Surf. 2021;19:336–40.
30. Gns HS, Gr S, Murahari M, Krishnamurthy M. An update on drug repurposing: re-written saga of the drug's fate. Biomed Pharmacother. 2019;110:700–16.
31. Low ZY, Farouk IA, Lal SK. Drug repositioning: new approaches and future prospects for life-debilitating diseases and the COVID-19 pandemic outbreak. Viruses. 2020;12(9):1058.
32. The Castleman Disease Collaborative Network. The ROADMAP Project. 2021. https://cdcn.org/roadmap/. Accessed on 7 July 2024.
33. Oprea TI, Mestres J. Drug repurposing: far beyond new targets for old drugs. AAPS J. 2012;14(4):759–63.

Methodologies Adopted in Drug Repurposing

2

Sateesh Kumar Vemula, Sunil Kumar Kadiri, Merugumolu Vijay Kumar, Nagarjuna Narala, Rajendra Kumar Jadi, Mounika Kuchukuntla, Sagar Narala, and Michael A. Repka

Abbreviations

AI	Artificial Intelligence
COVID-19	Coronavirus disease 2019
EGRs	Electronic health records

S. K. Vemula (✉)
Department of Pharmaceutics and Drug Delivery, School of Pharmacy, University of Mississippi, University, MS, USA

Department of Pharmaceutics, School of Pharmaceutical Sciences, Lovely Professional University, Phagwara, Punjab, India
e-mail: svemula@olemiss.edu

S. K. Kadiri · M. V. Kumar
Department of Pharmacology, College of Pharmaceutical Sciences, Dayananda Sagar University, Bengaluru, Karnataka, India

N. Narala
Department of Pharmaceutics and Drug Delivery, School of Pharmacy, University of Mississippi, University, MS, USA
e-mail: nnarala@go.olemiss.edu

R. K. Jadi · M. Kuchukuntla
Department of Pharmaceutics, School of Pharmacy, Anurag University, Hyderabad, Telangana, India

S. Narala
Pharmaceutical Technology, Ashland Specialty Ingredients, Wilmington, DE, USA
e-mail: Sagar.narala@ashland.com

M. A. Repka
Department of Pharmaceutics and Drug Delivery, School of Pharmacy, University of Mississippi, University, MS, USA

Pii Center for Pharmaceutical Technology, University of Mississippi, University, MS, USA

N. Chella et al. (eds.), *Drug Repurposing*,
https://doi.org/10.1007/978-981-97-5016-0_2

FDA	Food and Drug Administration
HCS	High content screening
HTS	High-throughput screening
NLP	Natural language processing
NSAIDs	Non-steroidal anti-inflammatory drugs
RNA	Ribonucleic acid
SARS-CoV-2	Severe acute respiratory syndrome coronavirus 2

Highlights

- Drug repurposing existing drugs present an opportunity to mitigate risks, expenses, and time constraints associated with the drug development process.
- Repurposing drugs allow for better use of therapeutic compounds and exploring the untapped potential of many molecules.
- Hybrid methodologies can expedite drug repositioning, leading to the discovery of novel treatments for various diseases.

2.1 Introduction

Drug repurposing is finding new therapeutic uses for existing drugs, including those already approved by regulatory agencies such as the US Food and Drug Administration, the European Medicines Agency, and other bodies or drugs being investigated. It also involves using newly developed drugs to treat diseases that differ from their original intended therapeutic use [1, 2]. Developing a novel active pharmaceutical ingredient is a tedious, time-consuming, and expensive process, with a success rate of only 2.0% [3–5]. Identifying new indications for licensed or well-established medications is critical to drug discovery [6]. This procedure can bypass certain preapproval assessments required for newly formulated drugs. Drug repurposing has numerous advantages in drug discovery, including reducing the likelihood of failure, lowering investment requirements, and shortening the development timeline. Since repurposed drugs have well-established safety profiles, streamlining the regulatory approval process is simple compared to new drug substances [7–11].

Repositioning is dependent upon two fundamental scientific foundations:

1. Understanding the human genome has facilitated the identification of shared biological targets across multiple diseases.
2. The idea of pleiotropic medicines.

2.2 Drug Repurposing Versus Traditional Drug Discovery

The traditional drug discovery process requires a considerable investment of time and resources and often involves setbacks and unsuccessful outcomes [12–14]. On the other hand, medication repositioning involves only the selection, procurement, development, and FDA post-market safety monitoring of a chemical [15]. Using bioinformatics, chemo-informatics techniques, and large biological and structural databases, medication development time, cost, and failure risk can be reduced through drug repositioning. Additionally, accidental discoveries have shown that repurposing existing medicines for unique therapeutic objectives can be effective [16–21]. This drug development method has clear advantages over conventional drug discovery programs [22].

Viagra, a phosphodiesterase-5 inhibitor developed by Pfizer in 1985 to treat heart disease, specifically angina, has been repurposed to treat erectile dysfunction. This repurposing strategy has potential benefits over traditional drug development strategies, including lower development expenses and shorter development time. For instance, oral metformin (Glucophage) is an established treatment for type 2 diabetes but is now in phase II/III clinical trials as a cancer therapy [11]. Drug repositioning has advantages over conventional drug discovery, significantly reducing research and development time. Developing a new drug using traditional approaches can take 10–16 years, whereas drug repurposing (DR) is expected to take 3–12 years [23]. Furthermore, identifying a novel medication using a drug repositioning strategy costs ~$1.6 billion, compared to ~$12 billion for traditional drug development [24–30].

2.3 Drug Repurposing Stages

Different stages in drug repurposing are as follows (given in Fig. 2.1):

(i) Compound identification (1.2 years): This stage involves selecting a potential therapeutic candidate to target a particular drug target precisely within the human body [32].
(ii) Acquisition of compounds throughout the early developmental stage (0–2 years): Once the potential therapeutic candidate has been identified, a license must be obtained for the new medication.

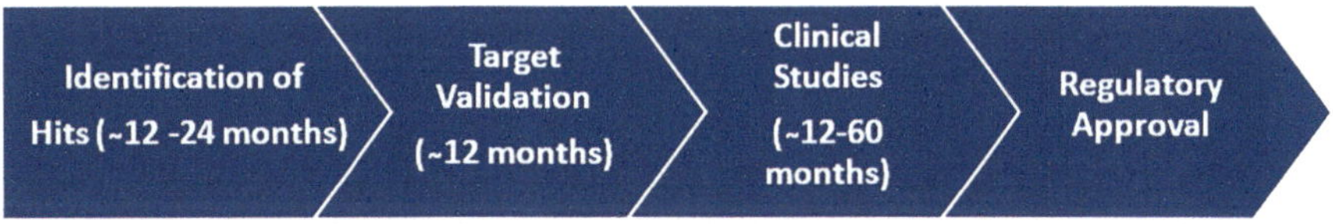

Fig. 2.1 Time taken along with methods and procedures for repositioning a current drug [31]

(iii) Development throughout 1 to 5 years: This stage can begin during the preclinical, phase 1, or phase 2 drug investigation. It involves thoroughly examining the available data to ensure the safety and efficacy of the new medication.
(iv) The post-market safety survey conducted by the Food and Drug Administration (FDA): The FDA monitors the safety of drugs and medical devices once made available to the general public.

2.3.1 Significance of Repurposing Drugs

Repurposing pre-existing medications presents a compelling opportunity to mitigate the risks, expenses, and time constraints associated with the drug development process. Exploring the possible effectiveness of previously licensed drugs for a specific therapeutic purpose in treating alternative diseases is a viable option. Furthermore, pharmaceuticals initially designed for a certain medical condition but considered inappropriate owing to adverse reactions might still have significance. This is because the observed undesirable effects may be linked to the drug's pharmacological properties, potentially beneficial in addressing a different ailment [33–40]. Considering the financial implications and efficacy of drug discovery and its crucial role in enhancing human well-being, it is essential to use methodologies that expedite and economize the drug development process [41]. These measures are necessary to facilitate the timely introduction of effective medications into the market.

2.3.2 Approaches to Drug Repurposing

The process of drug repurposing involves using analytical techniques such as homology modeling and molecular similarity for virtual screening of drug libraries. The goal is to identify suitable medicines and examine their interactions with target proteins [42]. Techniques such as binding free energy estimates and molecular docking are used to evaluate drug–receptor interactions and binding affinities [43, 44]. Recently, in silico techniques, artificial intelligence (AI) technologies, ligand-based drug design, and structure-based drug design have been utilized to expedite the drug repurposing process [45].

2.4 Computational Methods for Drug Reprofiling

Drug reprofiling is often regarded as the most promising method for identifying novel therapeutic uses for already approved or current medications, mostly due to its cost advantages [46]. Integrating metabolomics, proteomics, genomes, and transcriptomics has opened up several opportunities for drug discovery through a comprehensive approach. Researchers currently possess the most recent and advanced resources and information to research novel modes of action or pathways associated with target proteins/genes and/or distinctive biomarkers connected to disease progression. Using bioinformatics, data mining, and computational models to discover new applications for existing drugs based on their molecular structure, pharmacological properties, and known mechanisms of action [47–52]. Many computing techniques have been developed to accelerate and improve repurposing according to Rudrapal et al. [53]. These databases provide significant insights into the tridimensional structures of pharmaceutical compounds and their respective binding sites.

2.5 Methods for Experimentally Repositioning Drugs

The molecular methodologies employed in this context rely upon the comprehension of therapeutic efficacy and the pathophysiology of diseases [54]. These systems often rely on omics data, which includes genomic, transcriptomic, proteomic, pharmacological target, and chemical structure data. The stability and replicability of pharmaceutical and illness datasets have made transcriptomics and genomes the most common data for medication repurposing. Transcriptomics is a field of study that focuses on analyzing the expression levels of several genes. This is often achieved by quantifying RNA using techniques such as RNASeq or gene expression microarrays. One potential strategy for employing transcriptomics in medicine repurposing is the concept that reversing gene expression profiles may provide a favorable therapeutic outcome [55].

2.5.1 Approaches to the Use of Real-World Data

The real-world data strategy primarily identifies previously unknown and occasionally unexpected associations between pharmaceutical substances and medical conditions or their corresponding symptoms. The data in question are derived from individuals' health, habits, and behavior and have been obtained without any external influence or bias caused by the methods used for data collecting.

2.5.2 Process of Repurposing Based on Knowledge

Repurposing drugs might be predicated on medical professionals' and researchers' expertise in interpreting observations. In certain other situations, conducting a study may lead to serendipitous discoveries. One of the main causes of the early instances of repurposing was coincidence. Next, the drug's original intended disease was repurposed for its therapy to treat a group of disorders that shared altered pathways. Drug repositioning studies use bioinformatics or cheminformatics to include drug information, drug-target networks, chemical structures of targets and drugs, clinical trial data, FDA approval labels, signaling or metabolic pathways, etc. Blinded and target-based approaches may not reveal additional mechanisms beyond identified targets. Novel pharmacological targets, hidden drug–drug similarities, and new disease biomarkers are predicted using knowledge-based approaches. Knowledge-based drug repositioning strategies use well-known facts to increase prediction accuracy. By constructing models, this approach aims to anticipate unidentified targets, biomarkers, or disease-associated processes [56].

2.5.3 Concept of Repurposing Drugs Depending on Specific Targets

Target-based drug repurposing involves the use of proteins or biomarkers that are of interest. This process often involves the screening of drug compounds by high throughput (HTS/HCS) [57]. Consequently, this approach substantially enhances the probability of successful drug discovery. One of the key benefits of the target-based method is its capacity to effectively evaluate a wide range of pharmacological molecules that possess established chemical structures. Nevertheless, target-based approaches are limited in detecting and characterizing undiscovered processes that extend beyond the currently identified targets.

2.5.4 Concept of Pharmacological Repurposing Based on Pathway Analysis

Pathway-based drug repurposing compares diseases and treatments using metabolic, signaling, and protein–interaction networks. Researchers can recreate disease-specific networks using human or animal omics data to uncover new targets for repurposed medicines.

2.5.5 Concept of Drug Repurposing Using Target Mechanism-Based Approaches

The growing significance of precision medicine has served as a driving force behind the adoption of drug repurposing strategies. One notable benefit of these repurposing methodologies is their focus on elucidating the underlying mechanisms associated with illnesses or medications and the therapeutic interventions for certain diseases. The on-target technique investigates novel drug indications through the known target. Drugs with recognized pharmacological mechanisms work for new therapeutic indications. This method uses the same medication molecule biological target but a different ailment. It may work with the original drug's dose. Minoxidil (Rogaine) was repositioned using an on-target method since it has two therapeutic effects on the same target. Minoxidil became an anti-hair loss medicine from an antihypertensive vasodilator. Minoxidil, an antihypertensive vasodilator, widens blood vessels and opens potassium channels to deliver more oxygen, blood, and nutrients to hair follicles [58].

In the off-target technique, we find novel medication usage through an unexpected target and unknown mechanism of action. For new therapeutic indications, drugs target novel regions. Thus, the drug's target and indication are novel. The drug is not optimized for the target; thus, the dose is important. Docking and fingerprinting are options. Aspirin (Colsprin), an NSAID used to treat pain and inflammation, is an example of off-target repurposing. As an antiplatelet medication, it suppresses blood coagulation (clot formation) by reducing platelet function. It is used to treat heart attacks, strokes, and prostate cancer [59].

2.5.6 Signature-Based Repurposing

Signature-based repurposing involves the utilization of gene profiles derived from disease omics data to identify novel off-targets or disease processes. This technique compares gene expression patterns of medications and illnesses to find inverse relationships. Dudley et al. investigated inflammatory bowel disease drug-disease pairings [60]. The authors matched gene expression patterns from the gene expression omnibus database with the connectivity map's 164 pharmacological molecules. Consequently, several drug-disease combinations that were previously unidentified were identified, and one of these combinations was confirmed by experimentation on preclinical animals. The benefits of these techniques are their ability to discover novel mechanisms of action for pharmaceutical compounds. In contrast to knowledge-based approaches, these strategies incorporate molecular and/or genetic-level processes.

2.5.7 Process of Repurposing Depending on Phenotype

Phenotypic information has emerged as a novel avenue for medication repositioning. In recent years, there has been a growing utilization of this sort of data in systems-based methodologies for identifying genetic characteristics linked to human illnesses. The use of natural language processing (NLP) techniques in electronic health records (EHRs) has the potential to uncover previously unidentified adverse medication events that were not detected during the drug development process [61–63].

2.6 Mixed Approaches

This strategy, commonly referred to as "Mixed Approaches," involves using computational tools to generate predictions, which are subsequently confirmed by biological experiments and clinical trials [64]. Mixed methodologies provide researchers the potential to efficiently and expeditiously produce repositioned drugs.

2.7 Techniques for Repurposing Drugs

The approaches utilized in medication repurposing are often classified into three groups based on the availability and reliability of evidence pertaining to pharmacological, toxicological, and biological action.

(i) Drug-Oriented Methodology: Drug-based techniques anticipate therapeutic potentials and novel indications for current drugs using chemical, molecular, biological, pharmacological, and genetic data. Drug-based techniques are applied when drug data is available or there is a desire to research how pharmacological properties can contribute to drug repositioning [65]. The primary objective is to find compounds that exhibit biological effects through cell or animal experiments. This classification is rooted in conventional pharmacology and drug development practices, wherein investigations often focus on assessing the biological effectiveness of pharmacological compounds without prior knowledge of their specific biological targets [66–68].

(ii) Target-Oriented Methodology: The procedure uses in silico or virtual high-throughput screening to find medicines or compounds in drug libraries or databases. This is possible with ligand-based screening or molecular docking.

(iii) Disease/Therapy-Oriented Methodology: Disease-based techniques use phenotypic features, side effects, and indications to predict therapeutic potential and novel drug indications. Ailment-based techniques are utilized when drug data is limited or when investigating pharmacological properties can help drug reposition focused on a certain ailment. This approach is applicable when a substantial amount of information is available on the disease model. In such cases, medication repurposing can be led by the illness and/or therapy, considering the existing information regarding the disease process.

(iv) Genomics in Healthcare and Medicine: Genomics refers to the study of an organism's whole set of genes, including the analysis of disease-specific genetic information. Butte's group achieves a milestone toward holism and integration in genome-based therapeutic modeling by defining emergent genome-wide metrics for drug repositioning [69]. The intermediate genetic phenotypes they describe can be used to build new comprehensive models for complex systemic clinical diseases like diabetes mellitus type II that include more tissues. For targeted drug development, a genome-wide measure in various tissues and organs could combine the strengths of reductionism and integration. These conclusions are based on genome-wide metric assessments at the mRNA scale, which may be the best-characterized genomic intermediary states. Genomic metrics could also be used with clinical parameters for drug repurposing. From microRNA signatures to comparative genomic hybridization data, many intermediary genome-wide measures can be assessed at various scales. Thus, a conceptual framework for analyzing these scales' inputs and outputs is crucial for appropriate interpretation and application.

(v) Techniques and Approaches in Metabolomics Research: The field of metabolomics focuses on studying disease-specific metabolic pathways and patterns.

Metabolomics, a growing area that characterizes metabolites and metabolism in biological systems, helps scientists to understand biology. Complex diseases like cancer, heart disease, diabetes, obesity, infectious diseases, and others have metabolic abnormalities, according to research. Metabolomics helps us understand pathophysiology and physiology, disease biomarkers, and disease mechanisms. Metabolomics is essential for medication research and development due to its metabolic insights and sophisticated analytical and computational methodologies. Before drug development, basic drug research is essential. Metabolomics helps drug researchers to understand illness causes, therapeutic targets, and medication modes of action. Drug development, which includes preclinical studies, clinical development, and post-market evaluation, is expensive and time-consuming. Fortunately, metabolomics may accelerate drug development by assessing safety, effectiveness, toxicity, and adverse events.

2.8 Repurposing Drugs for COVID-19 and Other Diseases

Three strategies exist: computational, biological, experimental, and combined. Computational methodologies have emerged as novel and valuable tools in medication repositioning. This technique provides a concise overview of the interaction between severe acute respiratory syndrome coronavirus 1 and human host cells, specifically focusing on protein–protein interactions and potential therapeutic targets within the human system. This information is valuable for the discovery of repurposed drugs.

2.9 Examples of SARS-CoV19 Repurposed Antiviral Drugs

Remdesivir is an antiviral monophosphoramidate prodrug nucleoside analog. This drug was developed to treat Ebola initially. Later studies showed that it might fight parainfluenza type 3, measles, and Nipah. Remdesivir is a promising repurposed pharmacological treatment for COVID-19 infection due to its ability to slow SARS-CoV2 infection development [70]. Remdesivir is given intramuscularly over 10 days to COVID-19 patients in Italy at 200 mg every 12 h for loading and 100 mg for maintenance. Additional palliative medicines are usually used with this therapy. Hospitalized COVID-19 patients in Washington, USA, got intravenous remdesivir for 7 days. Remdesivir has FDA emergency use permission for severe COVID-19 patients. Japanese authorities allow remdesivir for COVID-19 treatment.

Japanese company Toyama Chemical Co. Ltd. created favipiravir. Japanese authorities license favipiravir, a nucleic acid purine base analog with the molecular name 6-fluoro-3-hydroxy-2-pyrazinecarboxamide. Most people use it to treat influenza. The medication inhibits RNA-dependent RNA polymerase. Hepatic favipiravir metabolism is primarily mediated by aldehyde oxidase. As a physiologically active metabolite of favipiravir, ribofuranosyl-5′-triphosphate mediates its pharmacological effects. Clinical data shows that COVID-19 patients without co-morbidities recover faster. Favipiravir was approved for COVID-19 therapy in China on February 15, 2020, and in Russia for hospitalized patients.

Sildenafil is a vasodilator that may be used as an antihypertensive medication. It inhibits the phosphodiesterase enzyme, hence inhibiting platelet aggregation. It was formerly employed as a potential therapy for angina because of these qualities. However, unanticipated adverse effects, such as penile erections, were reported during its clinical testing. Because of this adverse effect, Pfizer began marketing sildenafil as an erectile dysfunction medicine in 1998. Pfizer repurposed sildenafil in the treatment of pulmonary arterial hypertension in 2005.

2.10 Challenges and Opportunities

Unlike traditional drug discovery, drug repurposing is a safe, quicker, and less expensive approach. Recently, machine learning has been employed to increase the performance of drug repositioning. Discovering drug-disease linkages is better done through target protein-based screening, cell-based assay, animal model testing, and clinical testing than through computational techniques [71]. However, computational and experimental methods have been employed to uncover novel pharmacological indications, and computer methods can be validated using biological and clinical studies. A hybrid method of repositioning considers all possible options using improved datasets and technology [72, 73]. Medications that are repositioned require less research and development than new drugs, making them cost-effective [74–76]. This helps many pharmaceutical firms to make medicines available at a lower cost [77–81]. Hybrid methodologies may accelerate medication repositioning, discovering novel treatments for numerous diseases, which may positively impact the market demand and the economy. Repurposing pharmaceuticals for uncommon, neglected, orphan, or hard-to-treat illnesses is viable and may be lucrative for the pharmaceutical industry [82–87].

2.11 Summary

Drug repurposing is a promising approach to address unmet medical needs and make significant clinical advances. This can lead to long-term benefits for a sustainable healthcare system, both economically and socially. It is important to have motivation, financial support, and systematic study to fully unlock the potential of pre-existing medications to treat new and emerging diseases. There are still several illnesses for which effective treatment alternatives are lacking. Repurposing a medicine allows for better use of therapeutic compounds and exploring the unexplored potential of many molecules. Drug repositioning may need a better understanding and a combination of computational and experimental methods to increase success rates.

Author Contributions
Conceptualization: Sateesh Kumar Vemula and Michael A. Repka.
Writing—original draft:

- Introduction: Sagar Narala, Nagarjuna Narala, Sateesh Kumar Vemula.
- Methodologies and approaches: Sunil Kumar Kadiri, Merugumolu Vijay Kumar.
- Examples and challenges, Figures drawn: Rajendra Kumar Jadi, Mounika K.

Writing—review & editing: Sagar Narala, Sateesh Kumar Vemula, and Michael A. Repka.
Supervision: Sateesh Kumar Vemula and Michael A. Repka.

References

1. Hughes JP, Rees S, Kalindjian SB, Philpott KL. Principles of early drug discovery. Br J Pharmacol. 2011;162(6):1239–49. https://doi.org/10.1111/j.1476-5381.2010.01127.
2. Rudrapal M, Chetia D. Endoperoxide antimalarials: development, structural diversity and pharmacodynamic aspects with reference to 1,2,4-trioxane-based structural scaffold. Drug Des Devel Ther. 2016;10:3575–90. https://doi.org/10.2147/DDDT.S118116.
3. Feng LS, Xu Z, Chang L, Li C, Yan XF, Gao C, Ding C, Zhao F, Shi F, Wu X. Hybrid molecules with potential in vitro antiplasmodial and in vivo antimalarial activity against drug-resistant Plasmodium falciparum. Med Res Rev. 2020;40(3):931–71. https://doi.org/10.1002/med.21643.
4. Ferreira LG, Andricopulo AD. Drug repositioning approaches to parasitic diseases: a medicinal chemistry perspective. Drug Discov Today. 2016;21(10):1699–710. https://doi.org/10.1016/j.drudis.2016.06.021.
5. Kalita J, Chetia D, Rudrapal M. Design, synthesis, antimalarial activity, and docking study of 7-Chloro-4- (2-(substituted benzylidene) hydrazineyl) quinolines. Med Chem. 2020;16(7):928–37. https://doi.org/10.2174/1573406415666190806154722.
6. Allarakhia M. Open-source approaches for the repurposing of existing or failed candidate drugs: learning from and applying the lessons across diseases. Drug Des Devel Ther. 2013;7:753–66. https://doi.org/10.2147/DDDT.S46289.
7. Jin G, Wong ST. Toward better drug repositioning: prioritizing and integrating existing methods into efficient pipelines. Drug Discov Today. 2014;19(5):637–44. https://doi.org/10.1016/j.drudis.2013.11.005.
8. Kim TW. Drug repositioning approaches for the discovery of new therapeutics for Alzheimer's disease. Neurotherapeutics. 2015;12(1):132–42. https://doi.org/10.1007/s13311-014-0325-7.
9. Lionta E, Spyrou G, Vassilatis DK, Cournia Z. Structure-based virtual screening for drug discovery: principles, applications, and recent advances. Curr Topics Med Chem. 2014;14(16):1923–38. https://doi.org/10.2174/1568026614666140929124445.
10. Oprea TI, Overington JP. Computational and practical aspects of drug repositioning. Assay Drug Dev Technol. 2015;13(6):299–306. https://doi.org/10.1089/adt.2015.29011.Tiodrrr.
11. Talevi A, Bellera CL. Challenges and opportunities with drug repurposing: finding strategies to find alternative uses of therapeutics. Exp Opin Drug Discov. 2020;15(4):397–401. https://doi.org/10.1080/17460441.2020.1704729.
12. Koch U, Hamacher M, Nussbaumer P. Cheminformatics at the interface of medicinal chemistry and proteomics. Biochim Biophys Acta. 2014;1844(1 Pt A):156–61. https://doi.org/10.1016/j.bbapap.2013.05.010.
13. Rosa SGV, Santos WC. Clinical trials on drug repositioning for COVID-19 treatment. Rev Panam Salud Publica. 2020;44:e40. https://doi.org/10.26633/RPSP.2020.40.

14. Turanli B, Grøtli M, Boren J, Nielsen J, Uhlen M, Arga KY, Mardinoglu A. Drug repositioning for effective prostate cancer treatment. Front Physiol. 2018;9:500. https://doi.org/10.3389/fphys.2018.00500.
15. Napolitano F, Zhao Y, Moreira VM, Tagliaferri R, Kere J, D'Amato M, Greco D. Drug repositioning: a machine-learning approach through data integration. J Cheminform. 2013;5(1):30. https://doi.org/10.1186/1758-2946-5-30.
16. Appleby BS, Cummings JL. Discovering new treatments for Alzheimer's disease by repurposing approved medications. Curr Topics Med Chem. 2013;13(18):2306–27. https://doi.org/10.2174/15680266113136660162.
17. Chong CR, Chen X, Shi L, Liu JO, Sullivan DJ Jr. A clinical drug library screen identifies astemizole as an antimalarial agent. Nat Chem Biol. 2006;2(8):415–6. https://doi.org/10.1038/nchembio806.
18. Matthews H, Usman-Idris M, Khan F, Read M, Nirmalan N. Drug repositioning as a route to anti-malarial drug discovery: preliminary investigation of the in vitro anti-malarial efficacy of emetine dihydrochloride hydrate. Malaria J. 2013;12:359. https://doi.org/10.1186/1475-2875-12-359.
19. Serafin MB, Bottega A, Foletto VS, da Rosa TF, Hörner A, Hörner R. Drug repositioning is an alternative for the treatment of coronavirus COVID-19. Int J Antimicrob Agents. 2020;55(6):105969. https://doi.org/10.1016/j.ijantimicag.2020.105969.
20. Shah B, Modi P, Sagar SR. In silico studies on therapeutic agents for COVID-19: drug repurposing approach. Life Sci. 2020;252:117652. https://doi.org/10.1016/j.lfs.2020.117652.
21. Shivaprasad C, Kalra S. Bromocriptine in type 2 diabetes mellitus. Indian J Endocrinol Metab. 2011;15(Suppl. 1):S17–24. https://doi.org/10.4103/2230-8210.83058.
22. Shim JS, Liu JO. Recent advances in drug repositioning for the discovery of new anticancer drugs. Int J Biol Sci. 2014;10(7):654–63. https://doi.org/10.7150/ijbs.9224.
23. Pushpakom S, Iorio F, Eyers PA, Escott KJ, Hopper S, Wells A, Doig A, Guilliams T, Latimer J, McNamee C, Norris A, Sanseau P, Cavalla D, Pirmohamed M. Drug repurposing: progress, challenges and recommendations. Nat Rev Drug Discov. 2019;18(1):41–58. https://doi.org/10.1038/nrd.2018.168.
24. Adie EA, Adams RR, Evans KL, Porteous DJ, Pickard BS. Speeding disease gene discovery by sequence-based candidate prioritization. BMC Bioinform. 2005;6:55. https://doi.org/10.1186/1471-2105-6-55.
25. de Oliveira EAM, Lang KL. Drug repositioning: concept, classification, methodology, and importance in rare/orphans and neglected diseases. J Appl Pharm Sci. 2018;8(8):157–65.
26. Ekins S, Williams AJ, Krasowski MD, Freundlich JS. In silico repositioning of approved drugs for rare and neglected diseases. Drug Discov Today. 2011;16(7–8):298–310. https://doi.org/10.1016/j.drudis.2011.02.016.
27. Emig D, Ivliev A, Pustovalova O, Lancashire L, Bureeva S, Nikolsky Y, Bessarabova M. Drug target prediction and repositioning using an integrated network-based approach. PLoS One. 2013;8(4):e60618. https://doi.org/10.1371/journal.pone.0060618.
28. Köhler S, Bauer S, Horn D, Robinson PN. Walking the interactome for prioritization of candidate disease genes. Am J Hum Genet. 2008;82(4):949–58. https://doi.org/10.1016/j.ajhg.2008.02.013.
29. Mei JP, Kwoh CK, Yang P, Li XL, Zheng J. Drug-target interaction prediction by learning from local information and neighbors. Bioinformatics. 2013;29(2):238–45. https://doi.org/10.1093/bioinformatics/bts670.
30. Vanunu O, Magger O, Ruppin E, Shlomi T, Sharan R. Associating genes and protein complexes with disease via network propagation. PLoS Comput Biol. 2010;6(1):e1000641. https://doi.org/10.1371/journal.pcbi.1000641.
31. Selvaraj N, Swaroop AK, Kumar RR, Natarajan J, Selvaraj J. Network-based drug repurposing: a critical review. Curr Drug Res Rev. 2022;14(2):116–31.
32. Martínez V, Navarro C, Cano C, Fajardo W, Blanco A. DrugNet: network-based drug-disease prioritization by integrating heterogeneous data. Artif Intell Med. 2015;63(1):41–9. https://doi.org/10.1016/j.artmed.2014.11.003.

33. Fleuren WW, Alkema W. Application of text mining in the biomedical domain. Methods. 2015;74:97–106. https://doi.org/10.1016/j.ymeth.2015.01.015.
34. Gramatica R, Di Matteo T, Giorgetti S, Barbiani M, Bevec D, Aste T. Graph theory enables drug repurposing—how a mathematical model can drive the discovery of hidden mechanisms of action. PLoS One. 2014;9(1):e84912. https://doi.org/10.1371/journal.pone.0084912.
35. Jang G, Lee T, Lee BM, Yoon Y. Literature-based prediction of novel drug indications considering relationships between entities. Mol Biosyst. 2017;13(7):1399–405. https://doi.org/10.1039/c7mb00020k.
36. Krallinger M, Erhardt RA, Valencia A. Text-mining approaches in molecular biology and biomedicine. Drug Discov Today. 2005;10(6):439–45. https://doi.org/10.1016/S1359-6446(05)03376-3.
37. Kuusisto F, Steill J, Kuang Z, Thomson J, Page D, Stewart R. A simple text mining approach for ranking pairwise associations in biomedical applications. AMIA Jt Summits Transl Sci. 2017;2017:166–74.
38. Li J, Zhu X, Chen JY. Building disease-specific drug-protein connectivity maps from molecular interaction networks and PubMed abstracts. PLoS Comput Biol. 2009;5(7):e1000450. https://doi.org/10.1371/journal.pcbi.1000450.
39. Papanikolaou N, Pavlopoulos GA, Theodosiou T, Vizirianakis IS, Iliopoulos I. DrugQuest—a text mining workflow for drug association discovery. BMC Bioinform. 2016;17(Suppl. 5):182. https://doi.org/10.1186/s12859-016-1041-6.
40. Zhang M, Schmitt-Ulms G, Sato C, Xi Z, Zhang Y, Zhou Y, St George-Hyslop P, Rogaeva E. Drug repositioning for Alzheimer's disease based on systematic 'Omics' data mining. PLoS One. 2016;11(12):e0168812. https://doi.org/10.1371/journal.Pone.0168812.
41. Cheng D, Knox C, Young N, Stothard P, Damaraju S, Wishart DS. PolySearch: a web-based text mining system for extracting relationships between human diseases, genes, mutations, drugs, and metabolites. Nucleic Acids Res. 2008;36(Web Server issue):W399–405. https://doi.org/10.1093/nar/gkn296.
42. Cossarizza A, Chang HD, Radbruch A, Acs A, Adam D, Adam-Klages S, Agace WW, Aghaeepour N, Akdis M, Allez M, Almeida LN, Alvisi G, Anderson G, Andrä I, Annunziato F, Anselmo A, Bacher P, Baldari CT, Bari S, Barnaba V, et al. Guidelines for the use of flow cytometry and cell sorting in immunological studies (second edition). Eur J Immunol. 2019;49(10):1457–973. https://doi.org/10.1002/eji.201970107.
43. Cabot C, Darmoni S, Soualmia LF. Cimind: a phonetic-based tool for multilingual named entity recognition in biomedical texts. J Biomed Inform. 2019;94:103176. https://doi.org/10.1016/j.jbi.2019.103176.
44. Lee S, Kim D, Lee K, Choi J, Kim S, Jeon M, Lim S, Choi D, Kim S, Tan AC, Kang J. BEST: next-generation biomedical entity search tool for knowledge discovery from biomedical literature. PLoS One. 2016;11(10):e0164680. https://doi.org/10.1371/journal.Pone.0164680.
45. Plake C, Schiemann T, Pankalla M, Hakenberg J, Leser U. AliBaba: PubMed as a graph. Bioinformatics. 2006;22(19):2444–5. https://doi.org/10.1093/bioinformatics/btl408.
46. Fu G, Ding Y, Seal A, Chen B, Sun Y, Bolton E. Predicting drug target interactions using meta-path-based semantic network analysis. BMC Bioinform. 2016;17:160. https://doi.org/10.1186/s12859-016-1005-x.
47. Chen B, Ding Y, Wild DJ. Assessing drug target association using semantic linked data. PLoS Comput Biol. 2012;8(7):e1002574. https://doi.org/10.1371/journal.pcbi.1002574.
48. Hemnes AR, Beck GJ, Newman JH, Abidov A, Aldred MA, Barnard J, Berman Rosenzweig E, Borlaug BA, Chung WK, Comhair SAA, Erzurum SC, Frantz RP, Gray MP, Grunig G, Hassoun PM, Hill NS, Horn EM, Hu B, Lempel JK, Maron BA, et al. PVDOMICS: a multi-center study to improve understanding of pulmonary vascular disease through phenomics. Circ Res. 2017;121(10):1136–9. https://doi.org/10.1161/CIRCRESAHA.117.311737.
49. Mullen J, Cockell SJ, Woollard P, Wipat A. An integrated data driven approach to drug repositioning using gene-disease associations. PLoS One. 2016;11(5):e0155811. https://doi.org/10.1371/journal.pone.0155811.

50. Sun Y, Sheng Z, Ma C, Tang K, Zhu R, Wu Z, Shen R, Feng J, Wu D, Huang D, Huang D, Fei J, Liu Q, Cao Z. Combining genomic and network characteristics for extended capability in predicting synergistic drugs for cancer. Nat Commun. 2015;6:8481. https://doi.org/10.1038/ncomms9481.
51. Tao Y, Zhang Y, Jiang B. DBCSMOTE: a clustering-based oversampling technique for data-imbalanced warfarin dose prediction. BMC Med Genomics. 2020;13(Suppl 10):152. https://doi.org/10.1186/s12920-020-00781-2.
52. Zhu Q, Tao C, Shen F, Chute CG. Exploring the pharmacogenomics knowledge base (PharmGKB) for repositioning breast cancer drugs by leveraging Web ontology language (OWL) and cheminformatics approaches. Pac Symp Biocomput. 2014;2014:172–82.
53. Rudrapal M, Khairnar SJ, Jadhav AG. Drug repurposing (DR): an emerging approach in drug discovery. Drug repurposing-hypothesis, molecular aspects and therapeutic applications. 2020;10.
54. Ferreira LN, Vega-Oliveros DA, Cotacallapa M, Cardoso MF, Quiles MG, Zhao L, Macau EEN. Spatiotemporal data analysis with chronological networks. Nat Commun. 2020;11(1):4036. https://doi.org/10.1038/s41467-020-17634-2.
55. Xu R, Wunsch D 2nd. Survey of clustering algorithms. IEEE Trans Neural Netw. 2005;16(3):645–78. https://doi.org/10.1109/TNN.2005.845141.
56. Lu J, Chen L, Yin J, Huang T, Bi Y, Kong X, Zheng M, Cai YD. Identification of new candidate drugs for lung cancer using chemical-chemical interactions, chemical-protein interactions, and a K-means clustering algorithm. J Biomol Struct Dyn. 2016;34(4):906–17. https://doi.org/10.1080/07391102.2015.1060161.
57. Xu J, Wickramarathne TL, Chawla NV. Representing higher-order dependencies in networks. Sci Adv. 2016;2(5):e1600028. https://doi.org/10.1126/sciadv.1600028.
58. Rossi A, Cantisani C, Melis L, Iorio A, Scali E, Calvieri S. Minoxidil use in dermatology, side effects, and recent patents. Recent Pat Inflamm Allergy Drug Discov. 2012;6(2):130–6. https://doi.org/10.2174/187221312800166859.
59. Camilli M, Iannaccone G, La Vecchia G, Cappannoli L, Scacciavillani R, Minotti G, Massetti M, Crea F, Aspromonte N. Platelets: the point of interconnection among cancer, inflammation, and cardiovascular diseases. Exp Rev Hematol. 2021;14(6):537–46. https://doi.org/10.1080/17474086.2021.1943353.
60. Dudley JT, Sirota M, Shenoy M, Pai RK, Roedder S, Chiang AP, Morgan AA, Sarwal MM, Pasricha PJ, Butte AJ. Computational repositioning of the anticonvulsant topiramate for inflammatory bowel disease. Sci Transl Med. 2011;3(96):96ra76. https://doi.org/10.1126/scitranslmed.3002648.
61. Ai N, Wood RD, Welsh WJ. Identification of nitazoxanide as a group I metabotropic glutamate receptor negative modulator for the treatment of neuropathic pain: an in silico drug repositioning study. Pharm Res. 2015;32(8):2798–807. https://doi.org/10.1007/s11095-015-1665-7.
62. Hermeking H. The MYC oncogene as a cancer drug target. Curr Cancer Drug Targets. 2003;3(3):163–75. https://doi.org/10.2174/1568009033481949.
63. Wang W, Yang S, Zhang X, Li J. Drug repositioning by integrating target information through a heterogeneous network model. Bioinformatics. 2014;30(20):2923–30. https://doi.org/10.1093/bioinformatics/btu403.
64. King AD, Przulj N, Jurisica I. Protein complex prediction via cost-based clustering. Bioinformatics. 2004;20(17):3013–20. https://doi.org/10.1093/bioinformatics/bth351.
65. Macropol K, Can T, Singh AK. RRW: repeated random walks on genome-scale protein networks for local cluster discovery. BMC Bioinform. 2009;10:283. https://doi.org/10.1186/1471-2105-10-283.
66. Frey BJ, Dueck D. Clustering by passing messages between data points. Science. 2007;315(5814):972–6. https://doi.org/10.1126/science.1136800.
67. Wu Z, Wang Y, Chen L. Network-based drug repositioning. Mol Biosyst. 2013;9(6):1268–81. https://doi.org/10.1039/c3mb25382a.
68. Xiao W, Jing L, Xu Y, Zheng S, Gan Y, Wen C. Different data mining approaches based medical text data. J Healthc Eng. 2021;2021:1285167. https://doi.org/10.1155/2021/1285167.

69. Dudley JT, Deshpande T, Butte AJ. Exploiting drug-disease relationships for computational drug repositioning. Brief Bioinform. 2011;12(4):303–11. https://doi.org/10.1093/bib/bbr013.
70. Sun X, Chen W, Dai W, Xin H, Rahmand K, Wang Y, Zhang J, Zhang S, Xu L, Han T. Piper sarmentosum Roxb.: a review on its botany, traditional uses, phytochemistry, and pharmacological activities. J Ethnopharmacol. 2020;263:112897. https://doi.org/10.1016/j.jep.2020.112897.
71. Emon MA, Domingo-Fernández D, Hoyt CT, Hofmann-Apitius M. PS4DR: a multimodal workflow for identification and prioritization of drugs based on pathway signatures. BMC Bioinform. 2020;21(1):231. https://doi.org/10.1186/s12859-020-03568-5.
72. Lyne SB, Yamini B. An alternative pipeline for glioblastoma therapeutics: a systematic review of drug repurposing in glioblastoma. Cancers. 2021;13(8):1953. https://doi.org/10.3390/cancers13081953.
73. Nunes T, Campos D, Matos S, Oliveira JL. BeCAS: biomedical concept recognition services and visualization. Bioinformatics. 2013;29(15):1915–6. https://doi.org/10.1093/bioinformatics/btt317.
74. Ekins S, Bunin BA. The Collaborative Drug Discovery (CDD) database. Methods Mol Biol. 2013;993:139–54. https://doi.org/10.1007/978-1-62703-342-8_10.
75. Liu Y, Hu B, Fu C, Chen X. DCDB: drug combination database. Bioinformatics. 2010;26(4):587–8. https://doi.org/10.1093/bioinformatics/btp697.
76. Pacini C, Iorio F, Gonçalves E, Iskar M, Klabunde T, Bork P, Saez-Rodriguez J. DVD: an R/Cytoscape pipeline for drug repurposing using public repositories of gene expression data. Bioinformatics. 2013;29(1):132–4. https://doi.org/10.1093/bioinformatics/bts656.
77. Ashburn TT, Thor KB. Drug repositioning: identifying and developing new uses for existing drugs. Nat Rev Drug Discov. 2004;3(8):673–83. https://doi.org/10.1038/nrd1468.
78. Cuatrecasas P. Drug discovery in jeopardy. J Clin Invest. 2006;116(11):2837–42. https://doi.org/10.1172/JCI29999.
79. Delavan B, Roberts R, Huang R, Bao W, Tong W, Liu Z. Computational drug repositioning for rare diseases in the era of precision medicine. Drug Discov Today. 2018;23(2):382–94. https://doi.org/10.1016/j.drudis.2017.10.009.
80. Grinnan D, Trankle C, Andruska A, Bloom B, Spiekerkoetter E. Drug repositioning in pulmonary arterial hypertension: challenges and opportunities. Pulm Circ. 2019;9(1):2045894019832226. https://doi.org/10.1177/2045894019832226.
81. Sotiropoulou G, Zingkou E, Pampalakis G. Redirecting drug repositioning to discover innovative cosmeceuticals. Exp Dermatol. 2021;30(5):628–44. https://doi.org/10.1111/exd.14299.
82. Cong Y, Endo T. Multi-omics and artificial intelligence-guided drug repositioning: prospects, challenges, and lessons learned from COVID-19. Omics. 2022;26(7):361–71. https://doi.org/10.1089/omi.2022.0068.
83. Jegerlehner S, Suter-Riniker F, Jent P, Bittel P, Nagler M. Diagnostic accuracy of a SARS-CoV-2 rapid antigen test in real-life clinical settings. Int J Infect Dis. 2021;109:118–22. https://doi.org/10.1016/j.ijid.2021.07.010.
84. Khalili H. Clinical simulation practice framework. Clin Teacher. 2015;12(1):32–6. https://doi.org/10.1111/tct.12291.
85. Lotfi Shahreza M, Ghadiri N, Mousavi SR, Varshosaz J, Green JR. A review of network-based approaches to drug repositioning. Brief Bioinform. 2018;19(5):878–92. https://doi.org/10.1093/bib/bbx017.
86. Nowak-Sliwinska P, Scapozza L, Ruiz i Altaba A. Drug repurposing in oncology: compounds, pathways, phenotypes and computational approaches for colorectal cancer. Biochim Biophys Acta Rev Cancer. 2019;1871(2):434–54. https://doi.org/10.1016/j.bbcan.2019.04.005.
87. Yetisen AK, Qu H, Manbachi A, Butt H, Dokmeci MR, Hinestroza JP, Skorobogatiy M, Khademhosseini A, Yun SH. Nanotechnology in textiles. ACS Nano. 2016;10(3):3042–68. https://doi.org/10.1021/acsnano.5b08176.

In Silico Methods Used in Drug Repurposing

3

Neelanjan Chowdhury and Satheesh Kumar Nanjappan

Learning Objectives

- Advantages of computational methods in drug repurposing
- Various methodologies used in silico
- Challenges of in silico methodologies in drug repurposing

3.1 Introduction

To maximize drug production and utilization, drug repositioning—also called drug repurposing, drug retasking, or therapeutic flipping—identifies new therapeutic indications for already approved drugs. De novo drug development is still a labor-intensive process as well as an expensive procedure, despite advancements in biological and informational technologies. This is due to high failure rates. Contrarily, drug repositioning has recently gained popularity due to several success stories [1]. The FDA-approved medications currently on the market have well-established dosage regimens and pharmacokinetic (PK) and pharmacodynamic (PD) properties demonstrated in humans, so repositioning the medication could significantly reduce the time and expense required to bring a new medication to market. Since the bulk of drug failures occur in clinical safety and/or associated toxicity, it is less likely that new targets for existing treatments will fail in subsequent clinical trials. Because creating novel drug–target interaction profiles for FDA-approved medications takes less time and money than are already on the market and has established

N. Chowdhury · S. K. Nanjappan (✉)
Department of Natural Products, National Institute of Pharmaceutical Education & Research (NIPER) Kolkata, Kolkata, West Bengal, India
e-mail: satheesh.niperk@nic.in

N. Chella et al. (eds.), *Drug Repurposing*,
https://doi.org/10.1007/978-981-97-5016-0_3

formulations and manufacturing techniques, drug repositioning can be a potential option for drug development [2].

3.1.1 Significance of Drug Repositioning

Drug repositioning is a tactic for finding novel applications for authorized or experimental medications that fall outside the initial therapeutic recommendation range. Comparing this strategy to creating a brand-new medication specifically for certain indication reveals many benefits. The first and possibly most significant benefit is a decreased risk of failure. If early-stage trials have been completed, the repurposed drug has already been shown to be adequately safe in preclinical models and people, reducing the likelihood of failure in further effectiveness trials, at least in terms of safety. Second, there is a reduced time frame for drug development since most preclinical testing, safety assessment, and, in some cases, formulation development will have already been completed.

3.1.2 An Ideal Candidate for Drug Repurposing

A drug that has undergone clinical drug development and is present in the market is a good candidate for repurposing. Clinical trials for a drug may be skipped if there is enough data to support and justify doing so and the drug has passed established safety and toxicity testing in prior clinical studies that regulatory authorities have approved. It is crucial to ascertain the drug's method of action. The best prospects for repurposing are medications that have undergone numerous phases of clinical testing but have failed for factors unrelated to safety. Clinical trials have seen the repurposing of several drugs, such as the well-known Viagra (sildenafil), which was previously prescribed for angina and hypertension. Clinical research revealed the novel function of treating erectile dysfunction.

Examples of abandoned medications whose toxicity recurred with new indications exist. When used to treat morning sickness in pregnant women, the vomiting-inducing drug thalidomide causes various congenital abnormalities. Its usage was outlawed or strictly regulated in many nations because of this catastrophic impact on prenatal development. Later, leprosy and multiple myeloma were treated with the same medication. The scientific community has, however, voiced concerns about this kind of repurposing. When repurposing cases, the benefit-to-risk ratio might be taken into account. To resolve the question "Is it worth the risk?" the regulatory authorities must make a logical decision. Or will a currently used therapy outperform the medicine that has been repurposed?

3.2 Computational Drug Repurposing Strategy

Worldwide interest has grown recently due to the accessibility of a wealth of information on the structure of proteins, pharmacophores, illness information, clinical trials, or gene expression patterns of medications. Additionally, the growth of public social networking sites and the accessibility of computational genetic data have greatly aided computational approaches in predicting novel signals. Thus, for medication repositioning from various chemical spaces, most pharmaceutical corporations use computational or advanced bioinformatics resources, as in Fig. 3.1. Every pharmaceutical company's ultimate goal is to take advantage of the potent in silico technology's benefits of faster speed and lower costs. Techniques such as chemoinformatics, bioinformatics, network biology, systems biology, or genomic information uncover previously undisclosed targets or processes of licensed drugs, which expedites the repositioning process [3].

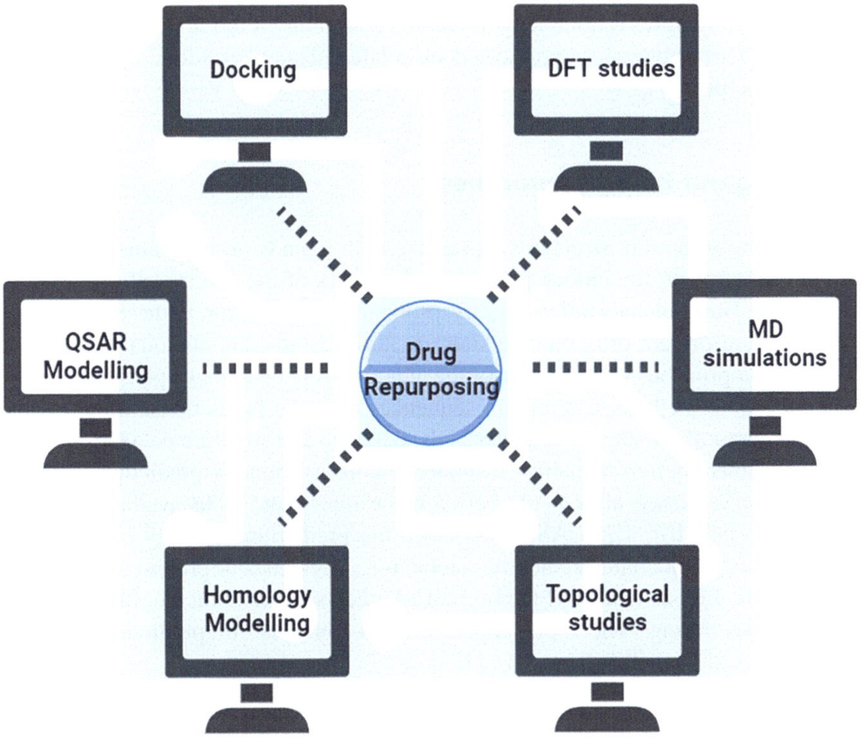

Fig. 3.1 A combined approach of homology modeling, molecular dynamics, and docking

3.2.1 Target-Based Approaches

Drugs have an affinity for therapeutic target proteins as well as additional proteins known as off-targets with which they interact. Pharmacological side effects are caused mainly by these off-target interactions. Following the identification of the interaction between sildenafil and phosphodiesterase 5 (PDE5), which was in charge of the erectile response, this renovation was completed. The ligand or receptor structure is the basis for the target-based drug repositioning techniques. These drug repositioning techniques include docking and/or pharmacophore models-based high-throughput in silico (virtual screening [VS]) drug screening employing chemical library medicines [4].

There are 142,807 total depositions in the protein data bank (PDB), available as an open source at https://www.rcsb.org (as of 27 February 2018). Retrieving a target protein structural information became simple with this database [5]. The three-dimensional structure of the protein can be predicted without experimental data using ab initio approaches, integrated methodologies, homology modeling, or protein threading/fold recognition [6]. The simplest method of foretelling three-dimensional structures is homology modeling, commonly referred to as comparative modeling. This approach is predicated on related sequences adopting essentially similar structures [7].

3.2.2 Ligand-Based Approaches

The majority of protein structures crystallize with ligands or drug molecules present, and as a result, the unique protein-binding sites of the co-crystallized ligands are present. But if ligands fail to solidify, problems start to occur. In these situations, predictions of protein-drug binding sites are made using computational techniques [8]. These approaches vary in terms of the data they rely on to make predictions. For instance, some methods use protein sequences to make predictions, while others rely on structural features or consider both sequence and structure data [9].

For instance, before focusing on sequence conservation for prediction, Firestar first creates sequence alignments between the query and PDB-available ligand-bound proteins [10]. The CASP10 assesses the prediction accuracy of computational methods that could predict the interaction of residues of protein amino acids with ligands. Firestar received the highest Matthews Correlation Coefficient score and accuracy rating. The top computational techniques for predicting protein-binding sites are listed in Table 3.1 [11].

We need the molecular makeup of the drug substances of interest to fully comprehend drug–protein interactions in silico. Numerous chemical databases provide a wide range of chemical data for effective virtual screening (VS). It has information on over 90 million molecules. Interestingly, it also keeps track of biological test data on chemicals, giving access to the drug's available experimental outcomes [12].

Table 3.1 In silico tools for protein-binding site prediction

Sr. No.	Program	Brief description
1.	FireStar	Utilizing the functionally significant residues in the FireDB database, predict functionally relevant residues
2.	SPAlign	The application optimizes the SP-score for alignment with the structure. Proteins sharing the same structural fold and DNA- or RNA-binding activity were found using SPalign
3.	Seok-server	Molecular docking is used to obtain information about the ligand binding site
4.	HHPredA	Utilizing databases like Pfam or SMART, the application compares profiles using an HMM-HMM algorithm
5.	IntFold3	Predictions are made using a multi-template modeling strategy driven by local and global quality assessments

Eleven thousand medication entries are available in the most recent version of DrugBank (5.0.11, released on 2017 December 20) [13]. A summary of some of the significant databases frequently utilized for VS is provided in Table 3.2.

The inverse docking paradigm is an addition to the traditional docking paradigm. In other words, this strategy entails docking a ligand against a range of biological targets, and the chosen targets are ranked according to their affinity for that specific ligand [14]. This is different from docking several ligands with a single target. Several researchers have effectively employed AutoDock Vina and other traditional docking tools as inverse docking tools to discover new indications [15].

Ligand profiling is essential for drug repurposing and is useful for predicting off-target interactions. One can embark on ligand profiling by comparing the chemical properties of unknown substances with ligand-based pharmacophores [16]. In chemoinformatics, chemical similarity is crucial and is frequently applied to forecast the structural and functional characteristics of the molecules. Similar substances are thought to elicit comparable biological effects (Fig. 3.2). Chemical fingerprints, which are abstract representations of the chemical characteristics of molecules, can be used to express molecular similarities. Binary representations of each attribute are used in 2D molecular fingerprints. The Tanimoto coefficient is the most popular method for determining how similar two chemicals are. The value is between 0 and 1, with 1 denoting the highest degree of similarity. When there is little to no substantial similarity between ligands or when there are various ligands, 3D similarity search methods can be used [17].

3.2.3 Knowledge-Based Approaches

Despite their success, target-based and blinded (unintentional, fortuitous findings) approaches cannot be employed to look into novel drug-target connections. Knowledge-based approaches foresee previously unknown processes, such as developing novel biomarkers, unknown drug-drug correlations, and previously

Table 3.2 List of key databases frequently used for virtual screening

Sr. No.	Program	Brief description
1.	ZINC	It's a free resource for buyable compounds that includes over 750 million chemicals that may be used for analog searches in addition to 230 million compounds that are ready to dock
2.	ChEMBL	A collection of bioactive substances with drug-like qualities. To help in converting genomic data into efficient novel pharmaceuticals, it combines chemical, bioactivity, and genetic data
3.	DrugBank	DrugBank integrates comprehensive drug target information with comprehensive drug data, including chemical, pharmacological, and pharmaceutical information, to be used as a cheminformatics and bioinformatics resource
4.	PubChem	It was established in 2004 as a part of the US National Institutes of Health's (NIH) Molecular Libraries Roadmap Initiatives. It is a public repository for data on chemical compounds and their biological activity
5.	BindingDB	A publicly available, web-accessible database called BindingDB has 1,439,799 binding records for 644,978 small compounds and 7042 protein targets
6.	ChemBridge Drug Library	A for-profit database containing almost 1.1 million chemicals that resemble drugs and lead
7.	ChemDiv's compound Libraries	The tool offers a 1.5 M varied chemicals library that may be purchased for virtual screening
8.	MolProt	Chemical compounds that can be bought are included in the MolProt dataset. Additional means to access them include through the MolPort Online portal
9.	SuperDrug2 Database	There are 4600 active pharmacological compounds in the SuperDRUG2 database. Additionally, it offers details on physicochemical characteristics and probable medication interactions
10.	SuperHapten	Currently, this database contains 24 carriers, 453 associated 7257 haptens, and commercially available antibodies
11.	Ligand Expo	It is an assemblage of Protein Data Bank entries for small compounds that crystallize with proteins

unidentified therapeutic targets for older drugs by using accessible knowledge about a drug [18]. Text analytics, cheminformatics, and bioinformatics are three categories of knowledge-based methods. Knowledge-based techniques, which use a huge amount of data for drug repurposing, increase prediction confidence. The most promising repurposing conclusion is produced by combining biological, chemical, and diagnostic elements. This approach can also aid in establishing a precise target and provide comprehensive knowledge of the drug's mechanism of action [19]. Table 3.3 displays a list of programs used for reverse docking and pharmacophore mapping in drug repositioning.

Several knowledge-based computational algorithms generate distinct disease pathways based on gene expression patterns for therapeutic repositioning. The

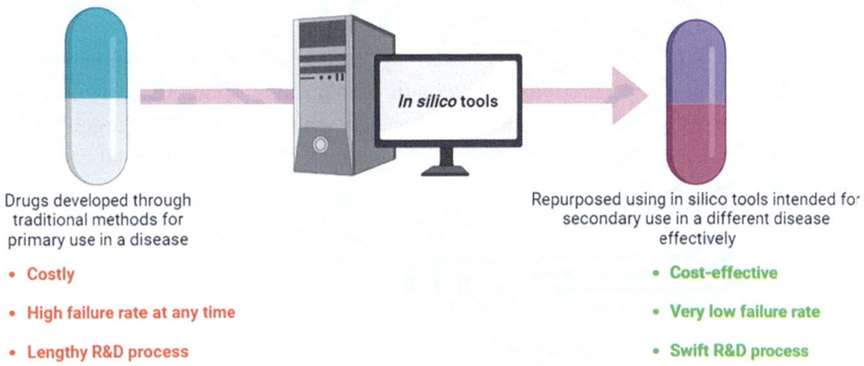

Fig. 3.2 Associations between drugs and molecular targets using various techniques

Table 3.3 List of programs used for pharmacophore mapping and reverse docking in drug repositioning

Sr. No.	Software	Description
1.	TarFisDock	It performs reverse ligand–protein docking using a database of potential medicinal targets. This is predicated on the DOCK 4.0 scoring mechanism
2.	MDock	The sphere-ligand matching strategy for possible ligand conformation and the ensemble docking approach for ligand docking creation are the foundations of MDock, an automated molecular docking program. It makes use of a knowledge-based scoring mechanism
3.	idTarget	It is an open-access web server that can identify biomolecular targets using a divide-and-conquer docking strategy
4.	INVDOCK	It was developed to find targets of tiny molecules that are proteins or nucleic acids (DNA or RNA). It makes use of a streamlined DOCK scoring mechanism without ligand pose optimization
5.	PharmMapper	For reverse pharmacophore mapping, it is frequently utilized. It uses various databases to describe the way that approved ligands attach to protein targets at their binding locations, including DrugBank, BindingDB, PDBBind, and PDTD. After that, LigandScout is used to map pharmacophores using the gathered structural data
6.	TargetHunter	By learning from the bioactive compound-target combinations that have been documented in the literature, it aids in target identification
7.	Similarity ensemble approach (SEA)	It generates cross-target similarity maps and establishes a set-wise association between proteins according to the molecular similarity of their ligands

underlying premise of these methods suggests that potential treatment targets could be the fundamental components of disease processes or proteins [20]. This approach provides valuable information for drug repurposing, such as the possibility of repurposing many drugs approved for Crohn's disease. It was successfully used to identify new applications for already-available medications [21].

Phenotypic information is being used by computational drug repositioning approaches to make predictions. Systematic evaluations show that phenotypic screening performs better for discovering new indications than target-based approaches [22]. New medication indications can now be suggested based on clinical side effects [23].

3.2.4 Signature-Based Approaches

Gene expression signatures, another name for gene signatures, are defined as one gene or a group of genes that exhibit a clearly identifiable pattern of gene expression as a result of a healthy or diseased state [24]. This data is used to assess the gene signatures of different diseases and therapeutic regimens, which forms the foundation of signature-based drug repositioning strategies [25]. Genome-wide expression profiles are crucial for drug repurposing because they clarify how a drug interacts with various cell processes [26]. By revealing the previously unknown mechanisms of action of pharmaceuticals, signature-based approaches give us a better understanding of pharmacological indications. It might be helpful to compare gene sets that are downregulated or unregulated in healthy or diseased states to identify a disease's signature [27].

Based on the type of probe utilized, DNA microarrays are divided into three main categories:

- cDNA microarrays: Spotted DNA arrays are another name for cDNA microarrays. These microarrays use complementary DNA to hybridize reverse-transcribed single-stranded RNA with complementary probes put on the array.
- Oligo DNA microarrays: Microarrays, also called oligo DNA microarrays, employ small DNA or RNA molecules called oligonucleotides, chemically manufactured, as probes.
- Single nucleotide polymorphism (SNP) microarrays: Polymorphisms within a population can be found using SNP microarrays.

Next-generation sequencing (NGS) is yet another essential tool for producing high-throughput gene expression data. A high-throughput DNA sequencing technique called NGS is used to quickly sequence DNA or RNA materials [28]. Numerous gene expression databases (Table 3.4) contain high-throughput gene expression data from the aforementioned methodologies. These databases act as storage facilities for quantifiable data and maintain a searchable index, enabling other applications to utilize the data for analyses and interpretations.

Table 3.4 A selection of important gene expression databases for drug repositioning predicted on signatures

Database	Notes
ImmGen database	An extensive microarray analysis of the regulation of gene expression in the mouse immune system is available in the public database ImmGen
GENEVESTIGATOR V3	Data on gene expression from RNAseq and microarray investigations can be found in GENEVESTIGATOR V3. Nearly 500 illnesses' expression data are included. This database can be used to research target identification, metabolisms, and gene function
Ensembl	EMBL-EBI's genome browser is available to study evolution, sequence variation, and comparative genomics. It includes information on gene expression for several organisms, including humans, mice, zebrafish, and others
ArrayExpress	EMBL-EBI's public repository, ArrayExpress, houses high-throughput genomic expression data

In signature-based drug repurposing approaches, the gene expression signature defines pharmacological action inside the biological milieu, independent of the chemical structure [29].

3.2.5 Network-Based Approaches

The two basic building blocks of biological networks, which are used to simulate the interactions between various biological elements, are typically nodes and edges [30]. Edges depict the connections between the numerous biological components, whereas nodes represent diverse biological elements like genes, proteins, or even multidimensional phenotypes like illnesses.

A bipartite interaction network is constructed in network-based drug–target interaction (DTI) models, where drugs and targets are shown as nodes. In contrast, connections between them are shown as edges. Later models have more robust construction than early models. Drug repositioning strategies based on similarities, commonly referred to as guilt by association (GBA) techniques, are becoming more popular [31].

3.2.6 Target Mechanism-Based Approaches

The target mechanism-based approaches combine treatment omics (genetic) data, accessible signaling circuit information, and protein interaction networks to identify the undiscovered mechanisms of the drug. Understanding therapeutic targets facilitates the development of more potent medication regimens that address issues like resistance, which is more prevalent in cancer. Targeted mechanism-based approaches (Table 3.5) that have yielded complex computer models to predict medication

Table 3.5 Target mechanism-based drug repurposing related omics databases

Database	Representation
KEGG	It is an open resource that provides details on intricate biological networks
WikiPathways	The scientific community maintains an open-source pathway database
Therapeutic Target Database (TTD)	This free resource contains information on the nucleic acid route and therapeutic proteins, among other topics
BioCyc Database	It includes information about pathways and genomes. It also offers the resources to examine them
PANTHER	Proteins and their genes are categorized in the database PANTHER (Protein Analysis Through Evolutionary Relationships) for high-throughput analysis
MiRTarBase	A database of microRNA–target interactions
BioCarta	This resource includes dynamic mapping for metabolic and signaling networks

effects and related targeted pathways have received less attention. This is because creating effective computational models is difficult [32].

3.3 Drug Safety Evaluation

After finding potential medication candidates, the next step is to evaluate their pharmacokinetic characteristics, such as absorption, distribution, metabolism, elimination, and toxicity (ADME-Tox). Thanks to improvements in machine learning algorithms and gathered datasets, ADME-Tox can also be predicted using computational methods. Preclinical testing on drug candidates is thought to result in the withdrawal of 40–60% of them due to ADME-Tox issues and various physiological barriers, including the blood–brain barrier and the gastrointestinal barrier.

3.4 Examples of Successful Drug Repositioning

Several medications have been successfully repositioned for innovative purposes using bioinformatics and experimental methodologies. Several of these repositioned molecules are currently undergoing clinical study. Due to insufficient efficacy in phase III clinical research, most drugs fail in the late stages of development. It would be ideal to look for new indications and drug repositioning for these safe, late-phase failed medications. Table 3.6 provides a list of a few instances of effective drug repositioning.

Table 3.6 Stories of effective drug repositioning

Drug	Original statement	New statement
Amphetamine	Stimulant	Children's hyperkinesis
Atomoxetine	Parkinson's condition	ADHD
Bupropion	Depression	Quitting smoking and obesity
Duloxetine	Depression	Chronic musculoskeletal discomfort, fibromyalgia, stress urine incontinence, and diabetic peripheral neuropathy
Everolimus	Immunosuppressant	Neuroendocrine tumors of the pancreas
Lumigan	Glaucoma	Uncomplicated hypotrichosis
Paclitaxel	Cancer	Restenosis
Phentolamine	Hypertension	Reversing agent for dental anesthesia
Naltrexone	Opioid dependence	Obesity

3.5 Findings and Challenges of In Silico Methods Used in Drug Repurposing

The conventional drug discovery method is changing as a result of computational drug repositioning. Drug repositioning is the quickest and least expensive technique to find new therapeutic indications since a plethora of information about the drug molecule is already available. Drug repurposing techniques are always changing; initially, drugs were examined by mistake against a variety of targets to discover new indications. The best illustration of accidental drug repurposing is thalidomide. The admission of computational tools has accelerated drug repurposing due to a better understanding of drug structure, interactions, and other properties. Repurposing pharmaceuticals lowers the cost of researching uncommon diseases and increases the success rate of treatment compared to traditional drug research. It is essential to include fields like bioinformatics, genomics, and molecular biology for effective drug repositioning. This partnership paves the way for new revenue models for repurposed medications and aids in discovering intuitive novel indications using highly qualitative quantitative data [33].

Although the academic and pharmaceutical sectors have recently shown great interest in computer-based techniques for drug repositioning, there are many limitations to the in silico methodologies used. In reality, most, if not all, computational methodologies require large amounts of data sets. Generally speaking, these methods are not helpful for drug repositioning, particularly for novel chemical or tartget combinations. The inverse docking approach has several disadvantages while being widely used for target-based medication repositioning and providing insights into possible new medical applications. These tactics don't yield the desired results because of the limitations of the target databases and the docking tools. We are aware that the docking program ranks the top hits by combining search algorithms with a scoring mechanism; this also has several disadvantages. Even though network pharmacology is expanding and shedding light on the myriad pharmacological side effects that are off-target, there are still certain inherent limits [34].

Similarly, several techniques have been created to clarify a relationship between the mechanisms of action of various pharmaceuticals by examining phenotypic data (indications and side effects). However, these methodologies are also insufficient to investigate drug repositioning because the side effect information for medications is ambiguous and sparse. A significant issue with phenotypic approaches is that the results depend on the person's genetic makeup, medical history, and features. Therefore, it is not very reliable and unlikely to succeed in directly comparing the phenotypic data of two medications to investigate their method of action.

3.6 Summary

The in silico discovery of disease-associated drug targets and therapeutic medicines has improved efficiency and accuracy over the past few decades. Due to recent rapid improvements in computational techniques and the accumulation of publicly available biological data, in silico drug discovery has increased. Still, careful observation is required to evaluate the above data under in vitro and in vivo conditions.

Acknowledgments The authors thank the Department of Pharmaceuticals and the Ministry of Chemicals and Fertilizers for their support.

References

1. Cha Y, Erez T, Reynolds I, Kumar D, Ross J, Koytiger G, et al. Drug repurposing from the perspective of pharmaceutical companies. Br J Pharmacol. 2018;175(2):168–80.
2. Wishart DS, Feunang YD, Guo AC, Lo EJ, Marcu A, Grant JR, et al. DrugBank 5.0: a major update to the DrugBank database for 2018. Nucleic Acids Res. 2018;46(D1):D1074–82.
3. Wishart DS, Knox C, Guo AC, Cheng D, Shrivastava S, Tzur D, et al. DrugBank: a knowledgebase of drugs, drug actions, and drug targets. Nucleic Acids Res. 2008;36(Suppl_1):D901–6.
4. Knox C, Law V, Jewison T, Liu P, Ly S, Frolkis A, et al. DrugBank 3.0: a comprehensive resource for 'omics' research on drugs. Nucleic Acids Res. 2010;39(Suppl_1):D1035–41.
5. Speck-Planche A, Kleandrova VV. Demystifying artificial neural networks as generators of new chemical knowledge: antimalarial drug discovery as a case study. In: Cartwright HM, editor. Machine learning in chemistry. Cambridge: Royal Society of Chemistry; 2020. p. 398–423.
6. Kumar S, Kovalenko S, Bhardwaj S, Sethi A, Gorobets NY, Desenko SM, Rathi B. Drug repurposing against SARS-CoV-2 using computational approaches. Drug Discov Today. 2022;27(7):2015–27.
7. Yella JK, Yaddanapudi S, Wang Y, Jegga AG. Changing trends in computational drug repositioning. Pharmaceuticals. 2018;11(2):57.
8. Li C, He W-Q. Global prediction of primary liver cancer incidences and mortality in 2040. J Hepatol. 2023;78(4):e144–6.
9. Jarada TN, Rokne JG, Alhajj R. A review of computational drug repositioning: strategies, approaches, opportunities, challenges, and directions. J Chem. 2020;12(1):1–23.
10. Menden MP, Iorio F, Garnett M, McDermott U, Benes CH, Ballester PJ, Saez-Rodriguez J. Machine learning prediction of cancer cell sensitivity to drugs based on genomic and chemical properties. PLoS One. 2013;8(4):e61318.
11. March-Vila E, Pinzi L, Sturm N, Tinivella A, Engkvist O, Chen H, Rastelli G. On the integration of in silico drug design methods for drug repurposing. Front Pharmacol. 2017;8:298.

12. Sardana D, Zhu C, Zhang M, Gudivada RC, Yang L, Jegga AG. Drug repositioning for orphan diseases. Brief Bioinform. 2011;12(4):346–56.
13. Ferreira LG, Dos Santos RN, Oliva G, Andricopulo AD. Molecular docking and structure-based drug design strategies. Molecules. 2015;20(7):13384–421.
14. Hu G, Agarwal P. Human disease-drug network based on genomic expression profiles. PLoS One. 2009;4(8):e6536.
15. Masoudi-Sobhanzadeh Y. Computational-based drug repurposing methods in COVID-19. Bioimpacts. 2020;10(3):205.
16. Jin G, Fu C, Zhao H, Cui K, Chang J, Wong ST. A novel method of transcriptional response analysis to facilitate drug repositioning for cancer therapy. Cancer Res. 2012;72(1):33–44.
17. Berman HM, Westbrook J, Feng Z, Gilliland G, Bhat TN, Weissig H, et al. The protein data bank. Nucleic Acids Res. 2000;28(1):235–42.
18. Wang X, Liu C, Wang J, Fan Y, Wang Z, Wang Y. Proton pump inhibitors increase the chemosensitivity of patients with advanced colorectal cancer. Oncotarget. 2017;8(35):58801.
19. Ehrt C, Brinkjost T, Koch O. Impact of binding site comparisons on medicinal chemistry and rational molecular design. J Med Chem. 2016;59(9):4121–51.
20. Manning G, Whyte DB, Martinez R, Hunter T, Sudarsanam S. The protein kinase complement of the human genome. Science. 2002;298(5600):1912–34.
21. Zhang Y-M, Cockerill S, Guntrip SB, Rusnak D, Smith K, Vanderwall D, et al. Synthesis and SAR of potent EGFR/erbB2 dual inhibitors. Bioorg Med Chem Lett. 2004;14(1):111–4.
22. Chen Y-C, Tolbert R, Aronov AM, McGaughey G, Walters WP, Meireles L. Prediction of protein pairs sharing common active ligands using protein sequence, structure, and ligand similarity. J Chem Inf Model. 2016;56(9):1734–45.
23. Hall DR, Kozakov D, Whitty A, Vajda S. Lessons from hot spot analysis for fragment-based drug discovery. Trends Pharmacol Sci. 2015;36(11):724–36.
24. Sharma PP, Bansal M, Sethi A, Pena L, Goel VK, Grishina M, et al. Computational methods directed towards drug repurposing for COVID-19: advantages and limitations. RSC Adv. 2021;11(57):36181–98.
25. Glicksberg BS, Li L, Chen R, Dudley J, Chen B. Leveraging big data to transform drug discovery. Methods Mol Biol. 2019;1939:91–118.
26. Kleandrova VV, Scotti MT, Speck-Planche A. Computational drug repurposing for antituberculosis therapy: discovery of multi-strain inhibitors. Antibiotics. 2021;10(8):1005.
27. Kadri H, Lambourne OA, Mehellou Y. Niclosamide, a drug with many (re) purposes. ChemMedChem. 2018;13(11):1088–91.
28. Chen B, Wei W, Ma L, Yang B, Gill RM, Chua M-S, et al. Computational discovery of niclosamide ethanolamine, a repurposed drug candidate that reduces the growth of hepatocellular carcinoma cells in vitro and mice by inhibiting cell division cycle 37 signaling. Gastroenterology. 2017;152(8):2022–36.
29. Zhang M, Schmitt-Ulms G, Sato C, Xi Z, Zhang Y, Zhou Y, et al. Drug repositioning for Alzheimer's disease based on systematic 'omics' data mining. PLoS One. 2016;11(12):e0168812.
30. Zhang M, Luo H, Xi Z, Rogaeva E. Drug repositioning for diabetes based on omics data mining. PLoS One. 2015;10(5):e0126082.
31. Koren G, Nordon G, Radinsky K, Shalev V. Identification of repurposable drugs with beneficial effects on glucose control in type 2 diabetes using machine learning. Pharmacol Res Perspect. 2019;7(6):e00529.
32. Park K. A review of computational drug repurposing. Transl Clin Pharmacol. 2019;27(2):59–63.
33. Brown AS, Patel CJ. A standard database for drug repositioning. Sci Data. 2017;4(1):1–7.
34. Gregori-Puigjané E, Mestres J. A ligand-based approach to mining the chemogenomic space of drugs. Comb Chem High Throughput Screen. 2008;11(8):669–76.

Role of Artificial Intelligence (AI) and Machine Learning (ML) in Drug Purposing

4

Neelanjan Chowdhury and Satheesh Kumar Nanjappan

Highlights

- Understanding the role of artificial intelligence in drug repurposing
- Understanding the concept of machine learning
- Understanding the application of machine learning in drug repurposing
- Challenges involved in using AI and ML in drug repurposing studies

4.1 Introduction

Artificial intelligence and machine learning fields have made enormous strides in recent years. However, more study is still required to thoroughly comprehend the information in biology and associated topics. Bioinformatics is a discipline of science that examines biological data using computational techniques, statistical tools, algorithms, and mathematical concepts [1]. Artificial intelligence, which can address several issues connected to human intellect, is the capacity to replicate human intelligence using computer hardware or software [2]. It involves machine learning, which, after receiving training, enables one to carry out all of these jobs. The bioinformatics tools are being used to develop novel or repurposed drugs to treat a wide range of ailments. These tools apply the principles of artificial intelligence and machine learning to specific tools [3]. Bioinformatics has been used to evaluate data and draw logical conclusions. The enormous amount of data generated by whole genome sequencing initiatives and bioinformatics is utilized to annotate biological data meaningfully.

N. Chowdhury · S. K. Nanjappan (✉)
Department of Natural Products, National Institute of Pharmaceutical Education & Research (NIPER) Kolkata, Kolkata, West Bengal, India
e-mail: satheesh@niperkolkata.ac.in; satheesh.niperk@nic.in

N. Chella et al. (eds.), *Drug Repurposing*,
https://doi.org/10.1007/978-981-97-5016-0_4

Gravey provided a brief overview of the history of artificial intelligence using three systems or paradigms: machine learning, which has been around since 2010; expert-based systems, which were in use from the 1980s to the 1990s; and Good Old-Fashioned AI (GOFAI), which dates back to the 1960s [4]. The first method, called GOFAI, was taken from general-purpose logical systems and is based on a heuristic strategy. The expert system switched from general intelligence to human chemists and physicians as its subject matter specialists, simulating decision-making procedures. As a result, significant artificial intelligence systems are now found in popular software like TurboTax and medical products like MYCIN [5].

Along with addressing the risks and difficulties involved, drug discovery processes have been revolutionized by artificial intelligence and computational methods [6]. Similarly, drug purposing is a method by which new medications are created from existing ones with medical applications. Comparing drug purposing to de novo synthesis, there are several benefits, which involves discovering and creating novel medication compounds [7]. On the other hand, purposing results in medication drugs manufactured far more quickly, because it contains elements that have undergone testing, it leads therapeutic molecules to the clinical phase with a substantially lower failure rate than novo synthesis. Additionally, drug purposing eliminates preclinical testing using people and other animal models, lowering the cost of developing drugs [8]. According to that information, the dosage will be compatible with the new indications if safety testing has been done. Random testing and exploration led to the accidental discovery of drug purposing. For instance, sildenafil citrate was initially developed by Pfizer as a hypertensive medication before being repurposed to treat erectile dysfunction, which was proven through clinical studies conducted directly after its formulation [9]. This resulted in enormous sales as well as additional health advantages [10].

Following the phenotypic discovery of unintentional drug compounds by polypharmacological activity, drug purposing was initiated. We thought that the medication's binding to the BCRABL1 fusion protein was the reason behind malignancy. Surprising findings from the computational research demonstrated that this medication binds more strongly than the wild form to the muted version of the T315I-linked BCRABL1. There are also signs that the drug's effectiveness against ponatinib-resistant cells may be diminished when it attaches to other molecules and that its dosage is ineffective. This proved that these algorithms can also highlight side effects before the drug moves on to the next stage [11]. As a result, in addition to diagnostics, artificial learning is currently utilized to forecast tailored drug molecules by examining patients' genomic information. These AI-based training techniques outperform non-AI-trained software in terms of efficiency. In the discipline of bioinformatics, in addition to neural networks, they also use linear regression, logistic regression, and k-nearest neighbor analysis for computational effectiveness in areas such as pharmacology, immunology, other biological areas, etc. [12]. Now, computer approaches can screen more than 100,000 influenza viruses using various techniques to reach a reasonable conclusion [13].

4.2 Drug Purposing

Medication purposing, also known as medication repositioning, reprofiling, redirecting, and drug rediscovery, is a tactic for finding new therapeutic uses for medicines that go beyond their initial therapeutic usage in medical indications. Compared to the de novo development of wholly new medications, drug purposing has several benefits, including the potential to hasten the decrease in failure rates and discovery process during the clinical development and testing phases [14]. Particularly when it comes to drugs that have been repurposed, safety evaluation in preclinical models and humans can be avoided if safety testing for the original indication has already been finished and the drug shows dose compatibility with the new indication. This could lead to lower overall development costs [15].

Drug research is now more focused on bioinformatics tools than on conventional procedures due to the progress of artificial intelligence [16]. The use of computer-aided drug design approaches has shown to be quite beneficial in handling large volumes of biological data and creating effective algorithms [17]. Artificial intelligence allows more effective approaches to complicated problems linked with drug research and development through the use of machine learning. Technologies built on artificial intelligence help the pharmaceutical sector find more effective medications as in Fig. 4.1 [18].

4.3 Impact of AI and ML on Drug Development

Machine learning and artificial intelligence have advanced significantly in recent years. It has greatly raised the quality of life while decreasing human burden [19]. To improve the efficiency and accuracy of drug research and development, artificial intelligence and machine learning are used [20]. Because of the ability to conduct

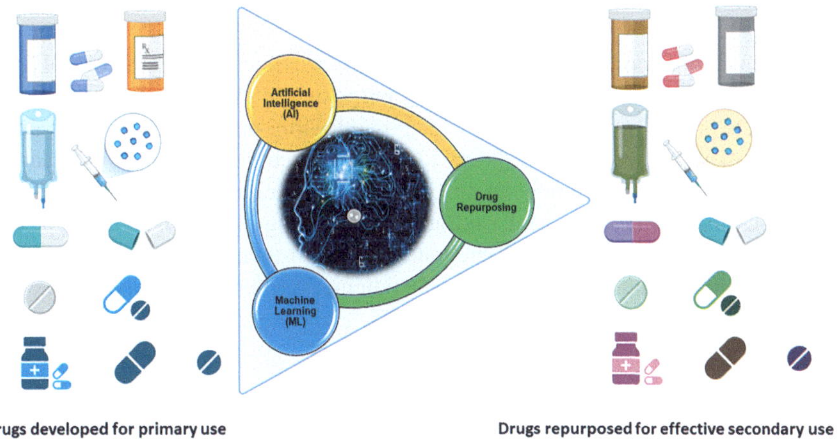

Fig. 4.1 AI & ML's role in accelerating drug discovery and purposing

simulations with these technologies, a body of work has been done to support the roles of machine learning and artificial intelligence in facilitating drug development and discovery processes [21]. This has included making these processes more cost-effective or completely eliminating the need for clinical trials. Additionally, they made it possible for scientists to thoroughly examine various chemicals without the need for trials [22]. The findings of this study show how commonplace machine learning and artificial intelligence techniques are in the field of drug discovery and point to a bright future for these technologies. As a result, scholars, students, and the pharmaceutical industry will be able to learn more about machine learning and artificial intelligence in the context of drug development and discovery [23].

4.4 AI in Drug Purposing

Medication purposing is the process of finding new therapeutic applications for medications that are currently on the market but have already received approval for a different condition or indication. Rather than starting from zero when creating new medications, scientists investigate how well-known medications might work for various ailments by taking use of their well-established pharmacokinetics, mechanisms of action, and safety profiles. This can drastically lower the risk, expense, and time involved in introducing a novel medication to the market. Additionally, purposing can reveal different ways to treat illnesses for which there are unmet medical requirements [24]. The ability of artificial intelligence (AI) to scan vast amounts of data and find new links and patterns has made it more and more valuable in the field of drug purposing. Three distinct approaches network-based, feature-based, and matrix-based are used by AI to assist in drug purposing [25].

4.4.1 Network-Oriented Approach

In order to identify connections between medications, illnesses, and biological pathways, the network-based approach to drug purposing by AI analyzes intricate biological networks. This method looks for possible drug purposing prospects using computer methods and network theory [26]. The detection of drug-disease connections that might not be immediately obvious using traditional methodologies is made possible by the network-based approach in drug purposing. This approach builds biological networks first, integrates data into them, analyzes the network once it is in place, and then identifies the medicine, verifies it, and conducts a clinical study [27].

4.4.2 Feature-Based Approach

In order to find possible purposing opportunities, the feature-based approach to drug purposing using AI analyzes and compares the traits or characteristics of pharmaceuticals and disorders. This method is centered on taking pertinent traits and applying them to various datasets in order to forecast the relationships between drugs and diseases. This approach involves first extracting features, integrating them into datasets, and then using machine learning and predictive modeling to help anticipate repositioned drugs, which are subsequently verified and clinically evaluated. The feature-based strategy makes use of AI methods to examine and take use of the innate qualities and traits of medications and illnesses [28]. This strategy allows for the identification of possible purposing possibilities based on the common traits or biological signatures of medications and diseases by taking into account pertinent aspects from various datasets [29].

4.4.3 Matrix-Based Approach

In order to uncover possible medication-disease connections, the matrix-based method to drug purposing by AI uses matrices to describe and analyze a variety of biological data. In order to extract patterns and relationships from the data, this method makes use of matrix operations and computational techniques. Potential drug-disease connections can be found by integrating and analyzing a variety of biological data using a matrix-based technique [30]. This strategy helps to uncover hidden patterns and associations from the data by using matrix operations and AI algorithms, which makes it possible to find attractive drug purposing opportunities.

4.5 Companies Employing AI for Drug Purposing

Many companies are starting to use these developments to accelerate the delivery of COVID-19 treatments and gain more insight into how the immune system combats the disease. GlaxoSmithKline (GSK) and Vir Biotechnology, two pharmaceutical companies, teamed together in April to use CRISPR and AI to expedite the development of coronavirus medicines. Furthermore, Google and Apple worked together to develop a Bluetooth-enabled contact-following app [31]. These methods might be effective in gathering a large amount of accurate data. The two best-known instances of this regarding COVID-19 to date are remdesivir (for treating Ebola) and hydroxychloroquine (recommended for treating malaria). Therefore, the effectiveness data set for these medications would be a good input for an AI model. The following is a list of companies employing AI to repurpose currently accessible COVID-19 medications [32].

BenevolentAI: The UK-based business is considered a titan in the study of drugs and artificial intelligence. The business has been purposing all its data using

AI-approved medications currently on the market against the new coronavirus. After a month, they reduced the number of options to six who showed the most promise. The most promising one was baricitinib, a medication used to treat rheumatoid joint discomfort [33].

- Innoplexus: The Indo-German organization analyzed patient data to assess the efficiency of medications like hydroxychloroquine and remdesivir against 2019-nCoV. Their AI stage suggested that a combination of chloroquine and tocilizumab (a drug used to treat arthritic joint pain), a third combination of hydroxychloroquine and remdesivir, as well as chloroquine and either plerixafor (an antiretroviral) or clarithromycin (an antibiotic) would have better viability [34].
- Deargen: The Korean organization, working with Dankook University, predicted how readily available antiviral drugs would react with Atazanavir, a medication used to treat HIV [35].
- Gero: Using its AI platform, the Singaporean company recognized nine drugs. Niclosamide and nitazoxanide, which contain broad-spectrum anti-parasitic and antiviral properties [36].
- Cyclica: On their AI-based medication purposing MatchMaker, a Canadian organization reported evaluating 6700 FDA-verified atoms or, at the very least, in phase I human preliminary tests. For in vitro and in vivo testing, they have teamed up with China's Institute of MateriaMedica [37].
- Healx: The UK-based company is using its foundation to offer licensed anti-infection drug bi- and tri-mixes. Despite focusing on rare illnesses, the organization is well-positioned to apply the understanding of why mortality rises when respiratory and cardiovascular system comorbidities occur [38].
- VantAI: The organization in New York employs a framework science approach to understand the interactions between over 500 human proteins and over 500 virus proteins throughout an illness [39].

4.6 ML in Drug Purposing

When it comes to computational drug purposing, we work with data from clinical trial reports, drug adverse event reports, gene expression, protein networks, and electronic health records. However, these high-dimensional, noisy, and complicated data provide significant difficulties for the development of computational methods. Using traditional statistical methods, finding a drug's molecular target among the many genomic, phenomic, and chemical databases is impossible. These methods can overcome the shortcomings of traditional statistical methods and their erroneous interpretations more effectively in the creation of such drug purposing techniques. Databases have facilitated the quick creation of a number of cutting-edge machine learning applications that use this data to repurpose drugs. The approaches are broken down into three main groups based on the fundamental machine learning methodology, and each category is covered in more detail in the subsequent subsections [40].

4.6.1 Networks-Based Approach

Because they have the ability to combine data from several sources, these are essential for the purposing of medications [41]. Network-based cluster techniques and network-based propagation approaches are the two general categories into which these approaches fall. The earlier methods aim to identify different sub-networks using cluster algorithms and concentrate on inexperienced drug–disease/target interactions. From the source node, preliminary data is transmitted to every network node and selected few sub-network nodes. These tactics are divided into two categories: local approaches and global approaches, based on the methods of dissemination [42].

Local propagation techniques only utilize a small amount of network information, and they can fail to produce precise forecasts [43]. The focus of a lot of current research is on global strategies for achieving exceptional performance. ProspectTR, PRINCE, and other network-based propagation techniques are examples [44]. Additional classification of the networks can be done into homogeneous and heterogeneous classes. Homogeneous networks use data from a single source to map out a disease route and identify therapeutic targets linked to several pathways [45]. Heterogeneous networks are typically integrated with different biological elements which help to improve the efficiency of previous models and provide tools for designing more stable and effective methods [46].

Cluster algorithms are used to find interesting components in order to identify biological modules. However, this approach has a flaw in that there are no criteria in place to evaluate the association between biological modules [47].

4.6.2 Text Mining-Oriented Approach

Text mining is a computer-assisted technique that has been widely used to identify novel, valuable biological entity correlations from a vast amount of biological and medical literature. It does this by using natural language processing to extract insights from unstructured text [48].

Appropriate information is culled from the literature during the information retrieval procedure [49]. The literature has certain concepts that are pointless, which filters the information. The next step in the biological name entity identification method is to map all pertinent words and phrases to a predefined set of categories in order to find concepts of biological interest in the text [50]. Afterwards, the pertinent data is extracted to find knowledge about biological concepts and create a knowledge graph during the biological information extraction and biological knowledge discovery processes [51].

4.6.3 Approaches Based on Semantics

Recently, semantic-based methods have been used for drug purposing. To build the semantic network, biological entity interactions are first retrieved from preliminary data in large medical databases, including chemical, pharmaceutical, biological, genomics, and biomedical databases [52].

The most significant challenge in creating a semantic network is merging data from multiple sources [53]. Building semantic networks that transport large amounts of medical data is therefore imperative. There have been several documented semantics-based strategies for medication purposing [54].

4.7 Prospective Difficulties and Potential Remedies

The pharmaceutical business currently faces two main obstacles when creating new drugs: rising costs and declining efficiency [55]. But there are a lot of chances to lower this cost, boost productivity, and save time throughout the drug research and development process thanks to ML techniques and recent advances in DL [56, 57]. The third wave of AI is being hinted at by developments in architectural hardware, big data accessibility, and AI algorithms, particularly in deep learning approaches [58]. Researchers' interest in AI methods to medication development has grown to the point that several pharmaceutical companies are working with AI companies [59]. In addition, the number of companies operating in this industry has increased, with 230 predicted by June 2020. Furthermore, DL techniques overcome the limitation of AI and ML approaches by integrating data at various levels through nonlinear models [60, 61]. However, the DL method is beneficial since it offers excellent accuracy and precision when data is integrated at numerous levels. Furthermore, DL offers a far more flexible architecture to build a neural network for a particular problem than AI and ML methods [62]. Artificial intelligence (AI) has surpassed humans in performance these days, making applications such as natural language processing, image and voice recognition easily achievable [63]. Therefore, it should come as no surprise that AI has a lot to offer in the drug discovery process [64]. AI is now utilized in drug discovery to optimize leads, find targets and hits, forecast ADMET, and plan clinical trials. Even with tremendous achievement, there are still many obstacles to overcome, such as obtaining high-quality data, which raises two important issues [65, 66]. First off, due to the complexity of drug action in biological systems, labeling cannot be binary; secondly, the quantity of data available for drug discovery pales in comparison to the vast amount of information already available [67]. As a result, a community that offers both quantity and quality of data is needed. Open data sharing is uncommon in the pharmaceutical sector; thus, the Pistoia alliance spearheaded a movement that has persuaded other businesses to share their data [68]. In addition, they want to create a standard data format, which is difficult to do technically [69]. Creating an algorithm that can handle sparse data is one way to approach this issue; Stanford University has created one such algorithm called "one-shot learning," which makes predictions about drugs' qualities based on heterogeneous data [70]. Furthermore, one can train an existing machine learning model by adjusting a large number of hyperparameters and optimizing it for good results, rather than developing new ones based on the accuracy and uncertainty of the experimental data [71]. Some studies have suggested that some reasonable parameters can be used to start the optimization process. Since molecular representation is one of the key elements in model construction, it presents additional difficulties [72]. Only a small number of newly created models improve the molecular representation to a standard by learning task-related properties from the raw data. Drug purposing was hitherto solely dependent on clinical findings [73, 74]. Nonetheless, the screening procedure might be enhanced by utilizing the vast quantity of data that is now available, which includes clinical trial outcomes, patents, and scientific publications [75]. Moreover, DL-based VS (virtual screening) can fully utilize the data and minimize false-positive rates that result from an imbalance between positive and negative data [76]. To design an effective medication with good ADMET characteristics and target activities, lead optimization is also a difficulty; yet, these criteria are separate and sometimes contradict one other. This issue can be resolved by fine-tuning each parameter independently and enhancing the model even further. It is difficult for pharmaceutical companies to get enough people for clinical studies. AI techniques will support patient recruitment and identification, as well as data management [53]. The main obstacle to neurodegenerative disease drug discovery is their unclear pathophysiology, which makes medication identification even more difficult. Another problem with machine learning models is that they are "black boxes," meaning that not even the most knowledgeable people can fully understand the biological process underlying a result or explain how the model arrived at it [77]. Furthermore, non-professionals are helpless since they cannot determine which machine learning model to use to address their problem due to the growing number of models and their claim to be the newest. Therefore, it would be preferable if users and developers agreed upon a standard objective evaluation before assessing the model's performance. It is also crucial to remember that the majority of nations do not grant patents for inventions made only with artificial intelligence. Additionally, in order to get patent rights, businesses that employ AI technology for medication research must go through a rigorous copyright process. Security is a big worry as well because AI-driven personalized treatment would need a person's genetic code, which will require personal data. In the end, processing large amounts of data would necessitate speedier calculation. It is predicted that quantum computers or other technologies may eventually replace today's supercomputers, completing tasks in minutes as opposed to hours. There hasn't been a single success story where a molecule created by AI made it to market for general usage, despite the fact that AI has provided numerous new targets and compounds for various ailments [78]. Recently, using AI-based technologies, a novel target and its innovative inhibitor have been proposed for the first time ever. Using AI-based technologies, the biotechnology company in silico medicine created a novel inhibitor from scratch and presented a novel target related to idiopathic pulmonary fibrosis. Both human cells and animal models have demonstrated the good efficiency of the discovered small molecule inhibitor. By early 2022, in silico hopes to have clinical trials underway after nominating their small molecule inhibitor for investigational new drug (IND) allowing investigations in December 2020. In the event that the trials are successful, this will be the first time ever that an approved new target and its inhibitor were suggested using AI-based technologies [79]. There is little doubt that in the near future, artificial intelligence (AI) will bring about revolutionary improvements in the drug research and development process, even though there are some inevitable impediments, and a great deal of effort needs to be done to incorporate AI tools in the drug discovery cycle.

4.8 Summary

As the field advances, scientists are searching for new drugs, cures, or therapies that work better than those that are already on the market. Understanding the fundamental mechanisms causing illness progression, the impact of currently developed drugs, and patients' genetic codes can help develop novel, precise medication therapies that will eventually improve their well-being and standard of living. It has been discovered that traditional methods for drug development are time-consuming and provide more false adverse outcomes. One can detect the medicine faster when employing computational methods that use AI. Numerous studies have demonstrated the effects of medications on various targets, and many of them are now being discovered. Purposing of pharmaceuticals offers novel molecular targets with characteristics that make them suitable for further exploitation and for the bioinformatics field's use of AI-based techniques to predict novel pharmacological compounds and to identify the targeted drug or protein. It is anticipated that bioinformatics will become increasingly important in the future to save time and money while analyzing data utilizing AI and ML techniques. It will also speed up developments in biological sciences, robotic surgery, and biomedicine.

References

1. Ezziane Z. Applications of artificial intelligence in bioinformatics: a review. Expert Syst Appl. 2006;30(1):2–10.
2. Narayanan A, Keedwell EC, Olsson B. Artificial intelligence techniques for bioinformatics. Appl Bioinforma. 2002;1:191–222.
3. Nicolas J. Artificial intelligence and bioinformatics. In: A guided tour of artificial intelligence research. Cham: Springer; 2020. p. 209–64.
4. Bülow RD, et al. How will artificial intelligence and bioinformatics change our understanding of IgA in the next decade? Semin Immunopathol. 2021;43:739–52.
5. Mitra S. Computational intelligence in bioinformatics. In: Transactions on rough sets III. Cham: Springer; 2005. p. 134–52.
6. Das S, Abraham A, Konar A. Swarm intelligence algorithms in bioinformatics. In: Computational intelligence in bioinformatics. Heidelberg: Springer; 2008. p. 113–47.
7. Kumar R, et al. Prediction of drug-plasma protein binding using artificial intelligence-based algorithms. Comb Chem High Throughput Screen. 2018;21(1):57–64.

8. Mellit A, Kalogirou SA. Artificial intelligence techniques for photovoltaic applications: a review. Prog Energy Combust Sci. 2008;34(5):574–632.
9. Chen SH, Jakeman AJ, Norton JP. Artificial intelligence techniques: an introduction to their use for modeling environmental systems. Math Comput Simul. 2008;78(2-3):379–400.
10. Laughton M. Artificial intelligence techniques in power systems. London: Institution of Electrical Engineers; 1997.
11. Fethi MD, Pasiouras F. Assessing bank efficiency and performance with operational research and artificial intelligence techniques: a survey. Eur J Oper Res. 2010;204(2):189–98.
12. Mellit A, et al. Artificial intelligence techniques for sizing photovoltaic systems: a review. Renew Sust Energ Rev. 2009;13(2):406–19.
13. Dilek S, Çakır H, Aydın M. Applications of artificial intelligence techniques to combating cyber crimes: a review. arXiv 1502.03552. 2015.
14. Elbadawi M, Gaisford S, Basit AW. Advanced machine-learning techniques in drug discovery. Drug Discov Today. 2021;26(3):769–77.
15. Vasant P. Handbook of research on artificial intelligence techniques and algorithms. Hershey: IGI Global; 2014.
16. McLaren BM, Scheuer O, Mikšátko J. Supporting collaborative learning and discussions using artificial intelligence techniques. Int J Artif Intell Educ. 2010;20(1):1–46.
17. Nawari N, Liang R, Nusairat J. Artificial intelligence techniques for the design and analysis of deep foundations. Electron J Geotech Eng. 1999;4:1–21.
18. Kohli A, et al. Concepts in US Food and Drug Administration regulation of artificial intelligence for medical imaging. Am J Roentgenol. 2019;213(4):886–8.
19. Shi F, et al. Review of artificial intelligence techniques in imaging data acquisition, segmentation and diagnosis for COVID-19. IEEE Rev Biomed Eng. 2021;14:4–15.
20. Rahmanifard H, Plaksina T. Application of artificial intelligence techniques in the petroleum industry: a review. Artif Intell Rev. 2019;52(4):2295–318.
21. Zhavoronkov A. Medicinal chemists versus machines challenge: what will it take to adopt and advance artificial intelligence for drug discovery? J Chem Inf Model. 2020;60(6):2657–9.
22. Sfetsos A, Coonick A. Univariate and multivariate forecasting of hourly solar radiation with artificial intelligence techniques. Sol Energy. 2000;68(2):169–78.
23. Zhou Y, et al. Artificial intelligence in COVID-19 drug purposing. Lancet Digit Health. 2021;3(1):e1.
24. Fleming N. How artificial intelligence is changing drug discovery. Nature. 2018;557(7706):S55.
25. Lamberti MJ, et al. A study on the application and use of artificial intelligence to support drug development. Clin Ther. 2019;41(8):1414–26.
26. Martinez-Romero M, et al. Artificial intelligence techniques for colorectal cancer drug metabolism: ontologies and complex networks. Curr Drug Metab. 2010;11(4):347–68.
27. Yang X, et al. Concepts of artificial intelligence for computer-assisted drug discovery. Chem Rev. 2019;119(18):10520–94.
28. Jiménez-Luna J, Grisoni F, Schneider G. Drug discovery with explainable artificial intelligence. Nat Mach Intell. 2020;2(10):573–84.
29. Díaz Ó, Dalton JA, Giraldo J. Artificial intelligence: a novel approach for drug discovery. Trends Pharmacol Sci. 2019;40(8):550–1.
30. Zhu H. Big data and artificial intelligence modeling for drug discovery. Annu Rev Pharmacol Toxicol. 2020;60:573–89.
31. Mohanty S, et al. Application of artificial intelligence in COVID-19 drug repurposing. Diabetes Metab Syndr. 2020;14(5):1027–31.
32. Barrett S, Langdon W. Advances in the application of machine learning techniques in drug discovery, design, and development. In: Applications of soft computing. Berlin: Springer; 2006. p. 99–110.
33. Zhavoronkov A. Artificial intelligence for drug discovery, biomarker development, and generation of novel chemistry. Mol Pharm. 2018;15(10):4311–3.
34. Zhong F, et al. Artificial intelligence in drug design. Sci China Life Sci. 2018;61(10):1191–204.

35. Paul D, et al. Artificial intelligence in drug discovery and development. Drug Discov Today. 2021;26(1):80–93.
36. Stephenson N, et al. Survey of machine learning techniques in drug discovery. Curr Drug Metab. 2019;20(3):185–93.
37. Torres-Carrión P, et al. Application of techniques based on artificial intelligence for predicting the consumption of drugs and substances. A systematic mapping review. In: Applied technologies. Cham: Springer; 2019. p. 39–52.
38. Chan HS, et al. Advancing drug discovery via artificial intelligence. Trends Pharmacol Sci. 2019;40(8):592–604.
39. Zhavoronkov A, Vanhaelen Q, Oprea TI. Will artificial intelligence for drug discovery impact clinical pharmacology? Clin Pharmacol Ther. 2020;107(4):780–5.
40. Weinstein JN, et al. Predictive statistics and artificial intelligence in the US National Cancer Institute's drug discovery program for cancer and AIDS. Stem Cells. 1994;12(1):13–22.
41. Kumar R, et al. Prediction of human intestinal absorption of compounds using artificial intelligence techniques. Curr Drug Discov Technol. 2017;14(4):244–54.
42. Koromina M, Pandi M-T, Patrinos GP. Rethinking drug repositioning and development with artificial intelligence, machine learning, and omics. Omics. 2019;23(11):539–48.
43. Bajorath JR, et al. Artificial intelligence in drug discovery: into the great wide open. J Med Chem. 2020;63(16):8651–2.
44. Levin JM, et al. Artificial intelligence, drug purposing, and peer review. Nat Biotechnol. 2020;38(10):1127–31.
45. Álvarez-Machancoses Ó, Fernández-Martínez JL. Using artificial intelligence methods to speed up drug discovery. Expert Opin Drug Discovery. 2019;14(8):769–77.
46. Wang L, et al. Artificial intelligence facilitates drug design in the big data era. Chemom Intell Lab Syst. 2019;194:103850.
47. Green CP, Engkvist O, Pairaudeau G. The convergence of artificial intelligence and chemistry for improved drug discovery. Future Med Chem. 2018;10(22):2573–6.
48. Bender A, Cortes-Ciriano I. Artificial intelligence in drug discovery: what is realistic, what are illusions? Part 2: a discussion of chemical and biological data used for AI in drug discovery. Drug Discov Today. 2021;26(4):1040–52.
49. Gertrudes J, et al. Machine learning techniques and drug design. Curr Med Chem. 2012;19(25):4289–97.
50. Lusci A, Pollastri G, Baldi P. Deep architectures and deep learning in chemoinformatics: the prediction of aqueous solubility for drug-like molecules. J Chem Inform Model. 2013;53(7):1563–75.
51. Duvenaud D, Maclaurin D, Aguilera-Iparraguirre J, Gómez-Bombarelli R, Hirzel T, Aspuru-Guzik A, Adams RP. Convolutional networks on graphs for learning molecular fingerprints. arXiv 1509.09292. 2015.
52. Gómez-Bombarelli R, Wei JN, Duvenaud D, Hernández-Lobato JM, Sánchez-Lengeling B, Sheberla D, Aguilera-Iparraguirre J, Hirzel TD, Adams RP, Aspuru-Guzik A. Automatic chemical design using a data-driven continuous representation of molecules. ACS Central Sci. 2018;4(2):268–76.
53. Weng H, Liu Z, Yan S, Fan M, Ou A, Chen D, Hao TA. Framework for automated knowledge graph construction towards traditional Chinese medicine. In: Health information science. Cham: Springer; 2017. p. 170–81. https://doi.org/10.1007/978-3-319-69182-4_18.
54. Jin W, Yang K, Barzilay R, Jaakkola T. Learning multimodal graph-to-graph translation for molecular optimization. arXiv 1812.01070. 2018.
55. Shi C, Xu M, Zhu Z, Zhang W, Zhang M, Tang J. GraphAF: a flow-based autoregressive model for molecular graph generation. arXiv 2001:09382. 2020.
56. Jin W, Barzilay R, Jaakkola T. Multi-objective molecule generation using interpretable substructures. Proc Mach Learn Res. 2020;119:4849–59.
57. Zang C, Wang F. MoFlow: an invertible flow model for generating molecular graphs. In: Proceedings of the 26th ACM SIGKDD international conference on knowledge discovery and data mining; 2020. p 617–626. https://doi.org/10.1145/3394486.3403104.

58. Bung N, Krishnan SR, Bulusu G, Roy A. De novo design of new chemical entities for sars-cov-2 using artificial intelligence. Fut Med Chem. 2021;13(06):575–85.
59. Zhavoronkov A, Zagribelnyy B, Zhebrak A, Aladinskiy V, Terentiev V, Vanhaelen Q, Bezrukov DS, Polykovskiy D, Shayakhmetov R, Filimonov A, Bishop M. Potential non-covalent SARS-CoV-2 3C-like protease inhibitors designed using generative deep learning approaches and reviewed by human medicinal chemist in virtual reality. ChemRxiv. 2020; https://doi.org/10.26434/chemrxiv.12301457.v1.
60. Cai Y, Zeng M, Chen YZ. The pharmacological mechanism of huashi baidu formula for the treatment of COVID-19 by combined network pharmacology and molecular docking. Ann Palliat Med. 2021;10(4):3864–95. https://doi.org/10.21037/apm-20-1759.
61. Xia QD, Xun Y, Lu JL, Lu YC, Yang YY, Zhou P, Hu J, Li C, Wang SG. Network pharmacology and molecular docking analyses on the Lianhua Qingwen capsule indicate that akt1 is a potential target to treat and prevent covid-19. Cell Proliferat. 2020;53(12):e12949.
62. Ren X, Shao XX, Li XX, Jia XH, Song T, Zhou WY, Wang P, Li Y, Wang XL, Cui QH, et al. Identifying potential treatments of covid-19 from Traditional Chinese Medicine (TCM) by using a data-driven approach. J Ethnopharmacol. 2020;258:112932.
63. Yan H, Zou C. Mechanism and material basis of Lianhua Qingwen capsule for improving clinical cure rate of covid-19: a study based on network pharmacology and molecular docking technology. J South Med Univ. 2021;41(1):20–30.
64. Guney E, Menche J, Vidal M, Barábasi AL. Network-based in silico drug efficacy screening. Nat Commun. 2016;7(1):1–13.
65. Wu Z, Wang Y, Chen L. Network-based drug repositioning. Mol BioSyst. 2013;9(6):1268–81.
66. Zeng X, Song X, Ma T, Pan X, Zhou Y, Hou Y, Zhang Z, Li K, Karypis G, Cheng F. Repurpose open data to discover therapeutics for covid-19 using deep learning. J Proteome Res. 2020;19(11):4624–36. https://doi.org/10.1021/acs.jproteome.0c00316.
67. Beck BR, Shin B, Choi Y, Park S, Kang K. Predicting commercially available antiviral drugs that may act on the novel coronavirus (sars-cov-2) through a drug-target interaction deep learning model. Comput Struct Biotechnol J. 2020;18:784–90.
68. Kim E, Choi AS, Nam H. Drug repositioning of herbal compounds via a machine-learning approach. BMC Bioinform. 2019;20(10):33–43.
69. Hooshmand SA, Ghobadi MZ, Hooshmand SE, Jamalkandi SA, Alavi SM, Masoudi-Nejad AA. A multimodal deep learning-based drug purposing approach for the treatment of covid-19. Mol Divers. 2020;25(3):1717–30. https://doi.org/10.1007/s11030-020-10144-9.
70. Huang K, Fu T, Xiao C, Glass L, Sun J. Deeppurpose: a deep learning based drug purposing toolkit. arXiv 2004.08919. 2020.
71. Belyaeva A, Cammarata L, Radhakrishnan A, Squires C, Dai Yang K, Shivashankar G, Uhler C. Causal network models of sars-cov-2 expression and aging to identify candidates for drug purposing. Nat Commun. 2021;12(1):1–13.
72. Liu R, Wei L, Zhang P. A deep learning framework for drug purposing via emulating clinical trials on real-world patient data. Nat Mach Intell. 2021;3(1):68–75.
73. Capodice JL, Chubak BM. Traditional Chinese herbal medicine-potential therapeutic application for the treatment of COVID-19. Chin Med. 2021;16(1):1–6.
74. Wang Z, Li L, Song M, Yan J, Shi J, Yao Y. Evaluating the traditional Chinese medicine (TCM) officially recommended in China for covid-19 using ontology-based side-effect prediction framework (OSPF) and deep learning. J Ethnopharmacol. 2021;272:113957.
75. Liao H, Wen G, Hu Y, Wang C. Convolutional herbal prescription building method from multi-scale facial features. Multimed Tools Appl. 2019;78(24):35665–88.
76. Guo F, Tang X, Zhang W, Wei J, Tang S, Wu H, Yang H. Exploration of the mechanism of traditional Chinese medicine by AI approach using unsupervised machine learning for cellular functional similarity of compounds in heterogeneous networks, XiaoErFuPi granules as an example. Pharmacol Res. 2020;160:105077.
77. Ruan C, Ma J, Wang Y, Zhang Y, Yang Y, Kraus S. Discovering regularities from traditional Chinese medicine prescriptions via bipartite embedding model. IJCAI. 2019;464:3346–52. https://doi.org/10.24963/ijcai.2019/464.

78. Wang Y, Jafari M, Tang Y, Tang J. Predicting meridian in Chinese traditional medicine using machine learning approaches. PLoS Comput Biol. 2019;15(11):e1007249.
79. Liu Z, Zheng Z, Guo X, Qi L, Gui J, Fu D, Yao Q, Jin L. Attentiveherb: a novel method for traditional medicine prescription generation. IEEE Access. 2019;7:139069–85.

Drug Repurposing in Cancer Therapy

5

Srijita Sen, Hemant Kushwah, and Om Prakash Ranjan

Abbreviations

CDK1	Cyclin-dependent kinase 1
CML	Chronic myeloid leukemia
CNS	Central nervous system
DDS	Drug delivery systems
DNA	Deoxyribonucleic acid
EGFR	Epidermal growth factor receptor
ELISA	Enzyme-linked immunosorbent assay
FDA	Food and Drug Administration
FITC	Fluorescein isothiocyanate
HSV	Herpes simplex virus
MCC	Multiple correlation clustering
MDR	Multidrug resistance
MOR	μ-opioid receptor
mTOR	Mammalian target of rapamycin
NFAT	Nuclear factor of activated T cells
NLC	Nanostructured lipid carriers
NSAIDs	Non-steroidal anti-inflammatory drugs
PCNA	Proliferating cell nuclear antigen
PLGA	Poly (lactic-co-glycolic acid)
R&D	Research and Development
SLN	Solid lipid nanoparticles
TKIs	Tyrosine kinase inhibitors

S. Sen · H. Kushwah · O. P. Ranjan (✉)
Department of Pharmaceutical Technology (Formulations), National Institute of Pharmaceutical Education and Research (NIPER) Guwahati, Guwahati, Assam, India
e-mail: omprakash@niperguwahati.in

N. Chella et al. (eds.), *Drug Repurposing*,
https://doi.org/10.1007/978-981-97-5016-0_5

TNBC	Triple-negative breast cancer
TNF	Tumor necrosis factor
TNIK	Traf2 and Nck-interacting protein kinase
TSC1	Tuberous sclerosis complex 1
VDAC	Voltage-dependent anion-selective channel
VEGF	Vascular endothelial growth factor
WHO	World Health Organization

Learning Objectives

- To identify the rationale behind the drug repurposing in cancer
- Summarize different category of drugs repurposed in different types of cancer
- Different in silico strategies adopted in the advancement of drug repurposing in cancer
- To understand the application of nanotechnology in the clinical advancement of repurposed drug(s) in cancer

5.1 Introduction

As per World Health Organization (WHO) reports, the number of deaths from cancer is about ten million worldwide, which is one in six persons. The concern is related to the precise approach to diagnosis and treatment. However, the major barrier is the higher cost of cancer treatment because the disease urges the development of novel strategies for molecular dissection at gene and protein levels to find appropriate targets and corresponding drugs/biological molecules. This ultimately demands an investment of a substantial amount of money, involvement of high-end techniques by pharmaceutical companies, and, most importantly, longer time duration. There are continuous efforts to shift the gears of cancer drug approval by adjusting the regulatory timelines, and there are some accelerated approvals. However, clinical data in the late-phase trials has often shown alarming survival conditions or toxic effects, although having very promising antitumor effects in the initial stage. As an appropriate exercise, a few anticancer drugs, after approval, have been withdrawn from the market on account of safety concerns. Sometimes, these got re-approval based on the additional data. This demands a more in-depth study with dose exploration for longer time duration and for the best possible acceptance by the regulatory authority in the new drug approval front. Therefore, an emphasis on the repurposing of old drugs in cancer has been given due to known toxicological profile. This will make the drug discovery and development pipeline somewhat faster by prioritizing the already approved drugs. This is more impactful in the case of some cancers where only very limited chemotherapeutic options are available. Despite considerable breakthroughs in a few cancer therapies, including breast cancer, the mortality rate is high because of chemoresistance to conventional therapeutics and formulations. The risk of quick resistance increased with enhanced dose requirement because of delayed detection at the late stage of disease progression, especially in developing countries. At higher doses and even in the required therapeutic dose, many approved anticancer drugs exhibit notable side effects by causing DNA damage. This challenge can be overcome by examining old drugs for repurposing [1]. Even the repurposed drug at non-cytotoxic concentration can act as a chemotherapy sensitizer. A good example is chloroquine, which inhibits autophagy-based survival in lung cancer cells (A549) and increases the apoptotic effect of topotecan [1]. In a molecular study, it has been revealed that exposure to amodiaquine leads to significant modulation of gene expression of a broad array of genes, which antagonizes the cell cycle progression and induces genotoxic, proteotoxic, and cytotoxic stress. The undermined autophagic lysosomal function in the malignant melanoma cells described the potentiality of its repurposing as a single or adjuvant chemotherapeutic agent [2]. Some drugs are extraordinary candidates to be repurposed because of their potent therapeutic effects in multiple cancers. Atovaquone is one such wonder molecule that has been found beneficial as mono or combination therapy in brain, lung, pancreas cancer, acute myeloid leukemia, etc., and even a few patents are there related to its repurposing [3]. Statin is another group of drugs (lipid-lowering agents) that have described antitumor effects in different cancers through mevalonate-dependent and independent pathways, including ovarian, prostate, endometrial, breast, and colorectal cancer [4]. Several studies have revealed the potentiality of vaccines available for infectious diseases that exert good antitumor effects after systemic or intra-tumor administration. Even some vaccines got the ticket for clinical trials to ascertain their safety, effectiveness, and immunological effects [5]. Opportunely, scientists are working in the same pipeline with multi-omics analysis and machine learning approaches, which can provide more precise results with lower financial expenditure [6]. Recent data are providing hope of getting personalized therapy based on repurposed drugs in cancer that can address genetic heterogenicity in the upcoming era. Here, the drug repurposing approach is a golden opportunity to decrease the pharmaceutical R&D challenges associated with screening and productivity, rising costs of new drugs, and overcoming the challenge of existing therapies and associated poor patient response rates. Several studies that have powerful conclusions for drug repurposing in cancer have been mentioned in the different sections of this book chapter.

5.2 Drug Repurposing in Different Cancers

5.2.1 Brain Cancer

Medulloblastoma is the most diagnosed pediatric malignant brain tumor. Drugs from several categories have been shown to have inherent repurposing ability in medulloblastoma. These are lipid-lowering agents (lovastatin, simvastatin, atorvastatin, fenofibrate) [7–9], cardiac glycosides (digoxin, proscillaridin, ouabain) [10, 11], drugs for cardiovascular diseases (verapamil) [12], anthelmintic drugs (nifurtimox, mebendazole) [13, 14], antimicrobial drugs (itraconazole) [15], antiviral agents (valganciclovir, abacavir) [16, 17], NSAIDs (celecoxib, diclofenac, tolfenamic acid, flurbiprofen) [18, 19], drugs for neurological disorders (valproic acid, sertraline, chlorpromazine) [20, 21], dermatological drugs (13-cis-retinoic acid, all-trans-retinoic acid) [22, 23], antidiabetic agents (phenformin) [24], anti-rheumatic drugs (sulfasalazine) [25], anti-alcoholic drugs (disulfiram) [26], etc. Notably, most of these drugs can cross the blood–brain barrier [27]. Among different brain tumors, glioblastoma is the most lethal form and involves surgical resection followed by radiation therapy and chemotherapy in most cases. However, there is no improvement in the quality of a patient's life after treatment. Glioblastoma patients (~20–60%) often suffer from epileptic seizures. So, antiepileptic drugs like valproate (mostly studied), levetiracetam, carbamazepine, and phenytoin have been found to elicit relief from seizure episodes. It is unfolded that abnormally activated histone deacetylases stimulate neoplasia. The histone deacetylase inhibitors (HDACi) of this enzyme can promote growth arrest by suppressing gene transcription and can induce cellular apoptosis. Valproic acid comes under this category and is known as an antineoplastic agent. Newer HDACi, i.e., topiramate, is an antiseizure drug, and it can reduce tumor angiogenesis (reducing VEGF expression) as well. Existing research also reported the antineoplastic activities of some antihypertensive agents, including losartan [28]. In the case of brain cancer metastasis, the cells undergo metabolic changes along with stress-induced autophagy regulation (either can help the cells survive in stress or allow programmed cell death). Therefore, therapeutic regulation of autophagy is a promising aptitude. In this context, antidepressants with reported anticancer activities demonstrated favorable results in glioma or glioblastoma either by inducing or inhibiting autophagy. New-generation antidepressants have yet to be explored for antineoplastic activity. Moreover, there is a need to evaluate the interactive effects of these drugs with the existing chemotherapeutic agents [29]. This research expansion will provide an effective manifesto to combat chemoresistance/radioresistance in brain cancer. Chloroquine (an antimalarial drug) inhibits autophagy and subsequently hinders the degradative function of autolysosome. This, in turn, sensitizes the glioblastoma cells to temozolomide because of the weakened protective mechanism of the cancer cells [30]. Regardless of the standard treatment of glioblastoma, it exerts a highly invasive and neurologically destructing phenotype. Actin-rich membranous structure formation and cellular matrix degradation are exhilarated, which is known as invadopodia. Three FDA-approved drugs, bortezomib, everolimus, and fludarabine, have been found after STRING analysis to target the invadopodia regulators (overexpressed genes). All three can reduce the basal levels of invadopodia-mediated FITC-gelatin degradation in LN229 cells. There are positive outputs in some clinical trials with these drugs as a single therapy or in combination with temozolomide/radiation therapy [31].

5.2.2 Oral Cancer

Global Cancer Observatory (GCO) data about the oral cancer incidence of 377,713 cases globally and about 40% predicted increment by 2040 with subsequent increased mortality is a real aspect to rethink about its therapy [32]. Fenofibrate, which was FDA approved to be used in hypertriglyceridemia and hypercholesterolemia, has been found to promote cytotoxicity in CAL27 oral cancer cells by a metabolic reprogramming process. The antitumor activity of fenofibrate has been observed to be linked to the adenosine 5′-monophosphate-activated protein kinase (AMPK) pathway and activation of the glycolytic pathway in both cell culture studies as well as in 4-NQO/arecoline mouse models. It caused activation of AMPK, which is correlated to the downregulation of mammalian target of rapamycin (mTOR) activity. This regulation of rapamycin complex 1 (mTORC1) signaling through TSC1/TSC2 leads to progressive metabolic shift in the cancer cells and simultaneously decreases mitochondrial respiration. The noticeable fact is that fenofibrate interrupts the interaction between hexokinase II and voltage-dependent anion channel (VDAC) by dissociating hexokinase from the mitochondria. Interestingly, it has the potential to suppress tumor lesion formation, including high-grade dysplastic lesions in squamous cell carcinoma, delaying tumor development and its multiplicity. However, this suppression is done by an AMPK-independent pathway through Akt/TSC2/Rheb signaling [33]. Selectivity is the primary need in chemotherapy to avoid adverse effects. One antifungal drug, Antimycin A, can selectively induce oxidative stress in oral cancer cells with a higher pan-caspase activity than in healthy cells. These lead to specific cancer cell apoptosis [34]. Although chemotherapy is important for a patient's life, it cannot improve the quality of life. The intolerable pain with reduced oral mobility is a real clinical problem. To date, cancer pain management is highly dependent upon the opioid analgesics. However, it cannot provide complete relief and gradually develop opioid dependency. There is experimental evidence that has shown that EGFR inhibitors can alleviate pain and have potential anticancer effects in the case of morphine tolerance. However, there is a crosstalk between MOR and EGFR, and prolonged morphine exposure may lead to adaptive changes in both signaling systems. Even longer EGFR inhibitor treatment can also raise resistance. Therefore, for oral cancer pain management, new-generation EGFR inhibitors, as single or in combination with other non-opioid (non-EGFR inhibitor) analgesics, should be evaluated [35].

5.2.3 Lung Cancer

One of the highly invasive types of cancer is lung cancer, with a poor prognosis against first-line therapy. Metformin, an antidiabetic drug, has been screened extensively for different cancers. It has been shown to inhibit mitochondrial metabolism, which decreases the ATP availability in cancer cells. This drug has undergone a few clinical trials to evaluate and observe the survival of advanced-stage patients suffering from non-small cell lung cancer (NSCLC) or lung adenocarcinoma without any toxicity when given as monotherapy (NCT02109549, NCT03086733) or in combination with lanreotide (NCT02823691), sintilimab (NCT03874000), nivolumab (NCT03048500), gefitinib (NCT01864681) [36]. Cyclooxygenase (COX) enzyme targeting has been found satisfactory in treating cancer as COX-1 and COX-2 play an important role in prostaglandin synthesis, which is involved in the growth and development of skin, breast, lung, and digestive cancer. Therefore, COX inhibitors can be screened to evaluate for anti-lung cancer effectivity. Celecoxib, being a cox-2 inhibitor, has emerged as an anti-arthritis drug for rheumatoid arthritis. However, a later study revealed that it can induce apoptosis in lung cancer cells, which might be associated with triggering stress in the endoplasmic reticulum via regulation of glucose-regulated protein 78 and caspase-4. Celecoxib got the ticket for clinical trials to investigate its efficacy for early-stage NSCLC and head and neck cancer, as chemoprotective in primary lung cancer, and in T-cell regulation for advanced NSCLC [37]. Another Cox inhibitor, indomethacin, has also been effective in cancer. Although indomethacin primarily acts through a COX pathway, one COX-independent pathway has been found to be interlinked for its anticancer effects in lung cancer, which is the downregulation of the CDK1-nucleolin regulatory axis. Consequently, decreased nucleolin levels in lung cancer cells increase the expression of SSAT-1 (spermidine/spermine N1-acetyltransferase) protein. This aids in decreasing the polyamine levels in the cancer cells, which regulates cellular growth and eventually induces cell cycle arrest. However, mammalian cells have the strategy to recover the decreased content of polyamines like spermidine and putrescine in the presence of an enzyme, namely polyamine oxidase (PAOX) [38]. In these circumstances, methoctramine (muscarinic antagonist) is beneficial to stop the recovery of polyamine homeostasis by specifically inhibiting PAOX. So, indomethacin and methoctramine can be repurposed as combination drugs for lung cancer. There are reports where indomethacin has been shown to potentiate the antimetastatic activity of non-specific immunotherapy and arsenic trioxide anticancer effects [39]. The spark in these findings has been added by nanotechnology, where liposomal loading of indomethacin has demonstrated superior effects in all lung cancer cell lines (H1299, A549, and H460), and the ex-vivo 3D-spheroid assay has shown more tumor reduction in comparison to free indomethacin [40]. Bedaquiline, a member of the diarylquinoline class of drugs that was FDA approved to be used as an anti-mycobacterial agent. It has shown anticancer effects at 40 μM concentration in H1299 non-small cell lung cancer cell line as a pure drug and its inclusion complex form (with β-CD) which is formulated for inhalation therapy [41].

5.2.4 Breast Cancer

There are some cancers in which the cellular machinery gets modified after a viral infection due to interference of molecular parasitism by-products or viral-oncoproteins. In the case of breast cancer, the human mammary tumor virus is the one infective agent that can drive the quiescent normal breast cells to a proliferative state after its genomic replication. This has shown the possible capability of antiviral agents to be repurposed in breast cancer or even in post-chemotherapeutic infection. Acyclovir, which is FDA approved for HSV encephalitis and genital herpes, has been shown to suppress the cellular invasion and colony formation in MCF7 breast cancer cells. This is related to the upregulation of E-cadherin secretion and downregulating of the aldehyde dehydrogenase activity, which is responsible for increasing the self-renewal capacity of the tumor cells [42]. A veterinary medicine, namely maduramicin (antiprotozoal agent), has also shown positive results as repurposed therapy in triple-negative breast cancer. Notably, when maduramicin treatment is given to mice in the form of nanoemulsion, the tumor size shrinks and represses the lung-liver metastasis [43]. An antiviral drug, gemcitabine (used against enteroviruses), was clinically approved for breast cancer in 2004 in combination with paclitaxel to be used in advanced and metastatic stages. Gemcitabine was previously approved for pancreatic cancer treatment in 1995. A randomized clinical study (phase 2) has shown that gemcitabine in combination with vinorelbine/cisplatin/capecitabine is able to improve the therapeutic response with good tolerability in metastatic breast cancer patients previously exposed to anthracyclines and/or taxanes [44], through mRNA and miRNA network analysis. One network analytical tool, namely RepCOOL, has been used to determine the structural similarity among the 10 FDA-approved anti-breast cancer drugs and repurposed drugs. After analysis, it has suggested trastuzumab, tamoxifen, paclitaxel, and doxorubicin for stage 2 breast cancer [45]. Using the same approach, a total of 1446 differentially expressed genes (847 significantly upregulated and 599 downregulated) have been identified in neo-adjuvant chemotherapy-resistant breast cancer patients. Later, in the downstream analysis, 413 genes were identified as most important in the development of drug resistance, which is the result of phenotypic changes due to cell malignancy. Among these, 330 genes (represented as turquoise modules in the co-expression modules) are involved in multiple biological processes and signaling pathways that are directors for drug resistance. Further implementing the MCC clustering algorithm, the top 10% of genes (known as hub genes) in the turquoise module have been identified, which are the key players in the highly connected complex network of genes and are involved in the chemoresistance process. Notably, ten hub genes among these 33 have been found as potential targets, and 19 drugs (including eight FDA-approved drugs) can be repurposed against these gene dysregulation, as shown in Fig. 5.1. Two signaling pathways, namely VEGF-VEGFR2 and tyrosine kinase receptor signaling, are found to be the most enriched ones that can be considered for drug repurposing in the future [46].

5.2.5 Gastric Cancer

The third leading cause of cancer-related mortality is due to gastric cancer because of cancer recurrence after failure of conventional chemotherapies and adverse effects after combination therapy. Doxycycline, an antibiotic, emerged as a repurposed potent inhibitor of the ERK/MAPK pathway in gastric cancer (GC). AGS cells treated with doxycycline exhibited significant downregulation in several gene

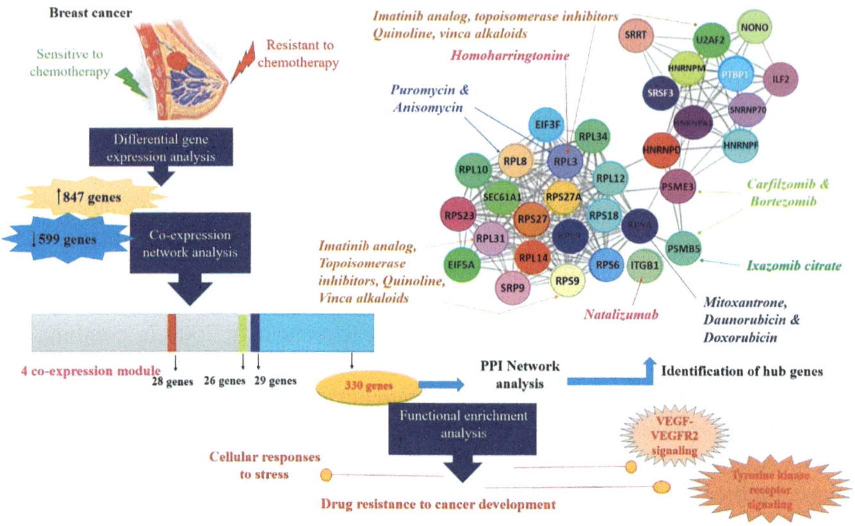

Fig. 5.1 Identification process of FDA-approved drugs (puromycin, anisomycin, imatinib, topoisomerase inhibitors, quinoline, vinca alkaloids, natalizumab, ixazomib citrate, mitoxantrone, daunorubicin, doxorubicin, carfilzomib, bortezomib) for breast cancer specific to disease associated hub genes by network analysis [46]. A part of the figure has been reproduced from [46] under the terms of the Creative Commons Attribution License (CC BY). Copyright © (2022)

families which are crucial for different cellular processes, including protein degradation (proteasomal pathway and ubiquitin ligation), DNA replication (DNA polymerases), and histone modification (histone genes) [47]. Another repurposed drug is disulfiram, an anti-alcoholic drug. This was tested on GES, MKN45, and SGC-7901 GC cells at varying concentrations. Disulfiram potently curbed cell proliferation in MKN-45 and SGC 7901, as evidenced by ELISA-confirmed PCNA downregulation, suggesting its potential as GC therapeutics [48]. In the case of resistant GC, it is difficult to rely on the existing therapeutic cocktail. Here, sertraline (an antidepressant) has been shown to enhance the sensitivity of cisplatin-resistant gastric cancer cells (specifically, SGC-7901). This suggests a promising avenue for its use in re-sensitizing drug-resistant GC, positioning it as a "new" adjunctive treatment in the therapeutic landscape [49]. Three drugs of different categories, namely benztropine (anticholinergic), natamycin (antibiotic), and nitazoxanide (antimicrobial—antiparasitic), have been screened in three GC cell lines (MKN28, MKN45, AGS). Nitazoxanide displayed the best effect in cell lines after 48 h of treatment and better efficacy as monotherapy than in combination with 5-Fluorouracil [49, 50].

5.2.6 Liver Cancer

In the context of liver cancer or hepatocellular carcinoma (HCC), PCSK9 (Proprotein Convertase Subtilisin/Kexin Type 9) is highly expressed in human tissues, which promotes tumor growth. A mechanistic study revealed that PCSK9 increases fatty acid synthase expression, thereby inhibiting apoptosis of HCC cells via the Bax/Bcl-2/Caspase9/Caspase3 pathway. Flubendazole, an anthelmintic drug, can downregulate the PCSK9 expression in the HCC cell line and have high prospects for repurposing. It even potentiates the effect of lenvatinib, which is approved in the US and in the European Union for HCC [51]. IL-6/GP130 signaling pathway also gets upregulated in HCC, which promotes survival and pro-inflammatory recruitment of mononuclear cells and inhibition of T cell apoptosis. Raloxifene, which is usually used in postmenopausal osteoporosis, can block this pathway in HCC [52]. Suppression of STAT-3 signaling in HCC cells is a good strategy to downregulate the transcription levels of downstream oncogenes. Pimozide, a neuroleptic agent, has been shown to exert STAT-3 suppression activity in HCC and can reverse the stem-like cell tumorigenic phenotypes induced by IL-6 treatment [53, 54]. Combination of disulfiram and copper has shown an interesting effect after repurposing. Disulfiram can inhibit the proliferation, migration, and invasion of liver cancer cells. However, Cu has no anticancer effects but can potentiate the activity when combined with disulfiram. In brief, disulfiram might work as a Cu ionophore that induces oxidative stress by promoting reactive oxygen species (ROS) generation and inhibiting NF-kB. This combination promotes ferroptosis (a recently discovered type of programmed cell death) by accelerating free iron accumulation and lipid peroxidation [55, 56]. In this context, a phase I clinical trial established the highest dose of Cu that could be administered with disulfiram in liver cancer patients without causing adverse effects. A daily dose of 250 mg of Cu gluconate is suitable. Another non-oncological drug, artesunate (antimalarial), can also induce ferroptosis in the liver cancer cell line (SMMC-7721) [57]. It can be used in sorafenib-resistant HCC patients. It also synergizes with sorafenib. On the other hand, artesunate primarily acts on lysosomes and synergizes with sorafenib to induce cathepsin B/L activation and ferritin degradation. These two processes are essential aspects to induce ferroptosis in cancer cells [58]. Antidepressant drug maprotiline can diminish cholesterol biosynthesis to inhibit the proliferation and metastasis of HCC cells via the ERK-SREBP2 signaling pathway by directly binding to CRABP1. It has been shown to enhance the sensitivity of HCC cells to sorafenib and can be repurposed as adjuvant therapy [59].

5.2.7 Colon Cancer

A number of non-oncological drugs from diverse pharmacological categories like anthelmintics (mebendazole, niclosamide), anti-retroviral drugs (zidovudine, tenofovir, indinavir, saquinavir, ritonavir), antibiotics (doxycycline, clarithromycin, gemifloxacin, azithromycin), antimalarial drugs (artesunate, mefloquine), antipsychotic (valproate), probiotic (butyrate), kidney transplant rejection inhibitor (sirolimus), antihyperglycemic agents (metformin, dapagliflozin), anti-inflammatory agent (celecoxib), antiplatelet drug (aspirin), etc. have been found satisfactory to be repurposed in colorectal cancer as shown in Fig. 5.2. However, in the repurposing story of colorectal cancer, metformin is a protagonist. It has been found to elicit anticancer effects in a multifaceted way in colorectal cancer. It blocks the G0/G1 phase in the life cycle of colorectal cancer cells (CRC). Metformin causes downregulation of IGF-1R, which decreases proliferation in colorectal neoplasia. This downregulation further reduces the expression of c-Myc, which activates underneath biological members and ultimately forces cancer cells to undergo apoptosis [60, 61]. Metformin can promote apoptosis in CRC cell lines via modulating oxidative stress and NF-κB inflammatory responses [60]. Apart from this, metformin also increases the activity of the sprouty two genes, which stops the cancer cell growth. Metformin exerts AMPK-mediated actions, which causes the suppression of the FAS gene that affects lipogenesis in cancer cells. This leads to decrement in cell proliferation by increasing fatty acid oxidation. This can be a panacea for those patients who have colorectal cancer along with diabetes mellitus. The histopathology of 86 CRC tissues of patients with type-2 diabetes revealed that metformin synergistically decreases poorly differentiated tumors in combination with 5-FU and reduces the resistance of 5-FU. Similar results can be observed with oxaliplatin in combination with 5-FU or metformin. This combination was found to show superior antitumor activity on chemo-resistant HT-29 or HCT-116 cells. Moreover, the association of metformin with the D3 vitamin demonstrated a chemo-preventive effect against 1,2-dimethylhydrazine (DMH)-induced CRC in rats and DMH-dextran sodium sulfate-induced colitis-associated CRC in mice [61]. Metformin

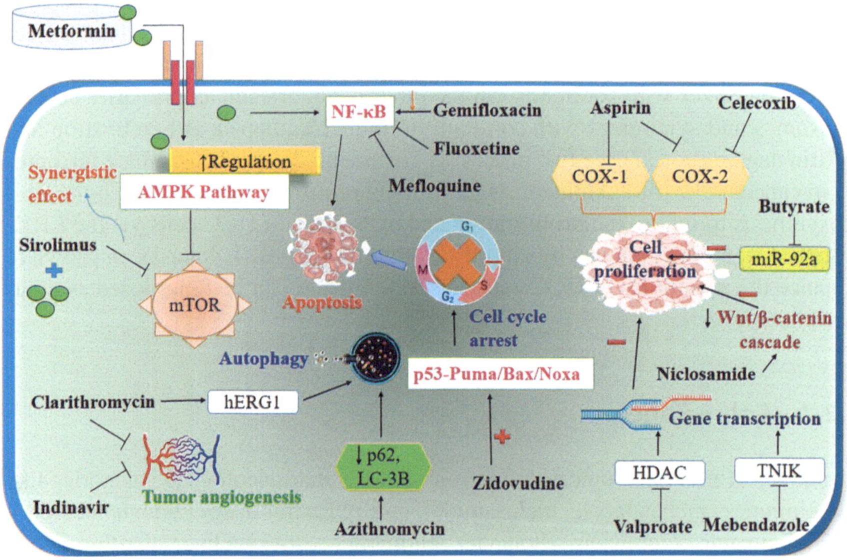

Fig. 5.2 Non-oncological drugs repurposing in colon cancer: different classes of drugs inducing cell apoptosis through inhibition, downregulation, or upregulation of cancer signaling pathways (NF-κB, AMPK, p53-Puma/Bax/Noxa, etc.), activation of cell cycle arrest and/or decreasing the cancer cell proliferation and invasion and through autophagy [60]

may boost the susceptibility of cancer cell lines to chemotherapeutic drugs such as paclitaxel and irinotecan [60].

5.2.8 Prostate Cancer

In the case of men, the second most common cancer is prostate cancer (PC), and more than 80% of cases are diagnosed at a late stage of 3 or 4, which increases the burden of treatment, especially the resistance to conventional treatments. Despite the initial improvement in the routine androgen-deprivation therapy with or without docetaxel, tumors relapse within 2–3 years. Therefore, studies have been conducted to find molecules that can be repurposed to synergize with docetaxel-mediated effects. A robust screening process has been designed where CP2 and SP1 cell lines have been selected, which are derived from highly aggressive tumors developed in Pbcre Ctnnb1(ex3Δ)/+ Ptenfl/+ and PbCre Spry2fl/+ Ptenfl/fl mice. Primary screening of 820 drugs revealed that anthelmintics (albendazole, mebendazole, flubendazole) belong to the top hits. Among these three hits, only mebendazole increased the cell growth inhibition in both cell lines and exerted higher combinatorial effects with docetaxel in a dose-dependent manner in mouse and human cell lines, including LNCaP, PC3, and CWR22, as shown in Fig. 5.3a. The synergistic anticancer effects obtained from the microtubule dynamics affected by both drugs. Docetaxel binds to the beta-tubulin and prevents microtubule depolymerization, whereas mebendazole binds to the tubulin-dimer and inhibits polymerization. As a result, the combination treatment resulted in increased multipolar divisions with higher fragmented nuclei. The detailed time-lapse microscopic investigation of mitotic timing in RPE1 GFP-tubulin cells has shown the formation of three abnormal daughter cells after combination treatment. The control cells and cells treated with mebendazole have shown proper mitotic division with correctly formed mitotic spindles and metaphase plates (Fig. 5.3b–d). When a genetically engineered mouse was subcutaneously injected with transferrin-conjugated liposome encapsulated docetaxel and mebendazole (dual-loaded), the prostate tumor size reduced drastically (Fig. 5.3e). Mebendazole is a good clinical repurposing candidate for resistant PC [63]. pH-dependent synergistic anticancer effects have been observed with very common drug combinations, i.e., pantoprazole and vitamin C in metastatic castration-resistant prostate cancer. Further evaluation revealed that cells pretreated with pantoprazole enhanced the uptake of vitamin C, and there was robust enhancement of apoptosis due to a simultaneous increase in ROS production. Moreover, the increased pH of the PC cells inhibits the exosome production. Altogether, the in vitro and in vivo data suggested that this repurposed drug combination is a good, cost-effective alternative to Lu-177, Ra-223, or Ac-225 therapy, which is costly and easily unavailable to patients [62]. Drugs from different categories, including steroids, have undergone clinical trials to validate their future clinical application as repurposed drugs in prostate cancer, including hormone-refractory prostate cancer, advanced stage (with or without mTOR mutation), stage IV castration sensitive PC, and/or metastatic PC. A few of them are metformin, suramin, (doxorubicin +

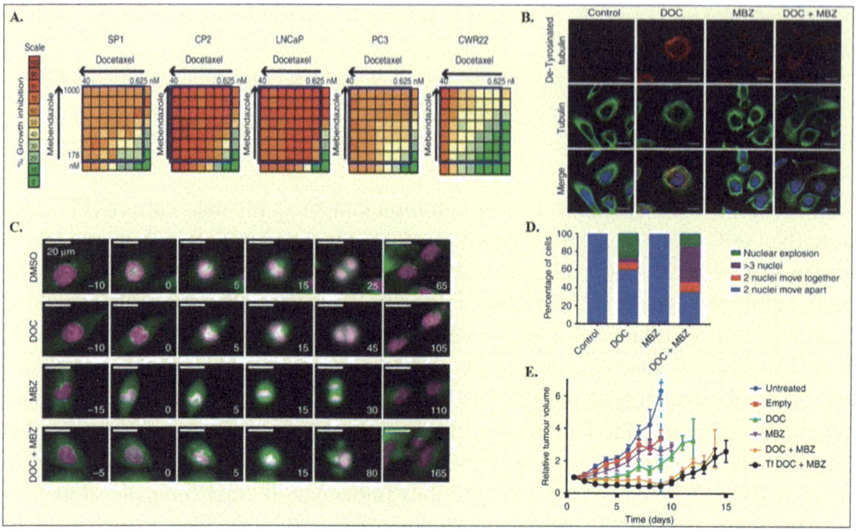

Fig. 5.3 (**a**) Combination of cell growth inhibition effect of docetaxel and mebendazole. Based on the combination index analysis, mebendazole convincingly synergized docetaxel-mediated growth inhibition in vitro, with the greatest synergism observed at low concentrations, suggesting an appealing therapeutic window. (**b–d**) Docetaxel and mebendazole combination reduced microtubule assembly and drastically impaired microtubule dynamics, resulting in aberrant cell division with frequent formation of multipolar spindles, aneuploid daughter cells, or arrest prometaphase. (**e**) Comparative antitumor study of individual and combination drug and their nanoencapsulated formulation in mouse tumor model. Reproduced from [63] under the terms of the Creative Commons Attribution License (CC BY). Copyright © (2020)

suramin), niclosamide in combination with enzalutamide, (abiraterone acetate + prednisone), (cabozantinib + docetaxel + prednisone), CTAP101 capsules (vitamin D analogue), zoledronic acid, (denosumab + zoledronic acid), octreotide acetate, dalteparin, nadroparin, fenretinide, and rapamycin [64].

5.2.9 Endocrine Cancer

There is a consistent pathophysiological relationship between metabolic syndrome and cancer. In this sense, many drugs that are used in metabolic-related pathologies have been screened in different cancers. Different antidiabetic drugs, gliflozins, metformin, and statins, seem to be a promising treatment in prostate, ovary, endometrial, pituitary, and endocrine-neuronal cancer [65]. Through a quantitative high-throughput analysis (qHTS), it has been found that more than 25% of drugs that belong to a class of anti-obesity and cardiotonic agents have profound anticancer effects in thyroid cancer, whereas only 12% of oncologic drugs can be repurposed. Bortezomib (reduction of approximately 50% cell growth at 6.6 nM and 90% cell death at 20 nM concentration), ouabain anticancer activity has been validated after qHTS screening. Therefore, bortezomib and ouabain, which have been previously approved in multiple myeloma and as cardiotonic agents, respectively, have been added to the repurposed drug library for the most common endocrine malignancy, i.e., thyroid cancer [66]. One preclinical study about the combination therapy using flavopiridol and carfilzomib has shown high efficiency against advanced adrenocortical carcinoma (ACC). There is increased expression of CDK1 and CDK2 mRNA in ACC, which is associated with the aggressiveness of the cancer and less survival of patients. In this case, flavopiridol has been selected because it is a potent inhibitor of CDK1 and CDK2. An-antiapoptotic protein XIAP level also gets increased. However, flavopiridol cannot significantly lower this level as monotherapy. Here, combining carfilzomib (FDA approved for refractory multiple myeloma) with flavopiridol has exerted synergy by targeting the dysregulated proteins, including XIAP, which will be potential in the advanced stage [67]. Despite the advancement of the cancer stage, chemoresistance is the primary cause of reproductive cancer. Moreover, the mortality rate is very high because of late diagnosis. Epithelial ovarian cancer is one of them. Therefore, efforts have been initiated to screen repurposed drugs in resistant ovarian cancer that can synergize with the frontline treatments. In this regard, 304 FDA-approved drugs with known effects in inflammation, metabolic disorders, and cancer have been selected from Selleck Chemical's database. These drugs have screened in OVCAR3, OVCAR4, OVCAR8, and A1847 cell lines in 2D and 3D culture models. The study confers 78 drugs as potential hits (15 hits in 3D, 38 in 2D, and 25 in both 2D and 3D model). Primary drug screening identified 15 hits including 47% anti-inflammatory drugs (pizotifen malate, licofelone, geniposide, cinepazide maleate, beclomethasone dipropionate, dexamethasone acetate, allylthiourea, glafenine), 26% metabolic modifier (phenformin HCl, mecarbinate, pioglitazone hydrochloride, risedronic acid), and 20% anticancer drugs (hydroxyurea, erlotinib, bergapten). A dose-dependent cytotoxic study across different cell lines has confirmed the anticancer activity of licofelone and glafenine against ovarian multicellular tumor spheroids to exhibit greater therapeutic resistance. Licofelone can reverse the stem-like properties of the tumors in a synergistic manner to paclitaxel by regulating mitochondrial metabolism. These promising findings show that further clinical experiments can be performed with extended evaluation for combination therapy with frontline drugs to overcome the clinical failure of available therapies in both primary and recurrent ovarian cancer [68].

5.2.10 Skin Cancer

Melanoma, which is a type of aggressive skin cancer that originates from melanocytes, often fails to respond to the targeted therapy by BRAF/MEK inhibitors. This is due to developed resistance. There is reported evidence that in the high-grade metastatic or resistant state of melanoma, the cells suffer from high inflammatory conditions, and inflammasomes serve as biomarker of the condition. Therefore, selective targeting of inflammasomes with existing drug repurposing may induce the activation of inflammasomes, which is a key regulator of pyroptosis (a cell death mechanism). In this context, temozolomide chemotherapy, along with an autophagy inhibitor chloroquine, has been found promising repurposed combinations in in vitro and in vivo studies. However, their translation to patient treatment is needed to evaluate their effectiveness for those having selective inflammasomes. Chloroquine and its different derivatives in combination with temozolomide have been subjected to clinical trials in melanoma. Several research studies showed that melatonin can potentiate the effects of BRAF-targeting agents. Oral treatment of melatonin in the murine metastatic melanoma model has been shown to significantly reduce the tumor size with reduced tumor angiogenesis via the ERK1/2 pathway. It can even augment the anti-melanoma activity of vemurafenib and inhibit the epithelial-to-mesenchymal transition [69, 70]. Impaired melanoma cell migration and cell invasion have also been observed upon treatment with niclosamide. Intraperitoneal niclosamide can potentially reduce the tumor volume by 62%, approximately at 20 mg/kg daily dose for 25 days in the A375 xenograft mice model. Most importantly, the number of metastatic nodules decreased after niclosamide therapy in the B16-F10 pulmonary metastasis model, probably by inhibiting myeloid-derived suppressor cells' immunosuppressive activity. These results confirmed that it is a promising repurposing candidate [71]. One antipsychotic drug, trifluoperazine, has shown a good adjuvant role in melanoma patients with lung, brain, and/or bone metastases [72]. Screening of 1280 small molecules from Prestwick Chemical Library has identified potent synergistic effects due to interaction between RAPTA-T (experimental ruthenium drug) and haloprogin (antifungal agent) at lower doses combination in breast (MD-MB231) and ovarian (A2780) cancer cell lines. Later, the synergistic anticancer effects were validated using murine B16F10 melanoma cells. However, further evaluations are needed to be done to conclude whether this combination can efficiently inhibit the metastatic spread from cutaneous melanoma cells [73].

5.2.11 Hematological Cancer

In chronic lymphocytic leukemia (CLL), the survival of the patients improved after treatment with ibrutinib, idelalisib, and venetoclax. However, remission-free life may not be achievable due to the occurrence of resistance to these small molecules by TNF family member B-cell activating factor (BAFF) over the entire spectrum of CLL. Therefore, an anti-BAFF molecule can significantly improve the situation. Belimumab, an antibody approved for the treatment of systemic lupus erythematosus (SLE), is a BAFF-neutralizing agent that re-sensitized the cells to all small molecule inhibitors irrespective of IbVH mutation or advanced stage of cancer [74]. Acute lymphoblastic leukemia (ALL), either originating from T-cell or B-cell progenitors, represents higher resistance to standard therapies and especially to glucocorticoids (GCs), which is one of the central components of the multi-treatment protocol. The resistance can arise due to multiple factors like overexpression of certain receptors or channels, activation or modulation of some signaling pathways directly regulating cell proliferation or mitochondrial activity, upregulation of anti-apoptotic proteins like Bcl-2, Mcl-1, inhibition of autophagy, decreased autoregulation of GC receptors, DNA mutation, and many more. In this regard, repurposing of FDA-approved established drugs has shown higher sensitivity toward the resistant-ALL and ability to reverse the GC-resistance (for example, azacitidine) in experimental preclinical models. The enhanced sensitivity results in the induction of autophagy, increased apoptosis, or decreased leukemic progression through targeting upregulated signaling axis like PI3K-Akt-mTOR, inhibition of efflux pump, and mitochondrial respiration/permeability to ions. Even the combination of the established anti-ALL drug(s) with the experimental repurposed candidate can perform synergistically and lower the toxicity of the other molecule (Fig. 5.4). Activated mTOR can increase glycolysis and provide fuel for leukemic cells to proliferate. Therefore, mTOR inhibitors (like dactolisib and rapamycin) are suitable for repurposing in ALL. Glycolysis can be reduced by blocking the glucose uptake. Niclosamide (anthelmintic) can block the GLUT-1 transporter and suppress the NFAT signaling, which reduces the leukemic progression (Fig. 5.4) [75]. One of the wonder molecules is metformin, which has exerted significant anticancer effects in different in vitro and in vivo experiments in all types of hematological cancers. Several clinical trials have been initiated and are going on for confirmative clinical

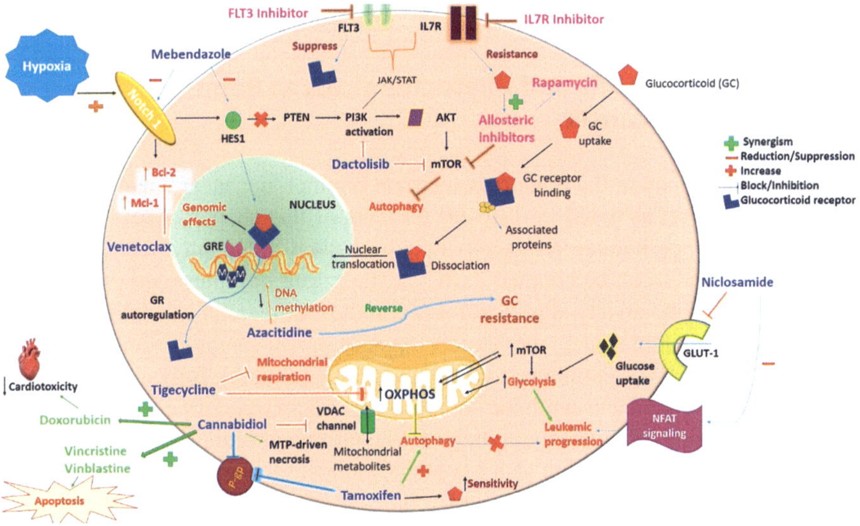

Fig. 5.4 Leukemic cell microenvironment in acute lymphoblastic leukemia (ALL) and biological targets for drug repurposing to overcome drug resistance. Tamoxifen increases the glucocorticoid sensitivity, induces autophagy, blocks the leukemic progression, and also inhibits the efflux protein pump (P-gp). Cannabidiol also inhibits P-gp and can exert synergistic apoptotic activity with vincristine and vinblastine. When it combined with doxorubicin to treat ALL, it can lower the cardiotoxic effect of doxorubicin. Cannabidiol targets one of the cell's key components, i.e., mitochondria, and induces MTP-driven necrosis. Tigecycline can block the mitochondrial respiratory process, which targets the OXPHOS, and its downregulation induces autophagy (**lower portion**). OXPHOS upregulation is also potentiated by mTOR activation. The mTOR activity can be reduced or inhibited indirectly by HES1, FLT3, and IL7R inhibitors or repurposing drugs that can downregulate Notch (**upper portion**) [75]

evidence in ALL, relapsed CLL, diffuse large B-cell lymphoma, relapsed/refractory ALL, multiple myeloma, and newly diagnosed ALL in adolescents and to study its preventive effects to stop the progression of monoclonal gammopathy of undetermined significance (MGUS), and smoldering multiple myeloma to active multiple myeloma. Metformin belongs to the biguanide family, and one drug, namely alexidine of the same family, which was first used as an oral disinfectant, has been shown to potentiate the conventional TKIs (of all three generations) treatment effects to eradicate CML leukemic stem cells. The strong synergism is due to enhanced caspase3/9 expression, PARP cleavage, reduced Bcl-xl related molecules, and, most importantly, blockade of the mitochondrial respiratory chain [76]. As per a recent study, one widely used anticonvulsant drug, valproic acid, has been found effective in one of the most common heterogeneous clonal disorders of hematopoietic cells, AML. It inhibits HDAC (histone deacetylase), which is abnormally assigned in the gene transcription process and is involved in AML1-ETO fusion protein-related oncogenesis. The apoptosis process associated with cell cycle arrest is induced in two ways by valproic acid. One involves acetylation of histone H3 level in the p21 promoter, which in turn increases the p21 expression. Another is the direct downregulation of cycline D1 & E1, CDK4/6. Both these events, retinoblastoma protein formation and successive biological events induce G0/G1 arrest [77].

5.3 Phytoceuticals and Phytochemicals as Repurposed Drugs in Cancer

Combination therapy in the non-cross-resistant regimen in cancer chemotherapy is in clinical practice, as noted in different portions of this chapter. Apart from synthetic drugs, other natural products, including plant-based phytoconstituent or herbal components, can be coupled to the existing drug or repurposed drug. Some natural products are already approved for use in the treatment of a spectrum of cancer, for example, vinblastine, vincristine, etoposide, taxol, camptothecin, topotecan, irinotecan, etc. In 1963, childhood leukemia was treated successfully with a cocktail combination of natural and synthetic chemotherapeutics (vincristine + amethopterin +6-mercaptopurine + prednisone). Such novel combinations are not only beneficial with respect to less toxicity but also lower the therapeutic burden due to affordability, especially for patients from low- and middle-income countries. Different classes of natural products like alkaloids, glycosides, flavonoids, isoflavonoids, tannins, plant acids, saponins, etc., have been reported for their excellent therapeutic efficacy and are known as natural medicinal agents. A literature review has concluded that different flavonoids, for example, kaempferol, apigenin, naringenin, luteolin, etc., are promising hits to be repurposed in lung cancer [78]. Many plant alkaloids have been reported to have anti-inflammatory, antioxidant, immunomodulatory, and antiproliferative properties. These should be considered for repurposing in cancer [79]. Therefore, different dietary components and phytochemicals must be systematically evaluated for repurposing for monotherapy or combination therapy in cancer. In a clinical study, curcumin and gemcitabine combination therapy has been found to be safe in pancreatic cancer and decrease fibrosis in the pancreatic stellate cells, which is one of the important factors in aiding cancer progression [80]. After performing molecular docking and dynamic simulation studies, a few polyphenols have been found to have higher binding affinities toward the targeted proteins in stomach cancer. Among 57 screening candidates, ten candidates, i.e., anthocyanin, curcumin, catechin, cyanidin, kaempferol, luteolin, naringenin, stilbene, theaflavin, and trolox, have been found most favorable to target the disease-causing protein linked to MET pathway [81].

However, there are several challenges that exist in phytochemical-based drug repurposing. The first is the low bioavailability of the compound. Detailed preformulation study and development of a suitable delivery vehicle can deal with the bioavailability issue. The second is the controversial effects exhibited by phytoconstituents in some cases. This suggests an extensive study of the tumor microenvironment. Third is the uncertain therapeutic window and failure in the in vivo study. A complete biodistribution study of the phytoconstituent, especially for drugs that undergo hepatic or renal clearance, must be performed. Even for combination therapy, the same study should be repeated with the conventional chemo-agent (s) for safety. To address the biological barrier, a nano DDS/microvector can be designed for simultaneous or co-delivery of the combination drugs [82]. The involvement of molecular docking and screening will allow faster exploration of a variety of phytochemicals and dietary constituents to expand the repurposing drug pool [83]. Already, efforts have been initiated to explore the multiplex natural drug repurposing based on computational chemistry and network-based analysis. A total of 384 terpenoids, 124 steroids, and 881 polyphenols were part of the analysis, and mulberroside C and terrestrosin D have been found promising in castration-resistant prostate cancer bone metastases [84].

5.4 Nanoformulation Landscape in Drug Repurposing in Cancer

Nanoformulations can provide a strong template to the old drugs for their faster clinical translation from bench side to bedside, improving biopharmaceutical, kinetic, and dynamic properties. The first medical need is safety and reduced adverse effects. This is dependent on the selective antitumor effects and cancer apoptosis. For example, chlorpromazine, an antipsychotic drug after loading into the PLGA nanoparticle, has shown improved antileukemic effects without accumulation in the brain after intravenous administration in zebrafish. This is excellent for AML therapy without affecting the central nervous system [85]. On the other hand, the poor solubility of the repurposed molecule is one of the major factors to look for in its nano-transformation or nanoencapsulation. Niclosamide, which is an antihelmintic drug, has been found to interrupt multiple oncogenic signaling networks, including NF-κB, Wnt, mTORc1, Notch, STAT3, and NF-κB pathways, and regulate the generation of ROS in several cancer cells. Interestingly, the niclosamide nanosuspension increases the glucose uptake in ovarian cancer cells (CP70 & SKOV3) with decreasing ATP synthesis, which shifts the cellular metabolism from oxidative phosphorylation to glycolysis. Oral treatment with niclosamide nanosuspension has been shown to reduce ovarian tumor size significantly [86]. Sometimes, the existing drug, which is originally administered for some disease condition, is not suitable to deliver through the same route or dosage form at the anticancer dose because of its increased toxicity. One such example is the β-blocker drug carvedilol, which, after systemic administration, may cause undesired cardiovascular effects. Therefore, to get the optimum anticancer benefits in melanoma, carvedilol has been loaded into a transferosome-based gel for topical application. A number of tumors with delayed incidence were observed after continual 25 weeks of ultraviolet-irradiation in a mice skin cancer model in the nano-gel treated group at only 10 μM dose. Even carvedilol was undetectable in plasma after topical nano-gel application at 100 μM dose, which confirms its reduced systemic effect with a very high skin deposition [87]. Nanoformulation is a well-known strategy for combination therapy. An outstanding result was obtained when suramin (antiparasitic agent) was co-loaded in glycol chitosan-based nanoparticles with doxorubicin. Suramin, which was abandoned due to its narrow therapeutic window, has been shown to exert synergistic effects on TNBC metastatic cells with reduced cardio-renal side effects [88]. The combination effect becomes stronger when any of the components of the nanoformulation has the capacity to reverse multidrug resistance. When rapamycin has been co-loaded in a PLGA nanoparticle with an alkaloid piperine, it has displayed enhanced cytotoxicity against MDR breast cancer because the piperine inhibits the P-gp mediated efflux of rapamycin [89]. In the case of itraconazole, one antifungal drug, the cancer cell death is obtained at a high concentration of about 97 μg/mL. Fabrication of itraconazole chitosan coated PLGA nanoparticles exerted better cytotoxicity at around 63 μg/mL due to rapid uptake by cancer cells. These itraconazole-nanoparticles have the potential to decrease sustained proliferation in comparison to itraconazole alone, which was evidenced by increased G0/G1 cells in the nanoformulation treated H12299 lung carcinoma cell line [90]. Many other examples have been mentioned in Table 5.1.

5.5 Emerging Technology in Drug Repurposing in Cancer

5.5.1 In Silico Computational-Based Drug Repositioning

It is complex to search out a drug, especially from a non-oncological therapeutics library, which can exert efficient anticancer effects. This has been made somewhat easier by the "in-silico" drug repurposing technology, which is a computational approach. Through a simple computational approach, encouraging data has been obtained where chlorhexidine, paromomycin, and deferoxamine have shown higher selectivity with stability at binding sites toward wild-type ABL-1 and chlorhexidine, ritonavir toward T315I mutant ABL-1 which are reasons behind anti-CML

Table 5.1 Drug delivery systems to improve drug repurposing in cancer

Sl. No.	Repurposed drug/drug combinations	Nanoformulation	Cancer type	Significance	References
1	Naproxen	Polyvinylpyrrolidone nanoparticle	Leukemia	Two-fold enhanced activity of nanoformulated naproxen than doxorubicin	[91]
2	Dasatinib	Poly(styrene-co-maleic acid) nano micelle	TNBC	Protection of dasatinib from extensive metabolic degradation and prolongation of its action	[92]
3	Ormeloxifene	PEGylated chitosan nanoparticles	Breast cancer	Conquered toxicities on kidney and spleen tissues than free drug due to sustained release from a nanoparticle, improved pharmacokinetic profile upon intravenous administration	[93]
4	Celecoxib	PEGylated SLN, NLC, and nanoemulsion	Breast cancer, acute promyelocytic leukemia	Controlled release of the repurposed drug, overcoming of resistance through inhibition of MK	[94]
5	Quinacrine	Inhalable cationic PLGA nanoparticles (modified with bovine serum albumin)	Lung cancer	Local drug accumulation, higher cellular uptake aided by albumin modification because of higher metabolic rate in cancer cells	[95]
6	Aceclofenac	Citronella oil-based nanoemulsion	Melanoma	Synergistic cytotoxic effect by aceclofenac and citronella oil, induction of cell cycle arrest at a significantly lower dose	[96]
7	Ketoconazole + disulfiram + tadalafil	Oral solid self-emulsifying drug delivery system	Hepatocellular carcinoma	Lower in vivo toxicity co-delivery with docetaxel allows adjuvant therapy with improved cytotoxicity	[97]
8	Pitavastatin	Bilosomes	Hepatocellular carcinoma	Achievement of active targeting after lactoferrin ligand modification, improved hepatocyte permeation	[98]
9	Quercetin + rapamycin	Hyaluronic acid-based nano hydrogel	Hormone-responsive breast cancer	Reduced inflammatory factors (IL-8, IL-6, IL-19, MMP2, and MMP9)	[99]

(continued)

Table 5.1 (continued)

Sl. No.	Repurposed drug/drug combinations	Nanoformulation	Cancer type	Significance	References
10	Chlorpromazine	PLGA nanoparticle	Acute myeloid leukemia	Prevent CNS affection, increased circulation time	[85]
11	Dihydroartemisinin	Biomimetic self-assembled nano platform	Apoptosis resistant glioblastoma	Selectively target lipoprotein receptor-related protein-1 (LRP1), able to cross the blood–brain barrier, induction of glioblastoma cell ferroptosis	[100]
12	Disulfiram	Folic acid modified metal organic framework (IRMOF3)-Zn^{2+}	Oral squamous cell carcinoma	Prevent disulfiram degradation during transportation, inhibition of cancer stem cells by Zn^{2+}	[101]

therapies [102]. Extensive virtual screening has been performed to recognize gefitinib as an allosteric inhibitor of Bcr-Abl oncogene (for CML) from 1424 approved drug molecules through multilevel processes including structure, pharmacophore and E-pharmacophore-based screening, Glide XP docking, Prime MM/GBSA rescoring, molecular dynamics simulation, steered molecular dynamics simulation, and finally cell line assay [103]. However, the conventional in silico approach fails to establish a good similarity or connection between the existing drugs and diseases due to complex relationships between different biological factors like genes, hormones, enzymes, etc., and drug pharmacology. In a drug repurposing study, applying systemic bioinformatics, tricyclic antidepressants, and G-protein coupled receptor inhibitors have been found to be potent apoptosis inducers in small cell lung cancer as well as in some high-grade neuroendocrine tumors. However, based on this preclinical result, when a clinical study was initiated [NCT01719861], satisfactory efficacy has not been achieved with desipramine [104].

5.5.2 Advanced Network-Based Computational Drug Repositioning

In contrast to the above limitation, the network-based computational approach has been found optimistic even in identifying drug(s) for different cancer subtypes (for example, ten subtypes of breast cancer) with the highest repurposability based on metabolomics, proteomics, genomics, and transcriptomics data. Even this novel approach permits screening a combination of drugs for the disease subtypes, which can triumph over a single drug [105]. Using protein-hub-based multi-omics technology, which integrates the epigenome and transcriptome network, has suggested a few drugs that can be repurposed in pituitary tumors. The suggested drug candidates have been confirmed for repurposing via in silico molecular docking. Linifanib, trametinib, niguldipine, palbociclib, eplerenone, and zuclopenthixol are the result of docking [106]. DeSign, a newly developed web-based bioinformatics tool, can build up a constructive gene expression profile from a large pool of cell line data (707 cell lines of both sensitive and resistant variants). This is better than connectivity map (CMap) analysis for drug repurposing, which can only provide a response based on pre- and post-therapeutic conditions of four cell lines. Whereas DeSign constructs higher order information considering cumulative gene expression factors of multiple genes. Still, there is a need to improve the information networking hierarchy by considering the post-translational modifications data (including phosphorylation, acylation, etc.). A recent study employing the DeSign tool has demonstrated the repurposing ability of bosutinib in a type of HNSCC (OSCC). There is a plan to upgrade the latest version of DeSign with expanded drug coverage with gene expression and drug response data from the Cancer Therapeutics Response Portal [107]. In CMap analysis, there is always a risk for errors in the output because it does not account for the drugs for specific cell lines. Therefore, in vitro or in vivo trials are mandatory for validation. In one repurposing candidate screening study for melanoma, researchers have used more than one method of approach for re-validation, which includes genetic variation analysis and gene expression profiling, biomedical associations, and metabolomics pathways. Thirty-five drug candidates, which are targetable to the melanoma-causing genes and proteins, have been extracted after analysis, as mentioned in Fig. 5.5. Moreover, these possess a great potential to be repurposed due to their similarity to the expression profile of known drugs in human melanoma cell lines [108].

5.5.3 Computational Repurposing in Rare Cancer

The computational approach has been found to yield significant valid results for drug repurposing for rare cancers like adrenocortical carcinoma (ACC). A research team has introduced one semi-supervised machine learning approach, namely Heter-LP, for this purpose and built a network link among the concerned genes for ACC, drugs, and their targets after the compilation of necessary data. Heter-LP has been predicted that cosyntropin is the drug that can be administered as an adjunct treatment to restimulate the adrenal gland after mitotane (the first treatment option in ACC) therapy. Cosyntropin has been found promising among the 11,140 drugs [109]. Seven drugs, including taxifolin, neratinib, nortriptyline, 5-fluorocytosine,

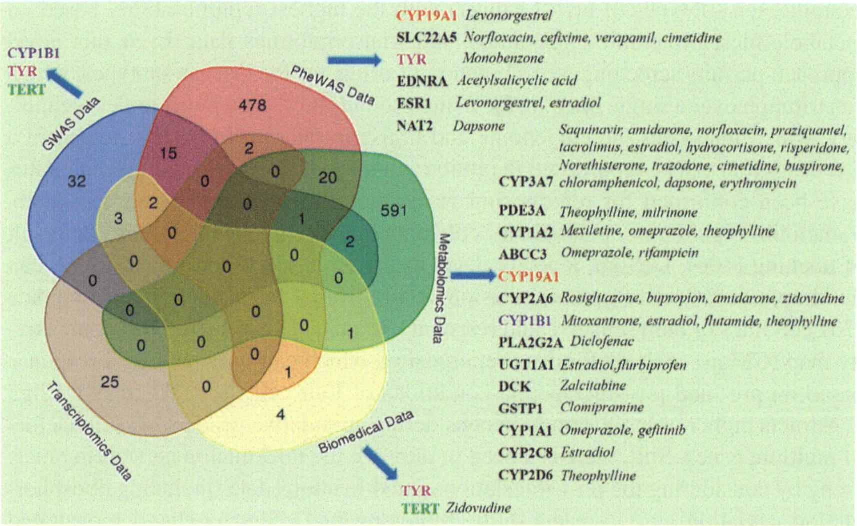

Fig. 5.5 Identification of precise druggable target (genes and proteins) for melanoma: Venn diagram representing the distribution of identified targets through different approaches (three targets from GWAS data, six targets from PheWAS data, 14 targets from metabolomics data, and two targets from biomedical data). A few druggable targets (TYR, TERT, CYP1A1, CYP1B1, and CYP19A1) have been fetched by more than one method. Thirty-five drugs with repurposing potential with the corresponding targets have been represented in italics. Few drugs can interact with multiple targets. [A part of the figure is reproduced from [108] under the terms of the Creative Commons Attribution License (CC BY). Copyright © (2019)]

vorinostat, puromycin, and zileuton, have been proposed to be effective in prolactinoma (lactotroph adenoma, benign tumor in pituitary) through a miRNA-mediated transcriptomic data based computational approach [110]. A study recruited an integrative GBM-based Biomedical Profile Network (GBPN) approach, which involved heterogeneous data for 92 glioblastoma-related rare diseases, associated genes, and drugs as well. Therefore, applying multiple network analyses, the data has been condensed into a small, manageable data space (Fig. 5.6). From that space, repurposing candidates have been screened through an optimization process. These include riluzole, cannabidiol, and VK-0214 [111].

5.5.4 Computational Repurposing in the Identification of Radiosensitizer

Hypoxic conditions in cancer cells contribute to resistance to radiotherapy. Therefore, the identification of genes associated with the hypoxic microenvironment is needed for drug repurposing. QUADrATIC, a connectivity mapping software, has been used to identify the connections between gene expression profiles of hypoxic prostate cancer and FDA-approved drug molecules. Menadione (vitamin K3) and gemcitabine have been selected as potential candidates to be repurposed. After in silico selection of these molecules when it has undergone downstream validation studies, menadione has been shown to sensitize cells to radiation treatment in hypoxic conditions, whereas the other drugs radiosensitized the cells in normoxic conditions (less sensitizer enhancement ratio-SER) [112].

5.5.5 Emerging Platforms to Repurpose Drugs for Cancer Precision Therapy

Few research companies are now providing the platform for precision drug repositioning based on patient stratification biomarkers to reduce development costs and risk, and maximize the revenue from R&D investment. Precision life is one

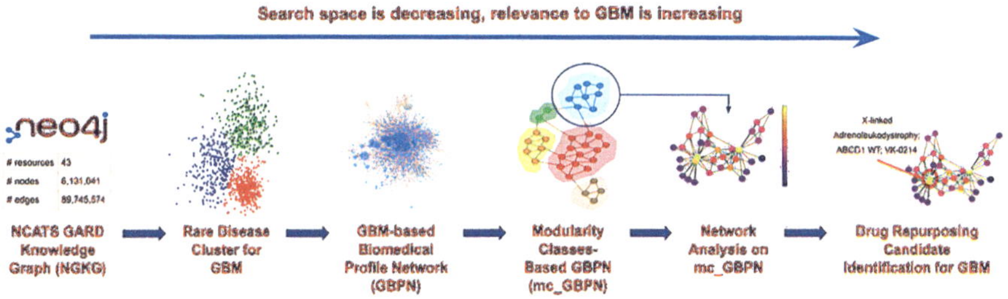

Fig. 5.6 Computational repurposing pipeline: identification of drug(s) for glioblastoma. Reproduced from [111] under the terms of the Creative Commons Attribution License (CC BY). Copyright © (2023)

such pioneering company that is providing licensing opportunities for old drugs for extended indications. This has conducted one systemic study where a detailed analysis of combinatorial features (genomic) of a breast cancer population has been done to identify combinations of single nucleotide-based polymorphism. This has allowed the identification of 175 gene targets of different breast cancer subpopulations. Twenty-three genes among these signatures have been found as novel targets for new and repurposed drugs. P4HA2 and TGM2 are found to be important one for breast cancer repurposing [113]. Advanced mass-spectroscopy-based bioinformatics approach allows comparative proteomic and phosphoproteomic profiling to accurately detect post-translational modifications in a cancer development or resistance. Nowadays, methodological improvements in this area allow simultaneous assessment of multiple samples in a single run and enable analysis of the whole proteome of an eukaryotic cell within a few hours [114]. One online accessible platform is DRepMel (an app based on machine and deep learning), which can provide rational drug combinations after analysis based on a multi-omics approach, whole exome sequencing, and RNA-sequence data [115].

5.6 Clinical Trials for Repurposed Approach

A successful clinical trial with a previously approved drug can give a second life to the drug. Few recent clinical trials have aimed to unfold the better pharmacologic action of the repurposed drug without or with traditional chemotherapy in combination (Table 5.2). However, prospective study designs are not always successful because of the poor study design. The main factor behind such unsatisfactory results is inappropriate dose. Statin is one such candidate with repurposed controversy for which most of the clinical trials have used a dose that is appropriate for hypercholesterolemia but cannot induce apoptosis because of lower plasma concentration. In the same way, the dosing schedule and dosing frequency of statin are the key to exerting anticancer effects. The short half-life of statins can effectively control the cholesterol level in the body when administered in the evening because the cholesterol synthesis level is higher at night. However, the same administration schedule may not be beneficial in anticancer therapy because geranylgeraniol synthesis in the tumor or malignant cells may not follow the same pattern as cholesterol [123]. Therefore, adequate design of clinical trials based on robust preclinical data is the gateway for repurposed drugs in cancer to achieve blockbuster status. Some drugs got regulatory approval for clinical repurposing in cancer, and a few of them have been presented in Table 5.3.

Table 5.2 Ongoing clinical trials for drug repurposing in cancer

Sl. No.	Repurposed drug (s)	Phase	Cancer type	MOA	Estimated completion year	Study reference	References
1	Ibrutinib	Phase 1b	Brain cancer in pediatric patients	Ibrutinib (Bruton's tyrosine Kinase inhibitor) will synergistically enhance the immune response in combined therapy with indoximod, cyclophosphamide, and etoposide	2026	NCT05106296	[116]
2	Metformin	Phase 2 (recruiting)	Lung cancer	Immune system reprogramming	2025	NCT04931017	[117]
3	Metformin + TKIs	Phase 3 (recruiting)	Lung tumors with EGFR mutations and LKB1 expression	Overcoming TKI resistance due to epithelial mesenchymal four transition	2025	NCT05445791	[118]
4	Decitabine	Phase 2	Advanced/metastatic, refractory (PDAC) pancreatic ductal adenocarcinoma (tumor which express KRAS-dependency signature)	Inhibit pyrimidine homeostasis	2025	NCT05360264	[119]
5	Metformin	Recruiting	Leukemia (as preventive drug)- preleukemia, myelodysplastic neoplasm, cytopenia		Primary completion by 2025	NCT04741945	[120]
6	Atovaquone	Recruiting (phase 2)	Ovarian cancer (platinum-resistant)	STAT 3 inhibition	2025	NCT05998135	[121]
7	ATRA + Gemcitabine + Nab-paclitaxel	Phase 2	PDAC	Stromal targeting	2024	NCT04241276	[122]

Table 5.3 Clinically approved repurposed drugs for cancer

Sl. No.	Drug name	Initially approved for	Repurposed in	Approval year	References
1	Fulvestrant + capivasertib	New drug combination for drug repurposing with fulvestrant which is used as antiestrogen	HR+ and HER2- locally advanced/metastatic breast cancer	2023	[124]
2	Fulvestrant	Antiestrogen	ER+, PR+, HER2-, advanced breast cancer	2017	[125]
3	Everolimus	Renal cancer (2009)	Neuroendocrine tumors of GI or lung-origin pancreatic cancer	2016	[126]
			Anastrozole- and letrozole-resistant metastatic breast cancer	2012	[127]
			Pancreatic cancer	2011	[128]
4	Capecitabine + docetaxel	Colon cancer	Metastatic and advanced breast cancer	2001	[129]
5	Paclitaxel	Ovarian cancer (1992)	Advanced and early-stage breast cancer	1994 and 1999	[130]
6	Doxorubicin	Antibiotic	Advanced breast cancer, lung, gastric, thyroid cancer, multiple myeloma, Hodgkin's and non-Hodgkin's lymphoma	1974	[131]

5.7 Summary

All the examples discussed in this chapter describe the potential to modulate multiple pathways simultaneously. Some repurposed candidates as single agents or in combination possess the ability to sweep the chemotherapeutic resistance to standard drugs and target two or even more structurally dissimilar proteins involved in the oncological signaling. Therefore, pathway-oriented repositioning of the drug can be more suitable for multi-targeted cancer. Even this strategy should be extended to personalized patient design, for example, breast cancer patients having the same molecular variations. However, more detailed molecular exploration by high-throughput analysis is needed to overcome the variations of response even within the same group of patients. Here, artificial intelligence, integrative bioinformatics, and network medicine will pave the new path for faster elucidation and validation of the specific drug repurposing based on the complex heterogeneity of cancer. Recently, efforts have been taken to examine traditional Chinese medicines and herbal products, including ayurvedic formulations, for repurposing in cancer. Some of the herbal extracts or phytoconstituents can act as epigenetic modifiers, i.e., act on genetic variables like DNA repairing. The immune-modulatory action of the repurposed candidate is also clinching the attention. Few candidates, like metformin, propranolol, chloroquine, losartan, etc., have shown better immune responses as antitumor agents and enhanced effects along with immune checkpoint inhibitors in preclinical studies. All these successful candidates in the preclinical evaluation can enter the market as clinically approved repurposed drugs only when the dosage is carefully evaluated, which may sometimes exceed the standard dose required for original indication.

5.8 Chapter Summary

The work in the domain of drug repurposing for cancer has been started in the twentieth century. Few of them, for example, paclitaxel and doxorubicin successfully reached to the bedside for repurposed therapy. The wheel of the journey speeded up after the introduction of computational and bioinformatics tool, network-based concept, multi-omics technology, phosphoproteomic profiling, and advanced new-generation analytical instruments including mass-spectrometry. These upgraded technologies like DeSign, QUADrATIC help in the precise identification of druggable targets for cancer which is important for therapeutic specificity and most importantly reduced the risk for clinical failure in the later phase. The multiple network analysis system helped to narrow down the huge heterogeneous data which found incredible for drug screening for resistance cancers as well as for rare type of cancers like adrenocortical cancer, pituitary cancer, etc. Machine learning-based software DRepMel can give rational output for drug combinations. Few companies have joined in this research which will help to get regulatory approvals for increased number of repurposed candidates. In the same pipeline, formulation science and nanotechnology are key drivers to reach to the intended goal for the repurposed drugs. This pipeline can be extended for phytopharmaceuticals and herbal components, where the initiatives already have started.

References

1. Wang Y, Peng R-Q, Li D-D, Ding Y, Wu X-Q, Zeng Y-X, Zhu X-F, Zhang X-S. Chloroquine enhances the cytotoxicity of topotecan by inhibiting autophagy in lung cancer cells. Chin J Cancer. 2011;30:690–700. https://doi.org/10.5732/cjc.011.10056.
2. Qiao S, Tao S, Rojo de la Vega M, Park SL, Vonderfecht AA, Jacobs SL, Zhang DD, Wondrak GT. The antimalarial amodiaquine causes autophagic-lysosomal and proliferative blockade sensitizing human melanoma cells to starvation- and chemotherapy-induced cell death. Autophagy. 2013;9:2087–102. https://doi.org/10.4161/auto.26506.
3. Villa-Ruano N, Anaya-Ruiz M, Villafaña-Diaz L, Barron-Villaverde D, Perez-Santos M. Drug repurposing of mito-atovaquone for cancer treatment. Pharm Pat Anal. 2023;12:143–9. https://doi.org/10.4155/ppa-2023-0015.
4. Joharatnam-Hogan N, Alexandre L, Yarmolinsky J, Lake B, Capps N, Martin RM, Ring A, Cafferty F, Langley RE. Statins as potential chemoprevention or therapeutic agents in cancer: a model for evaluating repurposed drugs. Curr Oncol Rep. 2021;23:29. https://doi.org/10.1007/s11912-021-01023-z.

5. Vandeborne L, Pantziarka P, Van Nuffel AMT, Bouche G. Repurposing infectious diseases vaccines against cancer. Front Oncol. 2021;11:688755. https://doi.org/10.3389/fonc.2021.688755.
6. Tanoli Z, Vähä-Koskela M, Aittokallio T. Artificial intelligence, machine learning, and drug repurposing in cancer. Expert Opin Drug Discov. 2021;16:977–89. https://doi.org/10.1080/17460441.2021.1883585.
7. Sheikholeslami K, Ali Sher A, Lockman S, Kroft D, Ganjibakhsh M, Nejati-Koshki K, Shojaei S, Ghavami S, Rastegar M. Simvastatin induces apoptosis in medulloblastoma brain tumor cells via mevalonate cascade prenylation substrates. Cancers (Basel). 2019;11:994. https://doi.org/10.3390/cancers11070994.
8. Takwi AAL, Li Y, Becker Buscaglia LE, Zhang J, Choudhury S, Park AK, Liu M, Young KH, Park W, Martin RCG, Li Y. A statin-regulated microRNA represses human c-Myc expression and function. EMBO Mol Med. 2012;4:896–909. https://doi.org/10.1002/emmm.201101045.
9. Urbanska K, Pannizzo P, Grabacka M, Croul S, Del Valle L, Khalili K, Reiss K. Activation of PPARα inhibits IGF-I-mediated growth and survival responses in medulloblastoma cell lines. Int J Cancer. 2008;123:1015–24. https://doi.org/10.1002/ijc.23588.
10. Huang L, Garrett Injac S, Cui K, Braun F, Lin Q, Du Y, Zhang H, Kogiso M, Lindsay H, Zhao S, Baxter P, Adekunle A, Man T-K, Zhao H, Li X-N, Lau CC, Wong STC. Systems biology-based drug repositioning identifies digoxin as a potential therapy for groups 3 and 4 medulloblastoma. Sci Transl Med. 2018;10:eaat0150. https://doi.org/10.1126/scitranslmed.aat0150.
11. Wolle D, Lee SJ, Li Z, Litan A, Barwe SP, Langhans SA. Inhibition of epidermal growth factor signaling by the cardiac glycoside ouabain in medulloblastoma. Cancer Med. 2014;3:1146–58. https://doi.org/10.1002/cam4.314.
12. Ingram WJ, Crowther LM, Little EB, Freeman R, Harliwong I, Veleva D, Hassall TE, Remke M, Taylor MD, Hallahan AR. ABC transporter activity linked to radiation resistance and molecular subtype in pediatric medulloblastoma. Exp Hematol Oncol. 2013;2:26. https://doi.org/10.1186/2162-3619-2-26.
13. De Witt M, Gamble A, Hanson D, Markowitz D, Powell C, Al Dimassi S, Atlas M, Boockvar J, Ruggieri R, Symons M. Repurposing mebendazole as a replacement for vincristine for the treatment of brain tumors. Mol Med. 2017;23:50–6. https://doi.org/10.2119/molmed.2017.00011.
14. Sholler S. Antitumor activity of nifurtimox is enhanced with tetrathiomolybdate in medulloblastoma. Int J Oncol. 2011;38:1329–41. https://doi.org/10.3892/ijo.2011.971.
15. Kim J, Tang JY, Gong R, Kim J, Lee JJ, Clemons KV, Chong CR, Chang KS, Fereshteh M, Gardner D, Reya T, Liu JO, Epstein EH, Stevens DA, Beachy PA. Itraconazole, a commonly used antifungal that inhibits hedgehog pathway activity and cancer growth. Cancer Cell. 2010;17:388–99. https://doi.org/10.1016/j.ccr.2010.02.027.
16. Baryawno N, Rahbar A, Wolmer-Solberg N, Taher C, Odeberg J, Darabi A, Khan Z, Sveinbjörnsson B, FuskevÅg O-M, Segerström L, Nordenskjöld M, Siesjö P, Kogner P, Johnsen JI, Söderberg-Nauclér C. Detection of human cytomegalovirus in medulloblastomas reveals a potential therapeutic target. J Clin Investig. 2011;121:4043–55. https://doi.org/10.1172/JCI57147.
17. Rossi A, Russo G, Puca A, La Montagna R, Caputo M, Mattioli E, Lopez M, Giordano A, Pentimalli F. The antiretroviral nucleoside analogue Abacavir reduces cell growth and promotes differentiation of human medulloblastoma cells. Int J Cancer. 2009;125:235–43. https://doi.org/10.1002/ijc.24331.
18. Eslin D, Lee C, Sankpal UT, Maliakal P, Sutphin RM, Abraham L, Basha R. Anticancer activity of tolfenamic acid in medulloblastoma: a preclinical study. Tumor Biol. 2013;34:2781–9. https://doi.org/10.1007/s13277-013-0836-6.
19. King JJG, Khalili K. Inhibition of human brain tumor cell growth by the anti-inflammatory drug, flurbiprofen. Oncogene. 2001;20:6864–70. https://doi.org/10.1038/sj.onc.1204907.
20. Kurita J, Hirao Y, Nakano H, Fukunishi Y, Nishimura Y. Sertraline, chlorprothixene, and chlorpromazine characteristically interact with the REST-binding site of the corepressor

mSin3, showing medulloblastoma cell growth inhibitory activities. Sci Rep. 2018;8:13763. https://doi.org/10.1038/s41598-018-31852-1.
21. Mascaro-Cordeiro B, Oliveira ID, Tesser-Gamba F, Pavon LF, Saba-Silva N, Cavalheiro S, Dastoli P, Toledo SRC. Valproic acid treatment response in vitro is determined by TP53 status in medulloblastoma. Childs Nerv Syst. 2018;34:1497–509. https://doi.org/10.1007/s00381-018-3817-7.
22. Gumireddy K, Sutton LN, Phillips PC, Reddy CD. All-trans-retinoic acid-induced apoptosis in human medulloblastoma: activation of caspase-3/poly(ADP-ribose) polymerase 1 pathway. Clin Cancer Res. 2003;9:4052–9.
23. Hallahan AR, Pritchard JI, Chandraratna RAS, Ellenbogen RG, Geyer JR, Overland RP, Strand AD, Tapscott SJ, Olson JM. BMP-2 mediates retinoid-induced apoptosis in medulloblastoma cells through a paracrine effect. Nat Med. 2003;9:1033–8. https://doi.org/10.1038/nm904.
24. Di Magno L, Manni S, Di Pastena F, Coni S, Macone A, Cairoli S, Sambucci M, Infante P, Moretti M, Petroni M, Nicoletti C, Capalbo C, De Smaele E, Di Marcotullio L, Giannini G, Battistini L, Goffredo BM, Iorio E, Agostinelli E, Maroder M, Canettieri G. Phenformin inhibits hedgehog-dependent tumor growth through a complex I-independent redox/corepressor module. Cell Rep. 2020;30:1735–1752.e7. https://doi.org/10.1016/j.celrep.2020.01.024.
25. Spiller SE, Logsdon NJ, Deckard LA, Sontheimer H. Inhibition of nuclear factor kappa-B signaling reduces growth in medulloblastoma in vivo. BMC Cancer. 2011;11:136. https://doi.org/10.1186/1471-2407-11-136.
26. Madala HR, Punganuru SR, Ali-Osman F, Zhang R, Srivenugopal KS. Brain- and brain tumor-penetrating disulfiram nanoparticles: sequence of cytotoxic events and efficacy in human glioma cell lines and intracranial xenografts. Oncotarget. 2018;9:3459–82. https://doi.org/10.18632/oncotarget.23320.
27. Hammoud H, Saker Z, Harati H, Fares Y, Bahmad HF, Nabha S. Drug repurposing in medulloblastoma: challenges and recommendations. Curr Treat Options in Oncol. 2021;22:6. https://doi.org/10.1007/s11864-020-00805-0.
28. Rundle-Thiele D, Head R, Cosgrove L, Martin JH. Repurposing some older drugs that cross the blood–brain barrier and have potential anticancer activity to provide new treatment options for glioblastoma. Br J Clin Pharmacol. 2016;81:199–209. https://doi.org/10.1111/bcp.12785.
29. Petrosyan E, Fares J, Cordero A, Rashidi A, Arrieta VA, Kanojia D, Lesniak MS. Repurposing autophagy regulators in brain tumors. Int J Cancer. 2022;151:167–80. https://doi.org/10.1002/ijc.33965.
30. Taylor MA, Das BC, Ray SK. Targeting autophagy for combating chemoresistance and radioresistance in glioblastoma. Apoptosis. 2018;23:563–75. https://doi.org/10.1007/s10495-018-1480-9.
31. Jones D, Whitehead CA, Dinevska M, Widodo SS, Furst LM, Morokoff AP, Kaye AH, Drummond KJ, Mantamadiotis T, Stylli SS. Repurposing FDA-approved drugs as inhibitors of therapy-induced invadopodia activity in glioblastoma cells. Mol Cell Biochem. 2023;478:1251–67. https://doi.org/10.1007/s11010-022-04584-0.
32. Anon. Oral cancer—the fight must go on against all odds…. Evid Based Dent. 2022;23:4–5. https://doi.org/10.1038/s41432-022-0243-1.
33. Jan C-I, Tsai M-H, Chiu C-F, Huang Y-P, Liu CJ, Chang NW. Fenofibrate suppresses oral tumorigenesis via reprogramming metabolic processes: potential drug repurposing for oral cancer. Int J Biol Sci. 2016;12:786–98. https://doi.org/10.7150/ijbs.13851.
34. Yu T, Hsieh C, Tang J, Lin L, Huang H, Wang H, Yeh Y, Chuang Y, Ou-Yang F, Chang H. Antimycin A shows selective antiproliferation to oral cancer cells by oxidative stress-mediated apoptosis and DNA damage. Environ Toxicol. 2020;35:1212–24. https://doi.org/10.1002/tox.22986.
35. Santi MD, Zhang M, Liu N, Viet CT, Xie T, Jensen DD, Amit M, Pan H, Ye Y. Repurposing EGFR inhibitors for oral cancer pain and opioid tolerance. Pharmaceuticals. 2023;16:1558. https://doi.org/10.3390/ph16111558.

36. ClinicalTrials.gov. https://clinicaltrials.gov/search?cond=lung%20cancer&term=metformin.
37. Kim B, Kim J, Kim YS. Celecoxib induces cell death on non-small cell lung cancer cells through endoplasmic reticulum stress. Anat Cell Biol. 2017;50:293. https://doi.org/10.5115/acb.2017.50.4.293.
38. Buelvas N, Ugarte-Vio I, Asencio-Leal L, Muñoz-Uribe M, Martin-Martin A, Rojas-Fernández A, Jara JA, Tapia JC, Arias ME, López-Muñoz RA. Indomethacin induces spermidine/spermine-N1-acetyltransferase-1 via the nucleolin-CDK1 Axis and synergizes with the polyamine oxidase inhibitor methoctramine in lung cancer cells. Biomol Ther. 2023;13:1383. https://doi.org/10.3390/biom13091383.
39. Schultz RM, Altom MG. Potentiation of nonspecific immunotherapy of experimental lung metastases by indomethacin. J Immunopharmacol. 1983;5:277–90. https://doi.org/10.3109/08923978309026449.
40. Sarvepalli S, Parvathaneni V, Chauhan G, Shukla SK, Gupta V. Inhaled indomethacin-loaded liposomes as potential therapeutics against non-small cell lung cancer (NSCLC). Pharm Res. 2022;39:2801–15. https://doi.org/10.1007/s11095-022-03392-x.
41. Parvathaneni V, Elbatanony RS, Goyal M, Chavan T, Vega N, Kolluru S, Muth A, Gupta V, Kunda NK. Repurposing bedaquiline for effective non-small cell lung cancer (NSCLC) therapy as inhalable cyclodextrin-based molecular inclusion complexes. Int J Mol Sci. 2021;22:4783. https://doi.org/10.3390/ijms22094783.
42. Shaimerdenova M, Karapina O, Mektepbayeva D, Alibek K, Akilbekova D. The effects of antiviral treatment on breast cancer cell line. Infect Agent Cancer. 2017;12:18. https://doi.org/10.1186/s13027-017-0128-7.
43. Song X, Zhang Y, Zuo R, Zhang J, Lin M, Wang J, Hu S, Ji H, Peng L, Lv Y, Gao X, Jiang S, Guo D. Repurposing maduramicin as a novel anticancer and anti-metastasis agent for triple-negative breast cancer as enhanced by nanoemulsion. Int J Pharm. 2022;625:122091. https://doi.org/10.1016/j.ijpharm.2022.122091.
44. Stemmler HJ, diGioia D, Freier W, Tessen HW, Gitsch G, Jonat W, Brugger W, Kettner E, Abenhardt W, Tesch H, Hurtz HJ, Rösel S, Brudler O, Heinemann V. Randomised phase II trial of gemcitabine plus vinorelbine vs gemcitabine plus cisplatin vs gemcitabine plus capecitabine in patients with pretreated metastatic breast cancer. Br J Cancer. 2011;104:1071–8. https://doi.org/10.1038/bjc.2011.86.
45. Fahimian G, Zahiri J, Arab SS, Sajedi RH. RepCOOL: computational drug repositioning via integrating heterogeneous biological networks. J Transl Med. 2020;18:375. https://doi.org/10.1186/s12967-020-02541-3.
46. Saha Detroja T, Detroja R, Mukherjee S, Samson AO. Identifying hub genes associated with neoadjuvant chemotherapy resistance in breast cancer and potential drug repurposing for the development of precision medicine. Int J Mol Sci. 2022;23:12628. https://doi.org/10.3390/ijms232012628.
47. Pandian J, Panneerpandian P, Devanandan HJ, Sekar BT, Balakrishnan K, Selvarasu K, Muthupandi K, Ganesan K. Identification of the targeted therapeutic potential of doxycycline for a subset of gastric cancer patients. Ann N Y Acad Sci. 2020;1467:94–111. https://doi.org/10.1111/nyas.14288.
48. Zhang J, Pu K, Bai S, Peng Y, Li F, Ji R, Guo Q, Sun W, Wang Y. The anti-alcohol dependency drug disulfiram inhibits the viability and progression of gastric cancer cells by regulating the Wnt and NF-κB pathways. J Int Med Res. 2020;48:030006052092599. https://doi.org/10.1177/0300060520925996.
49. Mu C, Peng R-K, Guo C-L, Li A, Yang X-M, Zeng R, Li Y-L, Gu J, Ouyang Q. Discovery of sertraline and its derivatives able to combat drug-resistant gastric cancer cell via inducing apoptosis. Bioorg Med Chem Lett. 2021;41:127997. https://doi.org/10.1016/j.bmcl.2021.127997.
50. Ribeiro E, Araújo D, Pereira M, Lopes B, Sousa P, Sousa AC, Coelho A, Rêma A, Alvites R, Faria F, Oliveira C, Porto B, Maurício AC, Amorim I, Vale N. Repurposing benztropine,

natamycin, and nitazoxanide using drug combination and characterization of gastric cancer cell lines. Biomedicine. 2023;11:799. https://doi.org/10.3390/biomedicines11030799.
51. Jin W, Yu J, Su Y, Lin H, Liu T, Chen J, Ge C, Zhao F, Geng Q, Mao L, Jiang S, Cui Y, Chen T, Jiang G, Li J, Miao C, Xiao X, Li H. Drug repurposing flubendazole to suppress tumorigenicity via PCSK9-dependent inhibition and potentiate lenvatinib therapy for hepatocellular carcinoma. Int J Biol Sci. 2023;19:2270–88. https://doi.org/10.7150/ijbs.81415.
52. Villarruel-Melquiades F, Mendoza-Garrido ME, García-Cuellar CM, Sánchez-Pérez Y, Pérez-Carreón JI, Camacho J. Current and novel approaches in the pharmacological treatment of hepatocellular carcinoma. World J Gastroenterol. 2023;29:2571–99. https://doi.org/10.3748/wjg.v29.i17.2571.
53. Chen J-J, Cai N, Chen G-Z, Jia C-C, Qiu D-B, Du C, Liu W, Yang Y, Long Z-J, Zhang Q. The neuroleptic drug pimozide inhibits stem-like cell maintenance and tumorigenicity in hepatocellular carcinoma. Oncotarget. 2017;8:17593–609. https://doi.org/10.18632/oncotarget.4307.
54. Chen J, Zhang L, Cai N, Zhang Z, Ji K. Antipsychotic agent pimozide promotes reversible proliferative suppression by inducing cellular quiescence in liver cancer. Oncol Rep. 2019;42(3):1101–9. https://doi.org/10.3892/or.2019.7229.
55. Li Y, Wang L, Zhang H, Wang Y, Liu S, Zhou W, Yuan X, Li T, Wu C, Yang J. Disulfiram combined with copper inhibits metastasis and epithelial–mesenchymal transition in hepatocellular carcinoma through the NF-κB and TGF-β pathways. J Cell Mol Med. 2018;22:439–51. https://doi.org/10.1111/jcmm.13334.
56. Ren X, Li Y, Zhou Y, Hu W, Yang C, Jing Q, Zhou C, Wang X, Hu J, Wang L, Yang J, Wang H, Xu H, Li H, Tong X, Wang Y, Du J. Overcoming the compensatory elevation of NRF2 renders hepatocellular carcinoma cells more vulnerable to disulfiram/copper-induced ferroptosis. Redox Biol. 2021;46:102122. https://doi.org/10.1016/j.redox.2021.102122.
57. Wang Q, Wu LM, Li AY, Zhao Y, Wang NP. Experimental studies of antitumor effect of artesunate on liver cancer. Zhongguo Zhong Yao Za Zhi. 2001;26:707–8, 720
58. Li Z, Dai H, Huang X, Feng J, Deng J, Wang Z, Yang X, Liu Y, Wu Y, Chen P, Shi H, Wang J, Zhou J, Lu G. Artesunate synergizes with sorafenib to induce ferroptosis in hepatocellular carcinoma. Acta Pharmacol Sin. 2021;42:301–10. https://doi.org/10.1038/s41401-020-0478-3.
59. Zheng C, Zhu Y, Liu Q, Luo T, Xu W. Maprotiline suppresses cholesterol biosynthesis and hepatocellular carcinoma progression through direct targeting of CRABP1. Front Pharmacol. 2021;12:689767. https://doi.org/10.3389/fphar.2021.689767.
60. El Zarif T, Yibirin M, De Oliveira-Gomes D, Machaalani M, Nawfal R, Bittar G, Bahmad HF, Bitar N. Overcoming therapy resistance in colon cancer by drug repurposing. Cancers (Basel). 2022;14:2105. https://doi.org/10.3390/cancers14092105.
61. Cunha Júnior AD, Bragagnoli AC, Costa FO, Carvalheira JBC. Repurposing metformin for the treatment of gastrointestinal cancer. World J Gastroenterol. 2021;27:1883–904. https://doi.org/10.3748/wjg.v27.i17.1883.
62. Li Z, He P, Long Y, Yuan G, Shen W, Chen Z, Zhang B, Wang Y, Yue D, Seidl C, Zhang X. Drug repurposing of pantoprazole and vitamin C targeting tumor microenvironment conditions improves anticancer effect in metastatic castration-resistant prostate cancer. Front Oncol. 2021;11:660320. https://doi.org/10.3389/fonc.2021.660320.
63. Rushworth LK, Hewit K, Munnings-Tomes S, Somani S, James D, Shanks E, Dufès C, Straube A, Patel R, Leung HY. Repurposing screen identifies mebendazole as a clinical candidate to synergise with docetaxel for prostate cancer treatment. Br J Cancer. 2020;122:517–27. https://doi.org/10.1038/s41416-019-0681-5.
64. Bahmad HF, Demus T, Moubarak MM, Daher D, Alvarez Moreno JC, Polit F, Lopez O, Merhe A, Abou-Kheir W, Nieder AM, Poppiti R, Omarzai Y. Overcoming drug resistance in advanced prostate cancer by drug repurposing. Med Sci. 2022;10:15. https://doi.org/10.3390/medsci10010015.

65. León-González AJ, Jiménez-Vacas JM, Fuentes-Fayos AC, Sarmento-Cabral A, Herrera-Martínez AD, Gahete MD, Luque RM. Role of metformin and other metabolic drugs in the prevention and therapy of endocrine-related cancers. Curr Opin Pharmacol. 2021;60:17–26. https://doi.org/10.1016/j.coph.2021.06.002.
66. Zhang L, He M, Zhang Y, Nilubol N, Shen M, Kebebew E. Quantitative high-throughput drug screening identifies novel classes of drugs with anticancer activity in thyroid cancer cells: opportunities for repurposing. J Clin Endocrinol Metab. 2012;97:E319–28. https://doi.org/10.1210/jc.2011-2671.
67. Nilubol N, Boufraqech M, Zhang L, Gaskins K, Shen M, Zhang Y-Q, Gara SK, Austin CP, Kebebew E. Synergistic combination of flavopiridol and carfilzomib targets commonly dysregulated pathways in adrenocortical carcinoma and has biomarkers of response. Oncotarget. 2018;9:33030–42. https://doi.org/10.18632/oncotarget.26050.
68. Hirst J, Pathak HB, Hyter S, Pessetto ZY, Ly T, Graw S, Koestler DC, Krieg AJ, Roby KF, Godwin AK. Licofelone enhances the efficacy of paclitaxel in ovarian cancer by reversing drug resistance and tumor stem-like properties. Cancer Res. 2018;78:4370–85. https://doi.org/10.1158/0008-5472.CAN-17-3993.
69. Pathipaka R, Thyagarajan A, Sahu RP. Melatonin as a repurposed drug for melanoma treatment. Med Sci. 2023;11:9. https://doi.org/10.3390/medsci11010009.
70. Ahmed F, Tseng H-Y, Ahn A, Gunatilake D, Alavi S, Eccles M, Rizos H, Gallagher SJ, Tiffen JC, Hersey P, Emran AA. Repurposing melanoma chemotherapy to activate inflammasomes in the treatment of BRAF/MAPK inhibitor resistant melanoma. J Invest Dermatol. 2022;142:1444–1455.e10. https://doi.org/10.1016/j.jid.2021.09.030.
71. Zhu Y, Zuo W, Chen L, Bian S, Jing J, Gan C, Wu X, Liu H, Su X, Hu W, Guo Y, Wang Y, Ye T. Repurposing of the anti-helminthic drug niclosamide to treat melanoma and pulmonary metastasis via the STAT3 signaling pathway. Biochem Pharmacol. 2019;169:113610. https://doi.org/10.1016/j.bcp.2019.08.012.
72. Xia Y, Xu F, Xiong M, Yang H, Lin W, Xie Y, Xi H, Xue Q, Ye T, Yu L. Repurposing of antipsychotic trifluoperazine for treating brain metastasis, lung metastasis and bone metastasis of melanoma by disrupting autophagy flux. Pharmacol Res. 2021;163:105295. https://doi.org/10.1016/j.phrs.2020.105295.
73. Riedel T, Demaria O, Zava O, Joncic A, Gilliet M, Dyson PJ. Drug repurposing approach identifies a synergistic drug combination of an antifungal agent and an experimental organometallic drug for melanoma treatment. Mol Pharm. 2018;15:116–26. https://doi.org/10.1021/acs.molpharmaceut.7b00764.
74. Tandler C, Schmidt M, Heitmann JS, Hierold J, Schmidt J, Schneider P, Dörfel D, Walz J, Salih HR. Neutralization of B-cell activating factor (BAFF) by belimumab reinforces small molecule inhibitor treatment in chronic lymphocytic leukemia. Cancers (Basel). 2020;12:2725. https://doi.org/10.3390/cancers12102725.
75. Olivas-Aguirre M, Torres-López L, Pottosin I, Dobrovinskaya O. Overcoming glucocorticoid resistance in acute lymphoblastic leukemia: repurposed drugs can improve the protocol. Front Oncol. 2021;11:617937. https://doi.org/10.3389/fonc.2021.617937.
76. Hu M, Chen Y, Ma T, Jing L. Repurposing metformin in hematologic tumor: state of art. Curr Probl Cancer. 2023;47:100972. https://doi.org/10.1016/j.currproblcancer.2023.100972.
77. Hao C, Liu P, Zhao L, Zhu C, Tian X, Wang L, Zhang Z. Inhibition of leukemic cells by valproic acid, an HDAC inhibitor, in xenograft tumors. Onco Targets Ther. 2013;6:733. https://doi.org/10.2147/OTT.S46135.
78. Yadav V, Mittal L, Paliwal S, Asthana S, Tonk RK. Repurposing of flavonoids as promising phytochemicals for the treatment of lung carcinoma. In: Drug repurposing for emerging infectious diseases and cancer. Singapore: Springer Nature; 2023. p. 89–110.
79. Efferth T, Oesch F. Repurposing of plant alkaloids for cancer therapy: pharmacology and toxicology. Semin Cancer Biol. 2021;68:143–63. https://doi.org/10.1016/j.semcancer.2019.12.010.
80. Ramakrishnan P, Loh WM, Gopinath SCB, Bonam SR, Fareez IM, Mac Guad R, Sim MS, Wu YS. Selective phytochemicals targeting pancreatic stellate cells as new anti-fibrotic

agents for chronic pancreatitis and pancreatic cancer. Acta Pharm Sin B. 2020;10:399–413. https://doi.org/10.1016/j.apsb.2019.11.008.
81. Ajmera H, Garge S, Suravajhala R, Malik B. Repurposing of polyphenolic compounds as novel drug targets in stomach cancer. AIP Conf Proc. 2023;2723:020023.
82. Zhang M, Chen X, Radacsi N. New tricks of old drugs: repurposing non-chemo drugs and dietary phytochemicals as adjuvants in anti-tumor therapies. J Control Release. 2021;329:96–120. https://doi.org/10.1016/j.jconrel.2020.11.047.
83. Manikkam R, Periyasamy V, Sabapathy I. Repurposing of phytocompounds-derived novel bioactive compounds possessing promising anticancer and cancer therapeutic efficacy through molecular docking, MD simulation, and drug-likeness/ADMET studies. In: Computational methods in drug discovery and repurposing for cancer therapy. London: Elsevier; 2023. p. 201–22.
84. Chang J, Jiang Z, Ma T, Li J, Chen J, Ye P, Feng L. Integrating transcriptomics and network analysis-based multiplexed drug repurposing to screen drug candidates for M2 macrophage-associated castration-resistant prostate cancer bone metastases. Front Immunol. 2022;13:989972. https://doi.org/10.3389/fimmu.2022.989972.
85. Gundersen ET, Førde J-L, Tislevoll BS, Leitch C, Barratt G, Gjertsen BT, Herfindal L. Repurposing chlorpromazine for anti-leukaemic therapy by nanoparticle encapsulation. Int J Pharm. 2022;612:121296. https://doi.org/10.1016/j.ijpharm.2021.121296.
86. Lin C-K, Bai M-Y, Hu T-M, Wang Y-C, Chao T-K, Weng S-J, Huang R-L, Su P-H, Lai H-C. Preclinical evaluation of a nanoformulated antihelminthic, niclosamide, in ovarian cancer. Oncotarget. 2016;7:8993–9006. https://doi.org/10.18632/oncotarget.7113.
87. Abdullah Shamim M, Yeung S, Shahid A, Chen M, Wang J, Desai P, Parsa C, Orlando R, Meyskens FL Jr, Kelly KM, Andresen BT, Huang Y. Topical carvedilol delivery prevents UV-induced skin cancer with negligible systemic absorption. Int J Pharm. 2022;611:121302. https://doi.org/10.1016/j.ijpharm.2021.121302.
88. Cheng B, Gao F, Maissy E, Xu P. Repurposing suramin for the treatment of breast cancer lung metastasis with glycol chitosan-based nanoparticles. Acta Biomater. 2019;84:378–90. https://doi.org/10.1016/j.actbio.2018.12.010.
89. Katiyar SS, Muntimadugu E, Rafeeqi TA, Domb AJ, Khan W. Co-delivery of rapamycin- and piperine-loaded polymeric nanoparticles for breast cancer treatment. Drug Deliv. 2016;23:2608–16. https://doi.org/10.3109/10717544.2015.1039667.
90. Alhakamy NA, Md S. Repurposing itraconazole loaded PLGA nanoparticles for improved antitumor efficacy in non-small cell lung cancers. Pharmaceutics. 2019;11:685. https://doi.org/10.3390/pharmaceutics11120685.
91. Kumar R, Siril PF, Javid F. Unusual anti-leukemia activity of nanoformulated naproxen and other non-steroidal anti-inflammatory drugs. Mater Sci Eng C. 2016;69:1335–44. https://doi.org/10.1016/j.msec.2016.08.024.
92. Bahman F, Pittalà V, Haider M, Greish K. Enhanced anticancer activity of nanoformulation of dasatinib against triple-negative breast cancer. J Pers Med. 2021;11:559. https://doi.org/10.3390/jpm11060559.
93. Agrawal S, Ahmad H, Dwivedi M, Shukla M, Arya A, Sharma K, Lal J, Dwivedi AK. PEGylated chitosan nanoparticles potentiate repurposing of ormeloxifene in breast cancer therapy. Nanomedicine. 2016;11:2147–69. https://doi.org/10.2217/nnm-2016-0095.
94. Üner M, Yener G, Ergüven M. Design of colloidal drug carriers of celecoxib for use in treatment of breast cancer and leukemia. Mater Sci Eng C. 2019;103:109874. https://doi.org/10.1016/j.msec.2019.109874.
95. Vaidya B, Kulkarni NS, Shukla SK, Parvathaneni V, Chauhan G, Damon JK, Sarode A, Garcia JV, Kunda N, Mitragotri S, Gupta V. Development of inhalable quinacrine loaded bovine serum albumin modified cationic nanoparticles: repurposing quinacrine for lung cancer therapeutics. Int J Pharm. 2020;577:118995. https://doi.org/10.1016/j.ijpharm.2019.118995.
96. Younis MK, Khalil IA, Younis NS, Fakhr Eldeen RR, Abdelnaby RM, Aldeeb RA, Taha AA, Hassan DH. Aceclofenac/citronellol oil nanoemulsion repurposing study: formulation, in vitro characterization, and in silico evaluation of their antiproliferative and pro-apoptotic

activity against melanoma cell line. Biomedicine. 2023;11:2531. https://doi.org/10.3390/biomedicines11092531.
97. Ardad RM, Manjappa AS, Dhawale SC, Kumbhar PS, Pore YV. Concurrent oral delivery of non-oncology drugs through solid self-emulsifying system for repurposing in hepatocellular carcinoma. Drug Dev Ind Pharm. 2023;49:377–91. https://doi.org/10.1080/03639045.2023.2216785.
98. Kharouba M, El-Kamel A, Mehanna R, Thabet E, Heikal L. Pitavastatin-loaded bilosomes for oral treatment of hepatocellular carcinoma: a repurposing approach. Drug Deliv. 2022;29:2925–44. https://doi.org/10.1080/10717544.2022.2120925.
99. Quagliariello V, Iaffaioli RV, Armenia E, Clemente O, Barbarisi M, Nasti G, Berretta M, Ottaiano A, Barbarisi A. Hyaluronic acid nanohydrogel loaded with quercetin alone or in combination to a macrolide derivative of rapamycin RAD001 (everolimus) as a new treatment for hormone-responsive human breast cancer. J Cell Physiol. 2017;232:2063–74. https://doi.org/10.1002/jcp.25587.
100. Liang J, Li L, Tian H, Wang Z, Liu G, Duan X, Guo M, Liu J, Zhang W, Nice EC, Huang C, He W, Zhang H, Li Q. Drug repurposing-based brain-targeting self-assembly nanoplatform using enhanced ferroptosis against glioblastoma. Small. 2023:19, e2303073. https://doi.org/10.1002/smll.202303073.
101. Cui J, Li W, Bu W, Liu J, Chen X, Li X, Liu C, Meng L, Chen M, Sun H, Wang J. Folic acid-modified disulfiram/Zn-IRMOF3 nanoparticles for oral cancer therapy by inhibiting ALDH1A1+ cancer stem cells. Biomater Adv. 2022;139:213038. https://doi.org/10.1016/j.bioadv.2022.213038.
102. Sohraby F, Bagheri M, Aliyar M, Aryapour H. In silico drug repurposing of FDA-approved drugs to predict new inhibitors for drug resistant T315I mutant and wild-type BCR-ABL1: a virtual screening and molecular dynamics study. J Mol Graph Model. 2017;74:234–40. https://doi.org/10.1016/j.jmgm.2017.04.005.
103. Singh VK, Chang H-H, Kuo C-C, Shiao H-Y, Hsieh H-P, Coumar MS. Drug repurposing for chronic myeloid leukemia: *in silico* and *in vitro* investigation of DrugBank database for allosteric Bcr-Abl inhibitors. J Biomol Struct Dyn. 2017;35:1833–48. https://doi.org/10.1080/07391102.2016.1196462.
104. Jahchan NS, Dudley JT, Mazur PK, Flores N, Yang D, Palmerton A, Zmoos A-F, Vaka D, Tran KQT, Zhou M, Krasinska K, Riess JW, Neal JW, Khatri P, Park KS, Butte AJ, Sage J. A drug repositioning approach identifies tricyclic antidepressants as inhibitors of small cell lung cancer and other neuroendocrine tumors. Cancer Discov. 2013;3:1364–77. https://doi.org/10.1158/2159-8290.CD-13-0183.
105. Firoozbakht F, Rezaeian I, Rueda L, Ngom A. Computationally repurposing drugs for breast cancer subtypes using a network-based approach. BMC Bioinform. 2022;23:143. https://doi.org/10.1186/s12859-022-04662-6.
106. Aydin B, Beklen H, Arga KY, Bayrakli F, Turanli B. Epigenomic and transcriptomic landscaping unraveled candidate repositioned therapeutics for non-functioning pituitary neuroendocrine tumors. J Endocrinol Investig. 2022;46:727–47. https://doi.org/10.1007/s40618-022-01923-2.
107. Lee BKB, Tiong KH, Chang JK, Liew CS, Abdul Rahman ZA, Tan AC, Khang TF, Cheong SC. DeSigN: connecting gene expression with therapeutics for drug repurposing and development. BMC Genomics. 2017;18:934. https://doi.org/10.1186/s12864-016-3260-7.
108. Khosravi A, Jayaram B, Goliaei B, Masoudi-Nejad A. Active repurposing of drug candidates for melanoma based on GWAS, PheWAS and a wide range of omics data. Mol Med. 2019;25:30. https://doi.org/10.1186/s10020-019-0098-x.
109. Lotfi Shahreza M, Ghadiri N, Green JR. A computational drug repositioning method applied to rare diseases: adrenocortical carcinoma. Sci Rep. 2020;10:8846. https://doi.org/10.1038/s41598-020-65658-x.
110. Aydin B, Arslan S, Bayraklı F, Karademir B, Arga KY. MicroRNA-mediated drug repurposing unveiled potential candidate drugs for prolactinoma treatment. Neuroendocrinology. 2022;112:161–73. https://doi.org/10.1159/000515801.

111. McGowan E, Sanjak J, Mathé EA, Zhu Q. Integrative rare disease biomedical profile based network supporting drug repurposing or repositioning, a case study of glioblastoma. Orphanet J Rare Dis. 2023;18:301. https://doi.org/10.1186/s13023-023-02876-2.
112. Bibby BAS, Thiruthaneeswaran N, Yang L, Pereira RR, More E, McArt DG, O'Reilly P, Bristow RG, Williams KJ, Choudhury A, West CML. Repurposing FDA approved drugs as radiosensitizers for treating hypoxic prostate cancer. BMC Urol. 2021;21:96. https://doi.org/10.1186/s12894-021-00856-x.
113. Taylor K, Das S, Pearson M, Kozubek J, Strivens M, Gardner S. Systematic drug repurposing to enable precision medicine: a case study in breast cancer. Digit Med. 2019;5:180. https://doi.org/10.4103/digm.digm_28_19.
114. Mavridou D, Psatha K, Aivaliotis M. Proteomics and drug repurposing in CLL towards precision medicine. Cancers (Basel). 2021;13:3391. https://doi.org/10.3390/cancers13143391.
115. Thompson ZJ, Teer JK, Li J, Chen Z, Welsh EA, Zhang Y, Ayoubi N, Eroglu Z, Tan AC, Smalley KSM, Chen YA. Drepmel—a multi-omics melanoma drug repurposing resource for prioritizing drug combinations and understanding tumor microenvironment. Cell. 2022;11:2894. https://doi.org/10.3390/cells11182894.
116. ClinicalTrials.gov. Chemo-immunotherapy using ibrutinib plus indoximod for patients with pediatric brain cancer. 2023. https://clinicaltrials.gov/study/NCT05106296?term=NCT05106296&rank=1. Accessed 15 Jan 2024.
117. ClinicalTrials.gov. Metformin for chemoprevention of lung cancer in overweight or obese individuals at high risk for lung cancer. 2024. https://clinicaltrials.gov/study/NCT04931017?term=NCT04931017&rank=1. Accessed 15 Jan 2024.
118. ClinicalTrials.gov. Metformin plus tyrosine kinase inhibitors for treatment of patients with non-small cell lung cancer with EGFR mutations (METLUNG). 2023. https://clinicaltrials.gov/study/NCT05445791?term=NCT05445791&rank=1. Accessed 15 Jan 2024.
119. ClinicalTrials.gov. tailOred dRug repurposIng of dEcitabine in KRAS-dependeNt refracTory pAncreaTic cancEr (ORIENTATE). 2023. https://clinicaltrials.gov/study/NCT05360264?term=NCT05360264&rank=1. Accessed 15 Jan 2024.
120. ClinicalTrials.gov. Repurposing metformin as a leukemia-preventive drug in CCUS and LR-MDS. 2023. https://clinicaltrials.gov/study/NCT04741945?term=NCT04741945&rank=1. Accessed 15 Jan 2024.
121. ClinicalTrials.gov. Repurposing atovaquone for the treatment of platinum-resistant ovarian cancer. 2024. https://clinicaltrials.gov/study/NCT05998135?term=NCT05998135&rank=1. Accessed 15 Jan 2024.
122. ClinicalTrials.gov. A randomised trial of ATRA in a novel drug combination for pancreatic cancer (STARPAC2). 2024. https://clinicaltrials.gov/study/NCT04241276?term=NCT04241276&rank=1. Accessed 15 Jan 2024.
123. Abdullah MI, de Wolf E, Jawad MJ, Richardson A. The poor design of clinical trials of statins in oncology may explain their failure—lessons for drug repurposing. Cancer Treat Rev. 2018;69:84–9. https://doi.org/10.1016/j.ctrv.2018.06.010.
124. U.S. Food and Drug Administration. FDA approves capivasertib with fulvestrant for breast cancer. https://www.fda.gov/drugs/resources-information-approved-drugs/fda-approves-capivasertib-fulvestrant-breast-cancer. Accessed 15 Jan 2024.
125. National Cancer Institute. FDA expands approval of fulvestrant for advanced breast cancer. 2017. https://www.cancer.gov/news-events/cancer-currents-blog/2017/fda-fulvestrant-breast-cancer. Accessed 15 Jan 2024.
126. U.S. Food and Drug Administration. Everolimus (Afinitor). https://www.fda.gov/drugs/resources-information-approved-drugs/everolimus-afinitor. Accessed 15 Jan 2024.
127. Royce ME, Osman D. Everolimus in the treatment of metastatic breast cancer. Breast Cancer. 2015;9:73–9. https://doi.org/10.4137/BCBCR.S29268.
128. Aggarwal S, Verma SS, Aggarwal S, Gupta SC. Drug repurposing for breast cancer therapy: old weapon for new battle. Semin Cancer Biol. 2021;68:8–20. https://doi.org/10.1016/j.semcancer.2019.09.012.

129. Gelmon K, Chan A, Harbeck N. The role of capecitabine in first-line treatment for patients with metastatic breast cancer. Oncologist. 2006;11:42–51. https://doi.org/10.1634/theoncologist.11-90001-42.
130. Alqahtani FY, Aleanizy FS, El Tahir E, Alkahtani HM, AlQuadeib BT. Paclitaxel. Profiles Drug Subst Excip Relat Methodol. 2019;44:205–38.
131. Drugbank. https://go.drugbank.com/drugs/DB00997. Accessed 15 Jan 2024.

Drug Repurposing in Inflammatory Disorders

6

Keshari Kumar Sriwastawa, Vaishnavi Sawkare,
Ayaan Ansari, and Ashutosh Kumar

Abbreviations

AD Atopic dermatitis
BSCE BaseSpace correlation engine
CD Crohn's disorder
CNS Central nervous system
CRP C- reactive protein
CVS Cardiovascular system
DM Diabetes mellitus
DPA Disproportionality analysis
GIT Gastrointestinal tract
IBD Inflammatory bowel disease
IL Interleukins
KEGG Kyoto encyclopedia of genes and genomes
MMP-3 Matrix metalloproteinase 3
RA Rheumatoid arthritis
SARS-CoV-2 Severe acute respiratory syndrome coronavirus 2
SLE Systemic lupus erythematosus
SSA Sequence symmetry analysis

K. K. Sriwastawa · A. Kumar (✉)
Department of Pharmacology and Toxicology, National Institute of Pharmaceutical Education and Research (NIPER) S.A.S Nagar, Mohali, Punjab, India
e-mail: ashutosh@niper.ac.in

V. Sawkare · A. Ansari
Department of Medical Devices, National Institute of Pharmaceutical Education and Research (NIPER) S.A.S Nagar, Mohali, Punjab, India

N. Chella et al. (eds.), *Drug Repurposing*,
https://doi.org/10.1007/978-981-97-5016-0_6

TNF Tumor necrosis factor
UC Ulcerative colitis
VEGF Vascular endothelial growth factor

Learning Objectives

- Inflammation and associated diseases
- Challenges with existing therapy
- Application of drug repurposing in inflammatory disorders
- Further challenges involved in drug repurposing against inflammatory disorders

6.1 Introduction

6.1.1 Drug Repurposing

Drug repurposing is the process of finding new therapeutic applications for drugs that have already received approval for use in other medical conditions. It is essential because it may significantly save the time, expense, and risk involved in creating new medications from scratch while offering new therapeutic alternatives to patients with unmet medical requirements [1]. There are two main types of drug repurposing: indication switch and indication expansion. Indication expansion refers to using a medication previously approved for one indication to treat another within the same therapeutic field. A medicine that has previously received approval for one indication may be used to treat another indication in a different therapeutic field. There are three different techniques for repurposing drugs: mixed approaches, biological experimental approaches, and computational approaches. Computational methods involve utilizing a blend of information and data. This enables a high-level comprehension of new signaling pathways and pharmacological actions. Biological experimental techniques involve carrying out tests to evaluate the effectiveness of existing medications against novel targets or indications. Effective drug repurposing uses mixed approaches, which combine computational and experimental approaches. High-throughput screening and literature-based methods are two approaches that fall under the mixed approach. Testing a wide variety of new chemical entities for efficacy against a particular target or disease is known as high-throughput screening. The purpose of literature-based techniques is to find possible new applications for currently available drugs by reviewing existing scientific literature [2]. Drug repurposing techniques differ from traditional strategies for drug development in several ways. First, rather than creating novel drugs from the start, drug repurposing focuses on finding new applications for existing medications. Second, since the safety and pharmacokinetic properties of currently available drugs are already well established, drug repurposing can frequently be faster and inexpensive than traditional drug development. Finally, since it permits the discovery of novel therapeutic applications for medications that have previously been licensed for other purposes, drug repurposing has the potential to offer new therapy alternatives for people with unmet medical needs [3].

6.1.2 Advantages and Recent Examples of Repurposing

Drug repurposing, the practice of utilizing existing drugs for new therapeutic purposes, offers several compelling advantages in pharmaceutical development. Firstly, it is notably cost-effective compared to the arduous process of creating novel drugs from scratch. Given that the safety and efficacy of the drug have already been established, repurposing can significantly reduce development expenses. Moreover, it expedites the drug development timeline by bypassing certain phases, such as preclinical testing, as the drug has already undergone rigorous evaluation. This acceleration of the process, in turn, mitigates risks associated with developing entirely new compounds. Another significant benefit is the heightened likelihood of success. Repurposing leverages drugs that have already received approval for one indication, increasing the chances that they will prove effective for a different medical use. This not only streamlines the development process but also addresses unmet medical needs by providing new therapeutic options for conditions with limited treatments. In doing so, it ultimately contributes to improved patient outcomes and well-being. Drug repurposing reduces regulatory burdens, benefiting patients and increasing revenue potential for pharmaceutical companies. It broadens a drug's market reach, driving reinvestment in research. Additionally, it is environmentally friendly, reducing animal testing and aligning with sustainability. Drug repurposing fosters collaboration, spurring innovation in medicine. Overall, it is a multifaceted strategy with far-reaching benefits for the pharmaceutical industry and global healthcare [4]. In recent times, drug repurposing has revealed exciting prospects in medicine. Baricitinib, originally a remedy for rheumatoid arthritis, was repurposed to combat severe inflammation in COVID-19 cases [5]. Clofazimine, traditionally used against leprosy, emerged as a potential glioblastoma treatment, promising improved therapies for brain cancer. Naltrexone, a cornerstone in addiction management, expanded its role to address conditions like fibromyalgia, Crohn's disease, and multiple sclerosis. These examples highlight the transformative potential of repurposing existing drugs, ushering hope and innovation in medical treatments [6, 7].

6.1.3 Inflammatory Disorders

Inflammation is desirable because it is an ordinary part of the body's defense against damage or infection. But inflammation can turn into the dark side of the sword when it affects healthy tissues or extends to the body for a long time, which is referred to as chronic inflammation. It results from many factors such as environmental chemicals, various injuries, genetic predisposition, and autoimmune conditions in which it plays a central role. It is involved in multiple disorders and affects almost all organ systems. Majorly involved are GIT, respiratory, CNS, CVS, etc. [8].

Different types of inflammatory disorders are as follows:

1. **Gastrointestinal disorders:** Anxiety and gastrointestinal (GI) disorders are prevalent and prompting clinical difficulties across the globe. Clinical traits of GI diseases can be complicated and are frequently linked to symptoms of anxiety. It is still completely unknown what causes GI illnesses and anxiety and the similar path mechanisms that underlie both conditions. This disorder involves inflammatory bowel disease (IBD), which includes Crohn's disorder (CD) and ulcerative colitis (UC), which is a severe progressive condition [9]. In COVID-19, the GI tracts get affected either by infections in ventilated patients or those who have received corticosteroids [10].
2. **Autoimmune diseases:** This disease is one in which the immune system unintentionally assaults the host. These illnesses induce the body's defenses to attack numerous organs in a systemic reaction, or they may only affect one organ locally, such as the skin. Every autoimmune illness impacts a significantly higher proportion of women. This illness comprises the following: Sjogren's syndrome, systemic sclerosis, rheumatoid arthritis, psoriasis, and systemic lupus erythematosus (SLE). Most cases of rheumatoid arthritis are marked by painful, swollen, stiffened joints, frequently accompanied by fever, exhaustion, and a sense of being weightless. Due to weakened synovium of the joints, it causes persistent inflammation in many joints. Various factors include the weather, poor dietary and lifestyle choices, inactivity, elevated stress levels, and/or inadequate sleep [11].
3. **Lung diseases:** Chronic airway inflammation has a role in the underlying causes of several respiratory illnesses, including pulmonary fibrosis, chronic obstructive disease, and asthma. This long-term inflammation brought on by an inappropriate immunity reaction toward non-pathogenic stimuli is the most common cause of inflammatory lung illnesses. The nose and tiny airways are merely two instances of the airways that this inflammation may affect. A dynamic interaction between cells and mediators results in inflammation. It majorly affects asthma triggered by inflammation in the respiratory system to generally harmless ambient inorganic and organic chemicals [12]. Patients with asthma were thought to be more susceptible than the general population to getting SARS-CoV-2 and the consequent, more severe COVID-19 disease, as proved and determined by studies with different respiratory viruses known to cause asthma flare-ups, such as the rhinovirus, influenza virus, and respiratory syncytial virus [13]. In a similar way, prior infections with rhinovirus suppress immunity and promote inflammation in asthma and COVID-19 [14].
4. **Neurodegenerative diseases:** Acute and chronic neurodegenerative conditions like Alzheimer's and Parkinson's disease, stroke, paraneoplastic condition, traumatic brain injury, and multiple sclerosis all exhibit neurodegeneration, which is the slow and progressive dysfunction and loss of neurons and axons in the central nervous system [15]. These detrimental cognitive effects are partially caused by neuroinflammation in the brain, especially in the hypothalamus and hippocampus, which is caused by obesity-associated systemic inflammation.

Intracellular insulin signaling is interfered with by high quantities of free fatty acids and inflammatory chemicals, which can result in a neuroinflammatory process [16].

5. **Metabolic diseases:** The concept of metabolic syndrome refers to an association of insulin resistance, diabetes, dyslipidemia, and hypertension corresponding to obesity, especially visceral adiposity [17]. High rates of death and disability are driven by these metabolic abnormalities, which also raise the risk of developing cardiovascular illness and type 2 diabetes mellitus (T2DM). Adipose tissue, liver muscle, and pancreas are sites of inflammation in the presence of obesity [18].
6. **Mental illnesses:** Severe mental illness (SMI), including major depressive disorder, bipolar disorder, and schizophrenia, is associated with increased inflammation. Systemic inflammation causes major changes in the functioning of the brain, and its impact on various disorders can be seen in its development. Tumor necrosis factor (TNF-α), a major proinflammatory molecule, changes the arrangement of CNS parenchyma from systemic blood circulation. Increased levels of proinflammatory molecules (IL-6, IL-1, TNF-α) are prone to anxious and depressed behavior [19].
7. **Cardiovascular (CVS) diseases:** Various CVS diseases now appear as a secondary effect of systemic inflammatory conditions. Along with conventional risk factors and normal mechanisms, inflammation accelerates plaque formation and poses problems of various artery blockages. The initiation of inflammation at the vascular and vessel injury site mediates and progresses the blockage conditions. Endothelial dysfunction is the central determinant in various autoimmune disorders and CVS diseases as it involves the recruitment of different types of proinflammatory molecules and cytokines, further aggravating the condition and finally disposing to a disease condition. The well-known biomarker C-reactive protein (CRP) is a mediator and marker of inflammation and CVS risk [20].
8. **Atopic dermatitis:** Common, chronic inflammatory dermatoses like atopic dermatitis (AD) are defined by persistent itching. Depending on the patient's age and the degree of skin dryness, the clinical presentation includes eczema-like eruptions, such as erythema, papules, and exudative lesions of an area in particular. The skin thickens and lichens as a result of persistent irritation and scratching. The pathogenesis of AD is based on aberrant inflammation in the TH2-produced mediators such as IL-4 and IL-13. T-helper (Th)1; Th22; and to a lesser extent, Th 17 cytokine-specific immune responses are strengthened during the acute stage of the illness. On the other hand, Th2 and Th22 responses grow more pronounced during the chronic stage, while Th1 responses become less pronounced. Around the world, AD affects 20% of children and 2–7% of adults, while regional prevalence varies [21].

The different types of drugs available for various inflammatory diseases have been mentioned in Table 6.1.

Table 6.1 Drugs available for inflammatory diseases

S. No.	Types of inflammatory disorder	Drugs	Drug type	References
1.	Gastrointestinal disorder (Inflammatory bowel diseases)	Mesalazine, olsalazine	Amino salicylates	[22]
		Prednisolone	Corticosteroid	
		Azathioprine Methotrexate	Thiopurines	
		Ciprofloxacin	Antibiotics	
		Infliximab, adalimumab	TNF blockers	
		Natalizumab Vedolizumab	Leucocyte adhesion inhibitors	
		Ustekinumab	Anti-Interleukin-12/13	
		Tofacitinib	JAK inhibitor	
2.	Autoimmune diseases (rheumatoid arthritis)	Methotrexate, azathioprine, cyclosporine	Immunosuppressants	[23]
		sulfasalazine, leflunomide	Others immunomodulators	
		etanercept, infliximab, adalimumab	TNF-alpha inhibitor	
		Anakinra, abatacept, rituximab	Other biologicals	
3.	Lung diseases (asthma)	Salbutamol, theophylline, ipratropium bromide, tiotropium bromide	Bronchodilators	[24]
		Omalizumab	Anti- IgE antibody	
		Montelukast, zafirlukast	Leukotriene antagonists	
		hydrocortisone, prednisolone, budesonide	Corticosteroids	
4.	Others (atopic dermatitis)	Tacrolimus, pimecrolimus	Calcineurin inhibitors	[25]
		Betamethasone dipropionate augmented, amcinonide, clobetasol, fluocinonide	Corticosteroids	

6.1.4 Unmet Needs in Therapeutics Available for Inflammatory Disorders

Even though there are various treatments for inflammatory illnesses, there are still a lot of unmet needs. Some individuals struggle to respond well to current medications, and many inflammatory diseases can be challenging to treat. Therapies with fewer adverse effects than considerable negative effects of many of the available therapies for inflammatory illnesses can restrict their use. Many existing therapies merely work to control the signs and symptoms. Treatments are required that can delay or stop the progression of the disease. Restoring eubiosis, restoring barrier function, and mucosal healing are a few examples of indicators for predicting the course of the illness, and predicting/monitoring the response to medication are some of the issues in IBD research. IBD patients have several complicated unmet needs, including managing medical problems, administering medicine with precision, and receiving behavioral and neuromodulatory therapy. Innovative and creative medicines are thus needed to satisfy these needs and improve the quality of life for IBD patients [26].

The best method to treat RA is to avoid getting it in its initial place because there is no known cure for it. However, there aren't any reliable preventive measures accessible right now. Major problems include drug-free remission, improved remission measurement methods, and novel taper therapy strategies. The prevention, treatment, and better care of patients in relapse are three main unmet needs for RA. More research is required to fulfill these needs and enhance the quality of life for RA patients [27].

Inadequate treatment compliance, improper use of inhalers, inability to access healthcare, insufficient diagnosis and care, comorbidities, and severity of the illness are all major needs for asthma disease. These unmet demands may significantly impact the lives of those who have asthma. They may result in worsened symptoms like wheezing, breathing difficulties, and tightness in the chest. Additionally, they may cause exacerbations to occur more frequently, necessitating hospitalization. Asthma can potentially be lethal in severe circumstances [28].

Plenty of individuals do not respond well to the medicines that have been approved thus far, necessitating the need for more potent alternative target therapies. The patient may not prefer subcutaneous injections, nor is it always the best option. The intricate link between AD severity, comorbidities, and options requires more investigation. A precise definition of remission for AD is not yet available: a specific therapeutic aim must be established. A further unmet need in AD management is the distinction between short-, medium-, and long-term goals. In conclusion, there is a need for atopic dermatitis treatments that are more potent, safer, and more suited to each patient's unique needs [29].

Despite recent advancements, there are still a lot of unmet necessities in the field of treating inflammatory conditions. Further study is required to create more efficient, safe, and cheap treatments for these ailments. The various unmet needs of different inflammatory diseases are shown in Fig. 6.1.

6.2 Drug Repurposing for Inflammatory Disorders

6.2.1 Drug Repurposing for Rheumatoid Arthritis (RA)

Rheumatoid arthritis (RA) is an autoimmune disease that causes inflammation in the joints, producing complications such as synovial inflammation, pannus

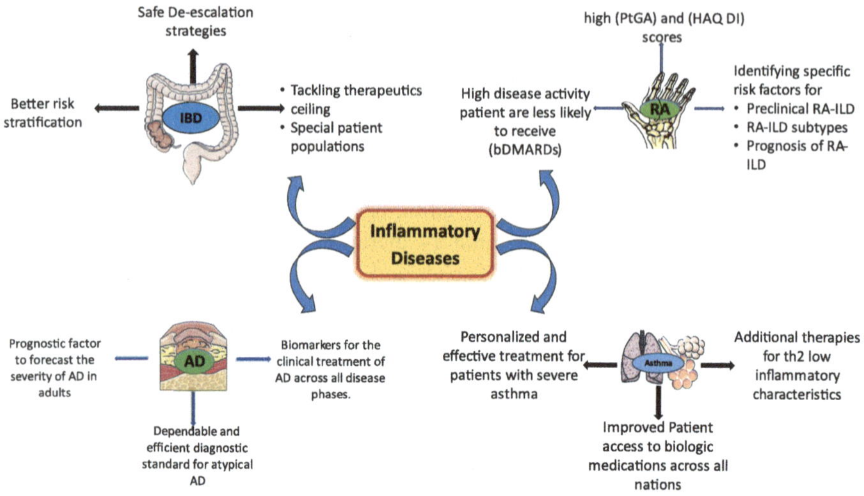

Fig. 6.1 (1) Inflammatory disease, therapeutics, and unmet needs in inflammatory bowel diseases [30]. (2) Unmet needs in rheumatoid arthritis [27]. (3) Unmet needs in atopic dermatitis [31]. (4) Unmet needs in asthma [32]. The artwork (Digestive system, Bone, Dermatology, Respiratory system) used in this figure has been taken from servier medical art. (http://servier.com/Powerpoint-image-bank). Servier medical art by Servier is licensed under a Creative Commons attribution 3.0 Unported license

hyperplasia, and damage to the joint cartilage. Drug repurposing for RA becomes essential because of the increased cost and precipitation of adverse drug reactions of the currently available therapies. Based on its ameliorative effects on the excessive production of fibroblast cells, recruitment of inflammatory cells, and collagen deposition, as revealed in in vivo research, pirfenidone, an anti-pulmonary fibrosis drug, was studied to repurpose its impact on RA. The pharmacological activity of pirfenidone against inflammation and angiogenesis makes it a liable candidate to be investigated against RA targeting synovial inflammation damage to the joints (Fig. 6.2a). It was reported that pirfenidone reduced synovial hyperplasia and inflammatory cell recruitment, joint inflammation, and damage in a collagen-induced arthritis rat model, along with decreasing the expression of Matrix Metalloproteinase 3 (MMP-3) and vascular endothelial growth factor (VEGF) in the articular chondrocytes and synovial cells. Moreover, the in vitro studies on TNF-α induced MH7A and EA.hy926 cell lines reported a decrease in the levels of cytokines such as interleukins (IL-1β, IL-8, and IL-6), MMP-1/3/2/9, VEGF and MMP-2/9 & VEGF, respectively [33] As per a study, antipsychotics possess the ability to affect several cytokines, which are also regulated by the available drugs for RA. Based on this ground, drug repurposing of haloperidol was performed by a method known as data mining. This method involves the use of big data. The chronology of this technique involves the use of signals known as disproportionality analysis (DPA) and sequence symmetry analysis (SSA) for screening the test drugs, followed by BSCE (BaseSpace Correlation Engine) software application to make a comparative analysis of the

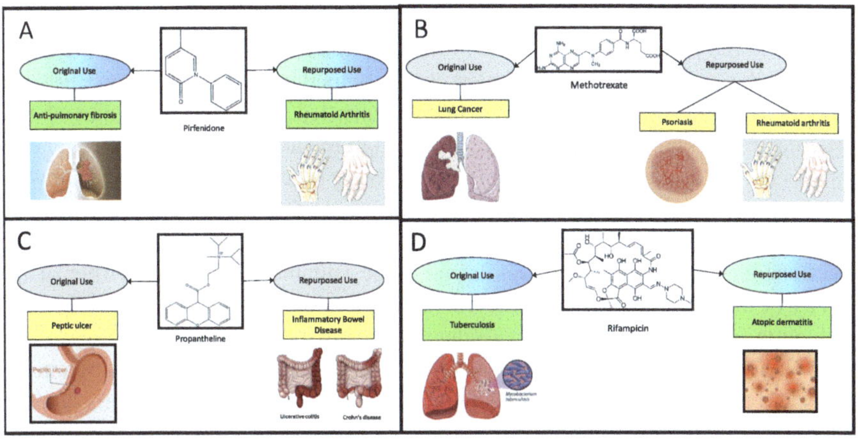

Fig. 6.2 Drugs repurposed for inflammatory diseases. (**a**) Pirfenidone repurposed for rheumatoid arthritis. (**b**) Methotrexate repurposed for psoriasis and rheumatoid arthritis. (**c**) Propantheline repurposed for inflammatory bowel disease. (**d**) Rifampicin repurposed for atopic dermatitis. The artwork (bones, lungs, and other organs) used in this figure has been taken from Servier Medical Art. (http://servier.com/Powerpoint-image-bank). Servier Medical Art by Servier is licensed under a Creative Commons Attribution 3.0 Unported License

investigational drugs signature with that of RA signatures, and finally interpretation of BSCE results to elucidate the molecular mechanism of action of the investigational drugs by the use of pathway databases, namely Reactome, Kyoto Encyclopedia of Genes and Genomes (KEGG), and ComPath. The results obtained from this study following these methods reported that haloperidol has the potential to act as an antirheumatic drug by repressing the pathogenesis of RA. BSCE bioinformatic analysis revealed that haloperidol exerts a significant pharmacological effect on RA via several immunological pathways, such as cytokine and chemokine signaling, major histocompatibility complex (MHC) class-II antigen presentation, and toll-like receptor (TLR) cascade pathway [34]. A study conducted to repurpose antidiabetic drugs for RA performed a 2-sample Mendelian randomization study to establish a relationship between the target genes of the antidiabetic drugs (insulin and its analogues, sulfonylurea and thiazolidinediones) and the pathogenesis of RA. Through assessment of genetic variations and genome-wide association studies, it was reported that thiazolidinediones possessed the least chance of causing RA and hence can be investigated for the mechanistic study [35].

6.2.2 Drug Repurposing for Inflammatory Bowel Disease (IBD)

Inflammatory Bowel Disease (IBD) includes a broad plethora of diseases, including ulcerative colitis and Crohn's disease, which target the intestine, leading to inflammation. Various studies have been conducted and are still being performed to illustrate the exact pathogenesis of this disease. A study focused on drug repurposing for

IBD through an in silico approach is based on the fact that genetic variations play a vital role in the pathogenesis of IBD. Following this approach, genome-wide association studies (GWAS) were conducted, which helped in the identification of single nucleotide polymorphisms (SNPs), which in turn led to the identification of the particular genes known as IBD risk genes through meta-analyses that act as genetic markers for the disease. Gene2D analysis involving several computational pathway databases and the IBD risk genes identified several test drugs, and their therapeutic reliability for IBD was analyzed. Cefadroxil, originally indicated for urinary tract infections and skin infections, was found to be suitable for IBD by virtue of its activity of inhibiting the enzyme alanine aminopeptidase, which is involved in the production of proinflammatory cytokines. Propantheline, an anticholinergic drug approved for the treatment of peptic ulcer and also used for enuresis and abdominal pain, can be repurposed for IBD because it exerts an analgesic effect on the pain arising due to inflammation in the intestines [22] (Fig. 6.2b). Leflunomide, an anti-rheumatic drug, has been explored for its therapeutic activity in IBD, which occurs due to a deficient gene known as tetratricopeptide repeat domain 7A (TTC7A). Any genetic variation in this gene leads to intestinal and immunological complications such as apoptotic enterocolitis, multiple intestinal atresia, and recurrent intestinal stenosis. In vitro and in vivo studies were conducted on TTC7A- knockout (TTC7A-KO) HAP1 (human haploid cells) and zebrafish. It was reported that TTC7A-KO HAP1 cells are subjected to death due to aberrant expression of caspase-3 and caspase-7, and leflunomide had an inhibitory effect on the expression of these caspases and decreased apoptosis. In zebrafish, leflunomide exerted ameliorative effects on reduced gut motility and intestinal dysfunctions by blocking the TTC7A gene [36]. A research team evaluated the therapeutic efficacy of topiramate, an anti-epileptic drug that enhances the activity of the neurotransmitter GABA (gamma-aminobutyric acid) in rats where IBD was induced using trinitrobenzene sulfonic acid, and it was found that topiramate decreased diarrhea, ameliorated the severe complications of the colon observed through biopsy and endoscopy [2].

6.2.3 Drug Repurposing for Atopic Dermatitis

Atopic dermatitis (AD) is a chronic inflammatory skin disorder caused due to hypersensitivity of the skin toward external allergens, which may also lead to asthma and allergic rhinitis [37]. Several drugs have been repurposed for AD, such as rifampicin, which is an antibiotic indicated for various diseases such as tuberculosis (TB), leprosy, and Legionnaires disease, exerts an inhibitory effect on the aggravated production of IgE and IL-4 in the serum, which is established hallmarks of this disease. Rifampicin possesses anti-inflammatory and anti-allergic activities pertaining to its pharmacological activity to reduce the production of histamine and β-hexosaminidase released from mast cells and basophils, respectively. Hence, it is proving to be a suitable drug to be repurposed for atopic dermatitis (Fig. 6.2d). Ethanolic extracts of Artemisia apiacea Hance have been reported to exert anti-inflammatory by suppressing the expression of proinflammatory mediators, namely

RANTES, IL-8, IL-6, IκBα, NF-κBp65, p38, and TARC [38]. Moreover, in an in vivo model of 2,4-dinitrochlorobenzene-induced atopic dermatitis, ethanolic extracts of Artemisia apiacea have been reported to show ameliorative effects on the enlarged ear and dorsal skin thickness in mice, proving to be an optimum candidate for the treatment of atopic dermatitis [39]. Another drug repurposing study evaluated the therapeutic efficacy of levocetirizine-loaded emulgel consisting of tamanu oil and sericin in atopic dermatitis. Incorporating tamanu oil and sericin into the emulgel enhanced the therapeutic efficacy of levocetirizine when applied through the topical route in the 2,4-dinitrochlorobenzene-induced atopic dermatitis rat model. This was further validated by the decrease in the pharmacodynamic parameters such as scratching frequency and erythema score in contrast to the typically formulated emulgels [40].

The different drugs that have been repurposed for various inflammatory diseases have been mentioned in Table 6.2.

6.3 Current Scenario and Future Scope

Drug repurposing or repositioning has been touted as a novel and useful strategy to mitigate various diseases as it involves the reuse of already approved drugs that have undergone clinical trials with an established safety profile. This method is cost-effective and time-saving, and various experimental data and studies report that drug repurposing should be done to bring out multiple therapeutic applications of the same drug while also keeping in mind to overcome the various administrative and technological challenges and unmet therapeutic needs associated with drug repurposing [2]. Inflammation is a common defensive action of the body in response to various harmful and obnoxious stimuli and has been reported to be an established

Table 6.2 Drugs repurposed for inflammatory disorders with their original use

Drug	Original use	Repurposed use	References
Baricitinib	Rheumatoid arthritis	COVID-19-related inflammation, systemic lupus erythematosus, psoriasis	[5]
Methotrexate	Cancer	Psoriasis, rheumatoid arthritis	[41]
Tofacitinib	Rheumatoid arthritis	Ulcerative colitis, Crohn's disease	[42]
Haloperidol	Psychosis	Rheumatoid arthritis	[34]
Minoxidil	High blood pressure	Alopecia areata	[43]
Montelukast	Asthma	Allergic rhinitis, chronic obstructive pulmonary disease (COPD), eosinophilic esophagitis	[44, 45]
Rifampicin	Tuberculosis	Atopic dermatitis	[46]
Pirfenidone	Anti-pulmonary fibrosis	Rheumatoid arthritis (RA)	[33]
Levocetirizine	Antihistamine	Atopic dermatitis	[47]
Leflunomide	Antirheumatic	Inflammatory bowel disease (IBD)	[36]

hallmark in almost every disease. The inability to control and modulate inflammation leads to chronic complications ranging from inflammatory dermatoses to diseases like cancer and diabetes. The pharmaceutical sector continues to suffer from a lack of efficiency in discovering novel medicines to regulate inflammation.

Furthermore, the traditional drug development approach has failed to provide high yields and quality treatments. Hence, it has become the need of the hour to repurpose drugs for inflammatory diseases [48]. Drug repurposing has been done for various inflammatory conditions such as RA, IBD, atopic dermatitis, sepsis, and asthma through in silico, in vitro, and in vivo approaches with an attempt to bring about a paradigm shift in the drug discovery process and providing patient compliance [22, 33–40]. A recent survey was done to identify the major obstacles and potentiators for drug repurposing revealed that out of 11,814 articles, 66% articles reported that deficient and unoptimized treatments, safety challenges, ineffective research architecture, complexity of the disease, unavailability of sufficient clinical and preclinical information, intellectual property, and regulatory issues are the major obstacles for drug repurposing. Collaborative studies, provision of access to scientific information about novel drugs through scientific databases and software, along with supplementation of funds by the government, can enhance drug repositioning and establish it as a proper, efficient, and validated therapeutic option to mitigate inflammatory as well as other diseases [49].

6.4 Summary

This book chapter discussed the definition, types, advantages, and need for drug repurposing. We have overviewed the different kinds of inflammatory diseases, such as gastrointestinal diseases, metabolic diseases, cardiovascular diseases, etc. It has been found from an etiological perspective that inflammation has a vital involvement in the pathogenesis of almost every disease since exaggerated inflammation surpasses its defensive role and precipitates other complications. Hence, creating more stable therapeutic strategies to mitigate inflammatory diseases in which drug repurposing comes into the frame becomes imperative. It is evident from the current scenario that drug repurposing is a very beneficial strategy to cure a broad spectrum of inflammatory and other associated diseases since it is cost-effective, bypasses the conventional time-consuming drug approval process, and is devoid of any safety concerns. We also discussed the unmet needs in the therapy of inflammatory diseases such as inflammatory bowel diseases, atopic dermatitis, rheumatoid arthritis, etc. We have also highlighted various studies done in the context of drug repurposing for various inflammatory diseases for the above-mentioned diseases. We have mentioned the recent cases of drug repurposing of different classes of drugs, such as antihistaminic, anticancer, antirheumatic, antitubercular, antiasthmatic, antihypertensives, etc., which have been repurposed for inflammatory diseases. Overall, it can be seen that unoptimized treatments, safety challenges, ineffective research architecture, complexity of the disease, unavailability of sufficient clinical and preclinical information, intellectual property, and regulatory issues are the major hurdles to drug repurposing. Collaborative studies, provision of access to scientific information about novel drugs through scientific databases and software, and supplementation of funds by the government can enhance drug repositioning and establish it as a proper, efficient, and validated therapeutic option to mitigate inflammatory and other diseases. Hence, it has become the need of the hour to accelerate the drug repurposing process, which can open new gates in the field of pharmaceutical research to establish the wide therapeutic applicability of the drugs to provide patient compliance, which can be a boon to human welfare.

References

1. Singh TU, Parida S, Lingaraju MC, Kesavan M, Kumar D, Singh RK. Drug repurposing approach to fight COVID-19. Pharmacol Rep. 2020;72:1479–508.
2. Pushpakom S, Iorio F, Eyers PA, Escott KJ, Hopper S, Wells A, et al. Drug repurposing: progress, challenges and recommendations. Nat Rev Drug Discov. 2018;18:41–58.
3. Dotolo S, Marabotti A, Facchiano A, Tagliaferri R. A review on drug repurposing applicable to COVID-19. Brief Bioinform. 2021;22:726–41.
4. Fetro C, Scherman D. Drug repurposing in rare diseases: myths and reality. Therapie. 2020;75(2):157–60.
5. Saber-Ayad M, Hammoudeh S, Abu-Gharbieh E, Hamoudi R, Tarazi H, Al-Tel TH, et al. Current status of baricitinib as a repurposed therapy for COVID-19. Pharmaceuticals. 2021;14(7):680.
6. Kučić N, Rački V, Šverko R, Vidović T, Grahovac I, Mršić-Pelčić J. Immunometabolic modulatory role of naltrexone in BV-2 microglia cells. Int J Mol Sci. 2021;22(16):8429.
7. Mulkearns-Hubert EE, Torre-Healy LA, Silver DJ, Eurich JT, Bayik D, Serbinowski E, et al. Development of a Cx46 targeting strategy for cancer stem cells. Cell Rep. 2019;27(4):1062–1072.e5.
8. Ahmed AU. An overview of inflammation: mechanism and consequences. Front Biol (Beijing). 2011;6(4):274.
9. Selvaraj DB, Vergil Andrews JF, Anusuyadevi M, Kandasamy M. Ranitidine alleviates anxiety-like behaviors and improves the density of pyramidal neurons upon deactivation of microglia in the CA3 region of the hippocampus in a cysteamine HCl-induced mouse model of gastrointestinal disorder. Brain Sci. 2023;13(2):266.
10. Cappell MS, Tobi M, Friedel DM. The impact of COVID-19 infection on miscellaneous inflammatory disorders of the gastrointestinal tract. Gastroenterol Clin N Am. 2023;52:115–38.
11. Angum F, Khan T, Kaler J, Siddiqui L, Hussain A. The prevalence of autoimmune disorders in women: a narrative review. Cureus. 2020;13:e8094.
12. Gillissen A, Paparoupa M. Inflammation and infections in asthma. Clin Respir J. 2015;9(3):257–69.
13. Assaf S, Stenberg H, Jesenak M, Tarasevych SP, Hanania NA, Diamant Z. Asthma in the era of COVID-19. Respir Med. 2023;218:107373.
14. Radzikowska U, Eljaszewicz A, Tan G, Stocker N, Heider A, Westermann P, et al. Rhinovirus-induced epithelial RIG-I inflammasome suppresses antiviral immunity and promotes inflammation in asthma and COVID-19. Nat Commun. 2023;14(1):2329.
15. Amor S, Puentes F, Baker D, van der Valk P. Inflammation in neurodegenerative diseases. Immunology. 2010;129(2):154–69.
16. Ana AP, Pedro PH, Frihling BEF, Souza E, Silva P, De Moraes LFRN, Migliolo L. Adipose tissue, systematic inflammation, and neurodegenerative diseases. Neural Regen Res. 2023;18:38–46.

17. Kumar A, Datusalia AK. Metabolic stress and inflammation: implication in treatment for neurological disorders. CNS Neurol Disord Drug Targets. 2018;17(9):642–3.
18. Esser N, Legrand-Poels S, Piette J, Scheen AJ, Paquot N. Inflammation as a link between obesity, metabolic syndrome and type 2 diabetes. Diabetes Res Clin Pract. 2014;105:141–50.
19. Firth J, Veronese N, Cotter J, Shivappa N, Hebert JR, Ee C, et al. What is the role of dietary inflammation in severe mental illness? A review of observational and experimental findings. Front Psychiatry. 2019;15:10.
20. Abou-Raya A, Abou-Raya S. Inflammation: a pivotal link between autoimmune diseases and atherosclerosis. Autoimmun Rev. 2006;5:331–7.
21. Mesjasz A, Zawadzka M, Chałubiński M, Trzeciak M. Is atopic dermatitis only a skin disease? Int J Mol Sci. 2023;24(1):837.
22. Grenier L, Hu P. Computational drug repurposing for inflammatory bowel disease using genetic information. Comput Struct Biotechnol J. 2019;17:127–35.
23. Ben Mrid R, Bouchmaa N, Ainani H, El Fatimy R, Malka G, Mazini L. Anti-rheumatoid drugs advancements: new insights into the molecular treatment of rheumatoid arthritis. Biomed Pharmacother. 2022;151:113126.
24. Bereda G (2022) Gudisa B. asthma [Internet]. www.austinpublishinggroup.com
25. Dattola A, Bennardo L, Silvestri M, Nisticò SP. What's new in the treatment of atopic dermatitis? Dermatol Ther. 2019;32(2):e12787.
26. Honig G, Larkin PB, Heller C, Hurtado-Lorenzo A. Research-based product innovation to address critical unmet needs of patients with inflammatory bowel diseases. Inflamm Bowel Dis. 2021;27(Supplement_2):S1–S16.
27. Mankia K, Di Matteo A, Emery P. Prevention and cure: the major unmet needs in the management of rheumatoid arthritis. J Autoimmun. 2020;110:102399.
28. Busse WW, Kraft M. Current unmet needs and potential solutions to uncontrolled asthma. Eur Respir Rev. 2022;31(163):210176.
29. Patrizi A, Costanzo A, Patruno C, Busà VM, Chiricozzi A, Girolomoni G. Unmet needs in atopic dermatitis management: an expert consensus. J Dermatolog Treat. 2022 Jul 4;33(5):2459–65.
30. Revés J, Ungaro RC, Torres J. Unmet needs in inflammatory bowel disease. Curr Res Pharmacol Drug Discov. 2021;2:100070.
31. Lobefaro F, Gualdi G, Di Nuzzo S, Amerio P. Atopic dermatitis: clinical aspects and unmet needs. Biomedicines. 2022;10:2927.
32. Caminati M, Vaia R, Furci F, Guarnieri G, Senna G. Uncontrolled asthma: unmet needs in the management of patients. J Asthma Allergy. 2021;14:457–66.
33. Gan D, Cheng W, Ke L, Sun AR, Jia Q, Chen J, et al. Repurposing of pirfenidone (anti-pulmonary fibrosis drug) for treatment of rheumatoid arthritis. Front Pharmacol. 2021;5:12.
34. Nakagawa C, Yokoyama S, Hosomi K, Takada M. Repurposing haloperidol for the treatment of rheumatoid arthritis: an integrative approach using data mining techniques. Ther Adv Musculoskelet Dis. 2021;13:1759720X2110470.
35. Qin C, Diaz-Gallo LM, Tang B, Wang Y, Nguyen TD, Harder A, et al. Repurposing antidiabetic drugs for rheumatoid arthritis: results from a two-sample mendelian randomization study. Eur J Epidemiol. 2023;38(7):809–19.
36. Jardine S, Anderson S, Babcock S, Leung G, Pan J, Dhingani N, et al. Drug screen identifies leflunomide for treatment of inflammatory bowel disease caused by TTC7A deficiency. Gastroenterology. 2020;158(4):1000–15.
37. Leung DYM, Boguniewicz M, Howell MD, Nomura I, Hamid QA. New insights into atopic dermatitis. J Clin Investig. 2004;113(5):651–7.
38. Prakash AV, Park JW, Seong JW, Kang TJ. Repositioned drugs for inflammatory diseases such as sepsis, asthma, and atopic dermatitis. Biomol Ther. 2020;28(3):222–9.
39. Yang JH, Lee E, Lee B, Cho WK, Ma J, Park KI. Ethanolic extracts of Artemisia apiacea Hance improved atopic dermatitis-like skin lesions in vivo and suppressed TNF-alpha/IFN-gamma–induced proinflammatory chemokine production in vitro. Nutrients. 2018;10(7):806.

40. Pal RR, Parashar P, Singh I, Saraf SA. Tamanu oil potentiated novel sericin emulgel of levocetirizine: repurposing for topical delivery against DNCB-induced atopic dermatitis, QbD based development and *in vivo* evaluation. J Microencapsul. 2019;36(5):432–46.
41. Misra DP, Gasparyan AY, Zimba O. Benefits and adverse effects of hydroxychloroquine, methotrexate and colchicine: searching for repurposable drug candidates. Rheumatol Int. 2020;40(11):1741–51.
42. Mishra S, Jena A, Kakadiya R, Sharma V, Ahuja V. Positioning of tofacitinib in treatment of ulcerative colitis: a global perspective. Expert Rev Gastroenterol Hepatol. 2022;16(8):737–52.
43. Abdallah MAER, Shareef R, Soltan MY. Efficacy of intradermal minoxidil 5% injections for treatment of patchy non-severe alopecia areata. J Dermatol Treat. 2022;33(2):1126–9.
44. Krishnamoorthy M, Mohd Noor N, Mat Lazim N, Abdullah B. Efficacy of montelukast in allergic rhinitis treatment: a systematic review and meta-analysis. Drugs. 2020;80(17):1831–51.
45. Chen Z, Qian X. Effect of montelukast/budesonide formoterol powder inhalation in chronic obstructive pulmonary disease. Trop J Pharm Res. 2022;21(5):1037–43.
46. Prakash AV, Park JW, Seong JW, Kang TJ. Repositioned drugs for inflammatory diseases such as sepsis, asthma, and atopic dermatitis. Biomol Ther (Seoul). 2020;28(3):222–9.
47. Pal RR, Maurya AK, Parashar P, Saraf SA. A comparative study of levocetirizine loaded vesicular and matrix type system for topical application: appraisal of therapeutic potential against atopic dermatitis. J Pharm Innov. 2021;16(3):469–80.
48. Hall CJ, Wicker SM, Chien AT, Tromp A, Lawrence LM, Sun X, et al. Repositioning drugs for inflammatory disease-fishing for new anti-inflammatory agents. Dis Model Mech. 2014;7(9):1069–81.
49. Krishnamurthy N, Grimshaw AA, Axson SA, Choe SH, Miller JE. Drug repurposing: a systematic review on root causes, barriers and facilitators. BMC Health Serv Res. 2022;22(1):970.

Drug Repurposing in CNS Disorders

7

Nazir Hussain, Snigdha Singh, and Amit Alexander

Abbreviations

AD	Alzheimer's disease
ADHD	Attention-deficit hyperactivity disorder
AEDs	Anti-epileptic medications
ALS	Amyotrophic lateral sclerosis
ASD	Autism spectrum disorder
AVM	Arteriovenous malformation
Aβ	Amyloid-beta
cGMP	Cyclic guanosine monophosphate
CNS	Central nervous system
COMT	Catechol- O-methyl transferase
CysLT1	Cysteinyl leukotriene
DF	Dimethyl fumarate
DM	Diabetes mellitus
DR	Drug repositioning
FTD	Frontotemporal dementia
GLP	Glucagon-like-peptide-1
HCV	Hepatitis C virus
HD	Huntington's disease
MAOI	Monoamine oxidase inhibitors
MPO	Myeloperoxidase
MS	Multiple sclerosis
NDD	Neurodevelopmental disorders

N. Hussain · S. Singh · A. Alexander (✉)
Department of Pharmaceutics, National Institute of Pharmaceutical Education and Research (NIPER), Guwahati, Assam, India
e-mail: amit@niperguwahati.in; amit.alexander@niperguwahati.ac.in

N. Chella et al. (eds.), *Drug Repurposing*,
https://doi.org/10.1007/978-981-97-5016-0_7

NFT	Neurofibrillary tangles
NMDA	N-methyl D-aspartate
PD	Parkinson disease
PDE-5	Phosphodiesterase-5
PK	Pharmacokinetic
PPARγ	Peroxisome-proliferated activated receptor γ
SCA	Sickle cell anemia
SCI	Spinal cord injury
SMA	Spinal Muscular Atrophy
SMA	Survival of motor neuron
SNRI	Serotonin and norepinephrine reuptake inhibitors
SSRI	Selective serotonin reuptake inhibitors
TBA	Traumatic brain injury

Learning Objectives

- To gain insight on the fundamentals of CNS disorders and their current treatment's availability
- To learn about the principles and strategies of drug repurposing for CNS disorders
- To gain insight on the case studies of some drugs that are repurposed for CNS disorders
- To learn about the challenges, it encountered during drug repurposing

7.1 Introduction

Central nervous system (CNS) disorders are uncommon, potentially fatal, or chronically devastating, caused by the pathogenic mutation of a particular gene or specific environmental insults (such as pesticides, metals, endotoxins, and prions) leading to a progressive loss of specific neuronal groups. Different regions of the central nervous system may be damaged, depending on the cause of the disease. This can lead to reduced motor and cognitive function, which has a substantial effect on people's quality of life [1]. The molecular, cellular, and neurological bases of CNS disorders are the subject of a great deal of research; nevertheless, there is a dearth of integration between these lines of inquiry and therapeutic repurposing. The Institute of Forum of Medicine on nervous system disorders has summarized the current obstacles to brain disorder drug research into two primary areas: knowledge discovery and regulatory affairs. The Food and Drug Administration (FDA) (2013) reported that just 2 (7%) of the 27 recently authorized drugs were indicated to be used in CNS disorders [2]. Drug development for these disorders is generally related to a high failure rate, an excessive time to market, and a financial burden that outweighs the need to treat a large amount of needy population. The etiology and pathophysiology of CNS disorders are poorly understood due to the complexity of the brain.

There is a lack of appropriate in vitro or in vivo models, biomarkers, and molecular targets to reproduce these disorders fully [3]. The genomic technology and the large data of "omics" consortia offer a chance to improve the strategy for drug repurposing as a treatment alternative for CNS disorders, particularly in psychiatric conditions. Drug repurposing may be defined as a process of locating and repurposing an established drug for a novel application. It is a cost-effective strategy because the repurposed drug has already received FDA approval, which lessens the financial burden of testing for safety, efficacy, tolerability, and side effects related to the phase I studies [4].

A great benefit of drug repurposing is that few of the cases have been shown to be suitably safe in pre-clinical trials. Additionally, the availability of clinical data along with the pre-clinical profile for approved drugs enhances the process of drug discovery. The discovery of novel drug–disease interactions is a key goal of drug repositioning. Several methods have been developed, including computational, biological, experimental techniques and a combination of the two [5]. This combination has accelerated the development of a computational technique with the availability of drug-related databases as an open resource. Simultaneously, the hybrid strategy that blends computational methods with biology significantly increases the efficacy of drug repositioning while cutting expenses.

7.2 Fundamentals of Central Nervous System Disorders

7.2.1 Overview of CNS Disorders

The CNS consists of the brain and spinal cord. The functionalities of the CNS include receiving, processing, and responding to sensory information [6]. However, the entangled network of neurons, neurotransmitters, and supporting structures also makes the CNS vulnerable to a myriad of disorders. The nervous tissues are very sensitive and delicate systems that may experience damage with little amount of force. Even though every damage is unique, every illness causes different levels of brain inflammation, and reactive microglial cells play a major role in this inflammation [7]. According to the World Health Organization (WHO) statistics, around one billion people in the globe are living with neurological disorders [8]. Some of the most frequent CNS ailments are traumatic brain injury (TBI), spinal cord injury (SCI), and stroke [9]. Likewise, Alzheimer's disease (AD) and Parkinson's disease (PD) are other CNS disorders where AD is characterized by a progressive neurodegenerative disorder associated with cognitive function declination and memory loss [10]. In comparison, Parkinson's disease (PD) results from the loss of dopamine-producing neurons and is characterized by motor symptoms like tremors and bradykinesia [11]. In this context, there were approximately 2.92 million cases of Alzheimer's disease and other dementias worldwide in 1990 and 7.24 million cases in 2019, representing a 147.95% increase in incidence between 1990 and 2019 [12]. Therefore, from these data, we can esteem that CNS disorder is a global burden.

7.2.2 Classification of CNS Disorders

CNS disorders are broadly classified into the following categories.

7.2.2.1 Neurodegenerative Disorders

As neurons are essential for communication, their existence is essential to the human brain's effective functioning [13, 14]. The majority of the neurons are produced during childhood only, but with aging in adulthood, their numbers are reduced. Specific pathological alterations in the brain, including extracellular protein deposits, cellular inclusions, and alterations in cell shape, are indicative of neurodegenerative conditions. Subsequently, neurodegenerations led to the dysfunctions of synapses, neural networks, and depositions of altered variants of proteins in the brain [15, 16]. Neurodegenerations are one of the central pathophysiology of many brain disorders [17] and are a significant health concern. AD, PD [18], prion disease, motor neuron disease, Huntington's disease (HD), spinal muscular atrophy [19], and spinocerebellar ataxia are among the most prevalent neurodegenerative disorders. The latest study on the pathophysiology of Alzheimer's disease has demonstrated that the build-up of tau and amyloid-beta (Aβ) proteins is essential to the disease's progression [20]. AD is characterized by the production of Aβ-containing plaques and NFTs made of hyperphosphorylated tau. These plaques and NFTs disrupt hippocampal circuitry, impairing short-term memory establishment into long-term traces. AD is characterized by vast neuronal damage, interrupted synaptic connections, and impairment in neurotransmitter systems crucial for brain functions, particularly memory [21]. Early-stage AD is primarily associated with selective memory impairment, affecting declarative episodic memory tied to the hippocampus and medial temporal lobe. Executive function impairment, judgment issues, and difficulties with problem-solving skills are also common clinical indices of AD and often emerge early in the disease progression [22]. In this context, it is estimated that more than 50 million people globally have AD, which is the most frequent cause of dementia and it is esteemed that this number will rise to 150 million by 2050 [12]. PD is a progressive brain disorder that causes problems like tremors, muscle stiffness, difficulty walking, and trouble with balance. It can be caused by both genetic factors and other things like age, excessive caffeine intake, smoking, and exposure to environmental toxins. Getting older is the main risk factor for Parkinson's [23, 24]. The exact reasons why some of these factors increase the risk are not entirely clear [25]. In PD, the brain changes atrophy in the frontal cortex and enlargement of the ventricles. However, the most noticeable change is the pigmentation loss in certain areas like the locus coeruleus and substantia nigra pars compacta. This happens because a type of neuron called dopaminergic neuromelanin-containing neurons die off. This cell loss affects the nigrostriatal pathway, leading to lower levels of dopamine in the striatum and causing the main motor symptoms of Parkinson's [26]. Apart from the areas mentioned, cell loss in other parts of the brain contributes to non-motor symptoms in Parkinson's. These areas include the nucleus basalis of Meynert, the raphe nuclei, the locus coeruleus, the pedunculopontine nucleus, the dorsal motor nucleus of the vagus nerve, the hypothalamus, and the olfactory bulb [27]. Parkinson's disease progresses because of various other factors, such as problems with α-synuclein, which is a protein that folds incorrectly and clumps together. Some other issues involve trouble with mitochondria (powerhouse of cells), problems with the systems that clear out unwanted proteins, issues with the ubiquitin-proteasome system, problems with the autophagy-lysosome system (ways the body cleans out waste), and inflammation in the brain. These all factors play a crucial role in developing Parkinson's disease [28, 29].

7.2.2.2 Psychiatric Disorders

Psychiatric disorders are a varied set of mental conditions characterized by abnormal mental or behavioral patterns that bring about distress or impairment for the individual. Psychological disorders, also known as mental illnesses or mental health conditions, impact one's thoughts, emotions, and actions, often influencing an individual's relationships and daily activities. Depression and anxiety are common examples of psychiatric disorders. These are associated with conditions affecting an individual's moods and emotional well-being [30]. Likewise, schizophrenia is another type of complex psychiatric disorder that involves disruptions of thought processes, emotions, and perceptions.

7.2.2.3 Neurodevelopmental Disorders

Neurodevelopmental disorders (NDDs) are conditions that are marked by deficiencies in cognition, communication, behavior, and motor skills due to atypical brain development. Intellectual debility, communication problems, autism spectrum disorder (ASD), and attention-deficit/hyperactivity disorder (ADHD) are common neurodevelopmental disorders [31]. ASD is a condition marked by challenges in social interaction and communication. Likewise, ADHD is characterized by difficulties in attention, hyperactivity, and impulsivity.

7.2.2.4 Neurovascular Disorders

The term "neurovascular disorders" refers to a wide spectrum of illnesses marked by complex interactions between the neurological and vascular systems, particularly the blood arteries supplying the brain [32]. These disorders can present in diverse ways, posing significant threats to neurological function. One common neurovascular disorder is stroke, where a disruption in blood flow to the brain causes damage to brain tissue, either through a clot (ischemic stroke) or bleeding (hemorrhagic stroke). Conditions like cerebral aneurysms and arteriovenous malformations (AVMs) increase the risk of hemorrhagic strokes by weakening the walls of blood vessels. Another noteworthy neurovascular disorder is cerebral venous thrombosis, marked by the formation of blood clots in the brain's venous sinuses, potentially leading to severe complications. Treatment strategies may involve medications to regulate blood clotting, surgical procedures, or endovascular interventions to address vascular abnormalities. Lifestyle adjustments, such as adopting a healthy diet and managing cardiovascular risk factors, also play a vital role in preventing and managing neurovascular disorders.

7.2.3 Current Treatment Approaches and Limitations

Treatment of CNS disorders is often disease-specific. Currently, several approaches are available for managing CNS disorders. Those are either the direct target specific to the pathogenesis or the indirect attempt to improve the symptoms experienced by the individuals. In subsequent sections, treatment approaches for a few CNS disorders are discussed.

7.2.3.1 Current Treatment Approach for AD

Managements of AD mainly involve the targeting of different pathways for disease progressions. US-FDA certified a few classes of drugs for the management of AD, which are described in the following section.

Cholinesterase Inhibitor

Several investigations have demonstrated that the elevated levels of acetylcholine in individuals with dementia mitigated their cognitive deterioration. The patient benefits from increased cholinergic activity when acetylcholine breakdown is limited by cholinesterase inhibitors. In AD, there is a declination level of cholinergic neurons in the cortical region of the brain. Therefore, in the current scenario, cholinesterase inhibitors are used as first-line medications for the treatment of AD [33]. Donepezil, rivastigmine, galantamine, and the *N*-methyl D-aspartate receptor antagonist memantine are some of the main drugs clinically approved by the FDA for the treatment of AD. Donepezil is prescribed for individuals with mild-to-moderate Alzheimer's disease, typically administered through oral tablets at doses of 5 or 10 mg/day. More recently, regulatory approval has been granted for using higher doses of donepezil, specifically 23 mg/day, either as a standalone treatment or in combination with memantine, for patients in the moderate-to-severe stages of the disease [34]. Rivastigmine, another acetylcholinesterase inhibitor for mild-to-moderate Alzheimer's disease, distinguishes itself by offering a transdermal patch option and targeting acetylcholinesterase and butyrylcholinesterase enzymes. Galantamine, a different class of cholinesterase inhibitors, has been approved for the same stages of AD at a daily dose ranging from 16 to 24 mg. In addition to its cholinesterase inhibitory effects, galantamine modulates nicotinic cholinergic receptors allosterically [35].

Amyloid-Beta (Aβ) Plaques Targeting

Biogen's aducanumab, a monoclonal antibody approved in June 2021 [36], represents a groundbreaking development as the first disease-modifying drug for Alzheimer's disease (AD). It is administered through intravenous infusion every 4 weeks. This IgG1 monoclonal antibody targets extracellular Aβ plaques in the brain, aiding in their clearance. While clinical data indicates a reduction in Aβ plaque load, its impact on cognitive function remains inconclusive. Despite conditional approval, the drug has generated enthusiasm among AD patients and advocacy groups, envisioning it as a pioneering step in altering the disease's pathology and paving the way for future therapies [37].

Glutamate Receptor Targeting
Glutamate is a key neurotransmitter in the brain responsible for excitatory signals. Excessive activation of glutamate receptors, especially through NMDA receptors, can lead to neuronal damage and cause neurodegeneration. However, completely blocking NMDA receptors causes severe side effects. To address this, memantine, an uncompetitive NMDA receptor antagonist, was developed. It benefits by regulating NMDA receptor activation and protecting against overstimulation [38]. Memantine was approved in 2003 for moderate-to-severe AD patients at 5–20 mg/day. Memantine has shown cognitive improvement over a placebo when used alone. Combining it with acetylcholinesterase inhibitors, donepezil in clinical trials demonstrated enhanced efficacy compared to monotherapy [39]. It is important to note that these drugs alleviate symptoms only but don't address the underlying AD pathology.

7.2.3.2 Current Treatment Approach for Parkinson's Disease

Parkinson's disease (PD) is the second most common neurodegenerative disorder, but effective therapies targeting the disease's underlying causes are lacking. Current options for the treatment of PD mainly focus on managing motor and non-motor symptoms separately for symptomatic relief.

The main approach for PD management involves restoring depleted dopamine levels in the brain's substantia nigra region. A common therapy of PD is a combination of levodopa, a dopamine precursor, with carbidopa, where levodopa helps to regain motor functions lost due to dopamine depletion. In contrast, carbidopa prevents the breakdown of levodopa outside the brain. Likewise, entacapone and tolcapone are also used to prevent levodopa breakdown through catechol-O-methyl transferase (COMT), preserving its effectiveness.

Also, dopaminergic agonists like apomorphine hydrochloride, pergolide, pramipexole dihydrochloride, ropinirole hydrochloride, and rotigotine mimic dopamine's effects and are used to treat PD. Monoamine oxidase inhibitors, such as selegiline and rasagiline, are other options to prevent dopamine breakdown in the brain [25]. Non-motor symptoms associated with Parkinson's disease include depression, psychosis, sleep difficulties, constipation, dementia, and olfactory abnormalities. These symptoms are addressed symptomatically. Rivastigmine, donepezil, and galantamine treat dementia symptoms, whereas clozapine, quetiapine, and pimavanserin treat psychotic symptoms. Melatonin or clonazepam can be used to alleviate sleep-related problems.

7.2.3.3 Current Treatment Approach for Depression and Anxiety

The symptoms of depression and anxiety are more often treated similarly. For this instance in some cases, therapy for this disorder can be tailored to an individual so that it helps to reduce the symptoms of both disorders. For the management of these disorders, several forms of psychotherapy can be effective. Subsequently, different types of antidepressant molecules are used for the treatment of depression. Antidepressants act on neurotransmitters; they play a vital role in regulating mood, cognition, and other activities like sleeping, behavior, etc. This antidepressant generally works by modulating the levels of neurotransmitters like dopamine, serotonin, and norepinephrine [40]. Different types of antidepressants are available, for example, selective serotonin reuptake inhibitors (SSRIs) such as fluoxetine, citalopram, sertraline, paroxetine, and escitalopram. SSRIs work by enhancing the levels of serotonin in the brain. Serotonin and norepinephrine reuptake inhibitors (SNRIs) are the groups of antidepressants that work by increasing the level of both serotonin and norepinephrine and help to regulate moods [41]; venlafaxine and duloxetine are examples of SNRIs. Atypical antidepressants are other groups of antidepressants that work differently than SSRIs and SNRIs. Bupropion, mirtazapine, and trazodone are common examples of these categories. Tricyclics, tetracyclic, and monoamine oxidase inhibitors (MAOIs) are the older group of antidepressants. Currently, this group of drugs is less often prescribed because of their deleterious side effects.

7.2.3.4 Current Treatment Strategies for Epilepsy and Seizures

Epilepsy is a chronic brain illness marked by recurring seizures. It affects 50 million individuals globally. Epilepsy involves alteration in the balance of excitatory and inhibitory neurotransmitters like GABA, glutamate, acetylcholine, and serotonin; modulations of these neurotransmitters and their receptors affect seizure activity and epileptogenesis. The treatment of epilepsy involves the use of anti-epileptic medications that are designed to manage and control seizures in individuals. The modern treatment of epilepsy begins after the introduction of phenobarbitone on the market in 1912 [42]. With the advent of more than 15 third-generation anti-epileptic medications (AEDs), the treatment of numerous forms of seizures has significantly improved during the past 30 years. Currently, the main treatments for epilepsies involve the use of anti-epileptic drugs (AEDs), but they only can remission of seizures in about 70–80% of the patients [43]. In addition, some patients have drug-resistant epilepsy (DRE) that does not respond to AEDs. There are no AEDs that can prevent the development of epilepsy from the first occurrence of a seizure [44].

7.3 Principles of Drug Repurposing

7.3.1 Definition and Concept of Drug Repurposing

The term "Drug repurposing," also known as drug repositioning, involves the identification and development of new therapeutic uses for existing drugs that were initially developed for different indications. The drug discovery and development for targeting CNS disorders are complicated due to its complex etiology, limited understanding of pathophysiology, lack of biomarkers and animal models, and promiscuity of drug–target interaction [45]. The approach of drug development by the repurposing of drugs offers several advantages over the conventional de novo approach, including reduced development costs, a faster path to developmental uses, and the potential to leverage existing safety data. In this context, when it comes to CNS disorders, the principle of drug repurposing is particularly more relevant. Some of the key principles of drug repurposing for CNS disorders are as follows:

(i) Target plausibility and mechanism of action: The identification of the drugs with known mechanisms of action that have the potential to modulate the targets relevant to the pathophysiology of the interested diseases. Moreover, it is also important to understand the molecular and cellular mechanisms underlying the disorder for selecting appropriate drugs for repurposing [46].
(ii) Biological pathways and network analysis: It is important to analyze the biological pathways and network associated with both the original indications of the drug and the new disorder of interest. Subsequently, identification of the point of intersection between the drugs' known mechanism of action and pathways following in the disorders of interest [47, 48].
(iii) Clinical and epidemiological observations: The considerations of the drug have shown some unexpected effects on some new disorders in some clinical observations or epidemiological studies. These serendipitous discoveries can provide valuable information about potential repurposing opportunities for drugs [49].
(iv) Potential biomarker identifications: The identifications of relevant biomarkers associated with both the drug's original indication and the new indications to facilitate monitoring and evaluations of treatment efficacy [50].
(v) Patient stratification: The tailoring of drug repurposing strategies based on patient subpopulations that may respond differently to treatment. The personalized medicine approach can improve the effectiveness of repurposing drugs in specific patient groups [51, 52].
(vi) Collaboration and data sharing: Integrating collaboration between academia, industry, and regulatory bodies to share data and expertise can provide valuable insight into the repurposing of drugs. Subsequently, collaborative networks can also accelerate the identification and validation of repurposing opportunities [53].

By considering the above principles, researchers and pharmaceutical companies can guide the complex landscape of disorders and identify promising opportunities for repurposing existing drugs, potentially offering faster and more cost-effective therapeutics for patients in need.

7.3.2 Advantages and Limitations of Drug Repurposing

7.3.2.1 Advantages of Drug Repurposing

(i) Cost efficiency: Repositioning existing drugs can be more cost-effective than developing new drugs by de novo approach.
(ii) Accelerated development timeline: As repurposed drugs have already undergone pre-clinical and clinical testing for their previous indications, the devel-

opment timeline of drugs can be significantly shortened, allowing for quicker availability of drugs in the market.

(iii) Risk reduction: Repurposing the drug mitigates some of the risks associated with drug development, as pharmacokinetic (PK), potential adverse effects, and toxicities are already known from previous studies.

(iv) Increased probability of success: The repurposed molecule has a higher probability of success in clinical trials when compared to entirely new compounds, as they have already demonstrated safety and efficacy in previous contexts.

(v) Expanded treatment option: Drug repositioning can offer new therapeutic options for different central nervous system disorders, as in cases when the traditional approach has limited success. This can help to broaden the range of therapeutics for the patients [54].

7.3.2.2 Limitations Associated with Repurposing of Drugs

Despite the enormous benefits of the repurposing of drugs, it possesses several limitations. A few of these limitations are mentioned below:

(i) Limited patent protection: In repurposed drugs, there may be limited or expired patent protection that reduces the financial incentives for pharmaceutical companies to invest in their development. Therefore, this can lead to a lack of investment in their development.

(ii) Off-level use challenges: The use of the drug for a purpose other than its originally approved indication may involve off-level prescribing in the market, which can pose challenges in terms of reimbursement, regularity compliance, and liability.

(iii) Potential lack of specificity: Drugs repurposed for central nervous system disorders may not be appropriately specific to the targeted pathways or mechanisms, leading to potential side effects or lack of efficacy in certain patient populations.

(iv) Heterogeneity of CNS disorders: The central nervous system disorders are often heterogeneous and bearing with different underlying causes and manifestations. A drug that is effective for one subtype of a disorder may not be as effective for others, which limits its overall impact.

(v) Unpredictable results: The success of drug repurposing can be unpredictable, as the efficacy observed in pre-clinical models or other clinical settings may not necessarily translate to success in the targeted disorders.

(vi) Regulatory challenges: The regulatory approval for new indications may still require extensive clinical trials, especially in cases where original safety and efficacy data are not directly applicable to new indications. In this case, there may be reductions in benefits of time and cost advantages.

Despite all of the challenges, drug repurposing remains an attractive option in CNS and other disorders, especially when we consider the urgent need for effective treatment and the potential benefits for the patients [55].

7.4 Strategies for Drug Repurposing

7.4.1 Computational and Bioinformatics in Drug Repurposing

During the last century, neurological disorders have contributed to significant challenges for the effective discovery of drugs. Around 50 lakhs of Americans have Alzheimer's disease and about five million of them are living with Parkinson's disease. Millions more have been diagnosed with other disease conditions such as Huntington's disease (HD), multiple sclerosis(MS), frontotemporal dementia (FTD), and amyotrophic lateral sclerosis (ALS) [56, 57], etc. The use of an in silico or computational drug repurposing approach anticipates getting the therapies for neurodegenerative disease to the clinic steadily [58, 59]. Conventional method of drug development involves clinical, pre-clinical research, and FDA post-market safety monitoring, which can take 15 years and a billion dollars to bring the drug to the market [60, 61]. On the other hand, several FDA-approved repurposed drugs find it a quicker and cheaper journey to the clinic [62]. The development of an omics-based approach as well as a high-throughput screening technique has propelled this computational method of repurposing as a cutting edge toward the drug discovery technique [5]. Also, computational screening possesses the capacity to reduce the cost as well as the time of drug development, and the economic and experimental challenges must be controlled [63].

Earlier, the accidental discoveries in the laboratory and clinic resulted in the success of repurposed drugs. The usage of zonisamide for treating Parkinson's disease is one such relevant illustration. Upon treating an epileptic patient, Murata, who had been already suffering from PD, the patient's PD symptoms also improved. In 2009, Japan approved zonisamide as an anti-PD therapy based on a serendipitous discovery. High-throughput structural, clinical, and molecular biology technologies, along with the development of large-scale computational capacity that is economically feasible, have recently given rise to a novel opportunity by using computational techniques rather than serendipitous discoveries [64, 65].

To identify the drugs that may alter the gene signature, one such group has led the way to the development of techniques based on transcriptomics for computational drug discovery [66]. These approaches compare the drug-gene expression profiles before and after drug treatment targeted toward disease gene expression signatures [67]. A resource developed at the Broad Institute, the Connectivity Map (CMap), is an early version of this hypothesis. It consists of a database having microarray gene expression signatures built before and after over and 1200 small molecular compounds were applied to the cell lines [68]. Pre-clinical identification of drugs such as parbendazole for conditions osteoporosis [68], metastatic colon cancer [69], and inflammatory bowel disease [70] are a few recent instances of the effectiveness of this approach. For the advancement of drug discovery, more recently developed techniques integrate genomic and gene expression data and make use of network-based techniques [71]. These methods allow for the high-throughput screening of already approved drugs without requiring the target molecule to be identified in advance. However, the availability of disease expression profiles specific to subtypes limits the use of transcriptome approaches. This is especially important when it comes to diverse neurodegenerative illnesses like PD and AD [72]. The presence of various genetic and demographic-related risk factors, such as apolipoprotein E(Apo E) ε4 allele [73], as well as vascular risk factors like hypertension and diabetes, has been linked to the differences in AD's pathology, disease risk, and biomarker profile according to the recent report. In creating biologically defined disease subtypes for AD, the 2018 National Institute of Aging (NIH) and Alzheimer's Association framework is a promising step forward. Large-scale datasets associated with neurodegenerative diseases have been compiled for easy access to research through the NIH-Accelerating Medicines Partnership-AD (NIH AMP-AD) portal. However, additional resources such as single-cell RNA-seq datasets for the targeted development are desperately needed, along with genetic information [74].

7.4.2 Targeted and Phenotypic Screening

Phenotypic screening is based on chemically probing a biological molecular target-agnostic chemical probing of a biological system related to disease. The following examples demonstrate how this empirical, biological approach offers tool molecules to connect therapeutic biology with unidentified signaling pathways, drug targets, and molecular mechanisms.

The hepatitis C virus (HCV), which affects 3% of the population and is thought to be responsible for 3 lakh deaths annually worldwide, is the cause of hepatitis C, a liver disease. The introduction of a combination of oral direct-acting antivirals (DAAs) that limit HCV replication and eradicate the virus in more than 90% of infected people has revolutionized HCV treatment during the past 10 years. One of the crucial elements of these DAA combinations is daclatasvir, which is a modulator of the HCV protein NS5A. Using an HCV replicon phenotypic screen, the significance of NS5A was first identified [75].

Type 1 Spinal Muscular Atrophy (SMA) is a rare neuromuscular disorder wherein 95% of children die before they turn 18 months old. The SMN1 gene, which codes for a protein called survival of motor neuron (SMN), which is essential in the development of neuromuscular junctions, is mutated to cause loss of function in sickle cell anemia (SMA). The SMN2 gene is closely linked to that of humans as well, yet a mutation affects its splicing resulting in the elimination of exon 7 and the formation of the unstable shorter variant of SMN. The phenotypic screening conducted by two different research groups has found small compounds that regulate the splicing of SMN2 pre-mRNA as well as enhance the level of SMN protein [76, 77].

7.5 Case Studies of Drug Repurposing in Specific CNS Disorders

7.5.1 Alzheimer's Disease

Alzheimer's disease (AD) is the most common cause of dementia among older adults. By estimates, 6.7 million Americans 65 years of age and older will have AD in 2023; by 2060, that number is expected to rise to 13.8 million [78]. The U.S. Food and Drug Administration (FDA) has approved only seven medications to treat AD that belong to amyloid β-directed monoclonal antibodies, cholinesterase inhibitors, glutamate regulators, and a combination of cholinesterase inhibitors and glutamate regulators. Drug repurposing is a promising approach to accelerate the therapeutic discovery process for Alzheimer's disease [79]. The researchers are exploring multiple ways to repurpose drugs alone or in combination with other therapeutics.

One of the strategies for repurposing drugs on AD is through modulations of nuclear factor-erythroid factor 2-related factor 2 (NRF2) neighborhood. In several conditions, including those in the neurodegenerative spectrum, NRF2 gives cytoprotection and is essential for the inflammatory response and cellular redox equilibrium. It is suggested that NRF2 and its associated proteins may be a novel, attractive therapeutic target against AD. In this context, Bourdakou et al. [80] conducted a study employing a systems biology approach to propose an in silico drug repurposing strategy for AD centered on the NRF2 interactome and regulome. Five candidate repurposed drugs were identified through a computational pipeline and analysis of publicly available data on NRF2-neighborhood expressions in AD. Then, a four-step computational meta-analysis validated these candidates, considering structural similarity, NRF2-diseasome evaluation, pathway relevance, and comparison with existing knowledge on AD and NRF2-related drugs. In cell culture experiments, the candidates activated NRF2 activity and increased NRF2 protein levels in hippocampus-derived cells. This research provides promising avenues for further investigation and clinical translation, offering potential therapeutic insights for AD [80]. In the following section, we discussed a few drugs that are repurposed against AD.

7.5.1.1 Montelukast

Montelukast is a $CysLT_1$ receptor antagonist marketed for treating asthma and seasonal allergies. It is a cysteinyl leukotriene (CysLT1) receptor antagonist, preventing leukotriene action and reducing inflammation. This drug has been repurposed for the treatment of AD; currently, it is in a phase II clinical trial [81]. In pre-clinical research, montelukast had shown improved learning and memory in a model of mild cognitive impairment and aging of the brain while also reducing neuroinflammation and elevating hippocampus neurogenesis [82]. However, this drug suffered from a poor bioavailability problem. Hence, the bioavailability of montelukast has been improved by formulating an oral film to evaluate the pharmacological properties in phase I clinical trials. It was observed to be safe and tolerable in healthy subjects with minimal fast pass effect and 52% higher bioavailability in CSF fluids compared to the already available formulation [83]. This type of novel formulation of montelukast may be an effective repurposed drug for the management of AD.

7.5.1.2 Zileuton

Some recent investigations suggested that zileuton, a 5-Lipoxygenase (5-LOX) inhibitor used to treat asthma, is useful in treating AD [84]. The overexpression of 5-LOX in AD patients was the hypothesis for its promising action against AD. Pre-clinical studies in mice models using zileuton have shown a reduction of amyloid deposition in the brain tissues [84].

7.5.1.3 Sildenafil and Tadalafil

These are phosphodiesterase-5 (PDE-5) inhibitors commonly used for the treatment of erectile dysfunctions. The role of the PDE-5 enzyme involves the regulation of cyclic guanosine monophosphate (cGMP). An increase in cGMP levels causes the brain's intracellular protein kinase to become active, phosphorylating several proteins that have a negative impact on tau hyperphosphorylation, Aβ building up, and neuroplasticity. In recent studies, it was observed that sildenafil has the potential to inhibit neuroinflammation and Aβ deposition in aged mice models [85]. Likewise, tadalafil had a higher blood–brain barrier penetration and could suppress the PDE-5 enzyme more effectively than sildenafil; it was more effective than sildenafil in terms of neuroprotection and memory improvement [86].

7.5.1.4 Imatinib

Imatinib is a tyrosine kinase inhibitor approved for the treatment of chronic leukemia. It has been claimed to be useful for treating AD because of its potential for neuroprotection as Aβ deposition gets diminished. Nevertheless, imatinib is easily effluxed by P-glycoprotein and fails to cross the blood–brain barrier efficiently. Tau proteins become less able to attach to microtubules as a result of hyperphosphorylation, which eventually causes NFTs to develop. In this context, such hyperphosphorylation has been demonstrated to be reduced by paclitaxel, a drug that is approved for non-small cell lung cancer, ovarian cancer, and breast cancer. In pre-clinical models, paclitaxel has also demonstrated effectiveness in reducing brain tauopathies.

7.5.1.5 Thalidomide

Another anticancer drug, thalidomide, is undergoing phase III clinical trials. It has been shown to modulate BACE-1 and decrease the proliferation of endothelial cells, angiogenesis, and the blood–brain barrier. It also shows potential in AD pathology. Additionally, it has demonstrated a decrease in hippocampus neuronal death by blocking TNFα. Furthermore, at a non-toxic dosage, carmustine, a tiny lipophilic, non-ionized nitrosourea chemical employed as an alkylating drug in brain cancer, improved tau pathology and significantly decreased Aβ formation in pre-clinical animals [86–88].

7.5.1.6 Pioglitazone

It is a peroxisome-proliferated activated receptor γ (PPARγ) agonist. It has been marketed for the treatment of type 2 diabetes mellitus (DM). The use of pioglitazone is linked to a decreased risk of dementia in people with DM, especially in those who have had a stroke or ischemic heart disease. This suggests that a customized strategy may be used when selecting pioglitazone to suppress dementia in DM patients [89]. Research has demonstrated that PPARγ receptor agonists may lessen inflammatory reactions in Aβ-induced brain microglial cells during AD [90]. Currently, pioglitazone has been repurposed for the management of AD.

7.5.2 Parkinson's Disease

Parkinson's disease (PD) is a serious healthcare burden on health systems and has a substantial negative impact on patient's quality of life. Drug repositioning, as an alternative to standard drug development methods, aims to provide novel pharmaceutical alternatives in the absence of safe and effective treatments for Parkinson's disease (PD). Some examples of drugs that are repurposed for the treatment of PD are discussed below.

7.5.2.1 Exenatide

It is a glucagon-like peptide-1 (GLP-1) receptor agonist, previously used for type 2 diabetes mellitus (T2DM); it has been repurposed for the management of PD. T2DM was identified as one of the possible risk factors for PD. Exenatide exhibited GLP-1 receptor-dependent neuroprotective effects in pre-clinical investigations. Exenatide's proof-of-concept and tolerance were demonstrated in an open-label experiment, and a subsequent randomized placebo-controlled trial found that some motor scores had improved clinically. In addition, phase III multi-center clinical trials are evaluating exenatide's neuroprotective benefits in people with PD who do not also have T2DM [91, 92].

7.5.2.2 Simvastatin

Statins block HMG-CoA reductase, a key enzyme in cholesterol manufacture, but they also have neuroprotective properties, particularly simvastatin [93]. It has also been suggested that simvastatin may reduce the incidence of progressive supranuclear palsy and postpone symptom onset [94]. With strong supporting results, simvastatin was selected at the first iLCT committee meeting in 2012, and it entered PD-STAT, a major statewide Phase II research in the United Kingdom. Involving 230 individuals with moderate PD, the study administered either simvastatin or a placebo daily for 24 months [95]. Preliminary data reported in September 2020 showed that simvastatin is safe in PD but does not change its course.

7.5.2.3 Salbutamol

Salbutamol is a beta2-adrenergic (β2AR) agonist, and it is commonly used for the treatment of bronchospasm. It has the potential to increase endogenous FGF20 production in the nigrostriatal tract and can potentially impact the survival of dopaminergic neurons. From the studies, it was found that β2AR is a regulator of the α-synuclein gene (*SNCA*). SNCA transcription is modulated by β2AR ligands by acetylating its promoter and enhancers on histone 3 lysine 27. It is also observed that patients treated with salbutamol β2AR agonists show neuroprotective activity, hence a lower risk of AD [96, 97].

7.5.3 Huntington's Disease

Huntington's disease (HD) is a neurodegenerative illness with a broad range of neuropsychiatric symptoms, including movement problems (mostly chorea), dementia, and behavioral or mental symptoms. The huntingtin gene (HTT), found on chromosome 4, increases the CAG (cytosine, adenine, and guanine) trinucleotide repeat, which causes HD, a polyglutamine disease. It has an autosomal dominant inheritance pattern [98, 99]. There is no cure for HD, but medication and treatment can help to relieve some symptoms. Some of the therapies for HD include the use of antipsychotic drugs to ease chorea and help control hallucinations, delusions, and violent outbursts, drugs to treat depression and anxiety, etc. For this instance, there is a wide scope of repurposing drugs for the management of Huntington's disease. Some of the drugs that are repurposed for the treatment of HD are mentioned.

7.5.3.1 Tetrabenazine (Xenazine®)

Initially, in 2008, the U.S. Food and Drug Administration (FDA) approved tetrabenazine for its antipsychotic properties repurposed for treating excessive involuntary movements (chorea) in HD patients. Tetrabenazine works as a reversible vesicular monoamine transporter 2 (VMAT2) inhibitor. VMAT2 is responsible for packaging neurotransmitters, including dopamine, into vesicles for release into the synapse. By inhibiting VMAT2, tetrabenazine reduces the uptake of cytosolic monoamines into vesicles and, consequently, decreases the release of dopamine from synaptic vesicles. This helps in managing the chorea observed in HD [100].

7.5.3.2 Deutetrabenazine

This drug belongs to the vesicular monoamine transporter 2 (VMAT2) inhibitors class. It shows its activity by altering certain natural substances in the brain that affect the nerves and muscles. In this context, after a successful randomized, double-blind, placebo-controlled trial, the FDA authorized AUSTEDO (deutetrabenazine) in April 2017 to treat chorea in HD patients [101].

7.5.4 Amyotrophic Lateral Sclerosis

Amyotrophic lateral sclerosis (ALS), or Motor neuron disease, is a debilitating neurological disorder characterized by progressive degeneration of the lower motor neuron in the area of the brainstem and spinal cord and the upper motor neuron of the motor cortex. With a prevalence of 6–9 per 1 lakh people, an incidence of 2 per lakh persons each year, and a lifetime risk of almost 1 in 350 individuals has been observed for ALS worldwide [102]. Some of the repurposed drugs are enlisted, such as:

7.5.4.1 Aldesleukin

An experimental design is being used in a phase II/III clinical research trial to investigate the use of aldesleukin in ALS. Clinigen is developing aldesleukin, a recombinant human IL-2 that is equivalent to endogenous IL-2 in terms of pharmacodynamic aspects, to treat ALS. Under the Proleukin brand, the FDA has approved it for treating metastatic melanoma and renal cell carcinoma. The immunoregulatory effects of aldesleukin were shown in vitro experiments using human cell lines. This also included the activation and multiplication of cytotoxic T lymphocytes, natural killer cells, and lymphokine-activated killer cells. It was also observed that low-dose IL-2 improved the function of T cells and that a major pathogenic characteristic of ALS is neuroinflammation [103, 104].

7.5.4.2 Verdiperstat

Biohaven Pharmaceuticals produced an inhibitor of myeloperoxidase (MPO), Verdiperstat, a small molecule that is administered orally. MPO exacerbates inflammation and oxidative stress in the brain, which are involved in several neurological diseases. Mitigating the pathogenic pathways linked to neuroinflammation may be possible by inhibiting MPO. Around 167 participants participated in a phase II/III placebo-controlled trial (NCT04297683, NCT04436510) as a part of the HEALEY ALS Platform trial to assess the safety and effectiveness of Verdiperstat. The trail did not yield positive outcomes in terms of primary outcomes as reported by the company that conducted trails [105].

7.5.5 Spinal Muscular Atrophy

Spinal muscular atrophy (SMA) is the most prevailing hereditary disease among newborns and young people, caused by a lack of a functional survival motor neuron 1 (SMN1) gene. To date, the Food and Drug Administration has approved only three drugs for the condition, Nusinersen-Spinraza; AVXS 101-Zolgensma; and Risdiplam—Evrysdi. Despite the promising results, the therapies have some important limitations, including high costs, side effects, and a failure to consider SMN as an independent target [106]. Following are the repurposed drugs for SMA:

7.5.5.1 Sodium Butyrate

Sodium butyrate, a salt of butyric acid and histone deacetylase inhibitor (HDAC) has anticancer properties. The researchers converted lymphocytes collected from SMA patients with the Epstein–Barr virus to create lines of immortalized lymphoid cells. They used quantitative reverse transcriptase to ascertain which drug led to an enhancement in the production of full-length SMN2 mRNA after exposing the cells to the potential treatment. Compound I enhanced the development of full-length SMN2 mRNA by increasing SMN2 transcription and the inclusion of exon 7 in SMN2 transcripts. A similar rise was seen with SMN protein [107].

7.5.5.2 Ceftriaxone

Although the β-lactam antibiotic, ceftriaxone has physicochemical features that are not often thought to be necessary for central nervous system penetration, it still crosses the blood–brain barrier via active transport [108]. Even though this compound has shown to enhance the glutamate reuptake and enhance the phenotype of ALS in the murine model, a clinical trial including ceftriaxone in ALS was unsuccessful. However, when tested in animals with severe SMA, this compound demonstrated a modestly longer lifespan [109]. The animals receiving treatment had a somewhat better neuromuscular phenotype and higher survival rate. Enhanced amounts of the transcription factor Nrf2, the glutamate transporter Glt, and the SMN protein appear to be the mechanism underlying the neuroprotective effect. This study proposed the first proof of β-lactam antibiotic as a possible benefit in treating SMA [107].

7.5.6 Epilepsy

Epilepsy stands as a classic example of a chronic ailment, profoundly impacting various facets of life due to its elusive nature and the complexities associated with its management. Its inherent unpredictability restricts routine activities such as driving, attending school, and pursuing employment. Despite relentless efforts in research and the application of diverse treatment approaches such as anti-epileptic drugs, specialized diets, vagal nerve stimulation, and surgery, approximately 30% of patients continue to grapple with uncontrolled epilepsy. Additionally, a substantial number experience adverse effects from medications, undermining their overall quality of life.

Developing new antiseizure drugs encounters significant impediments, marked by prolonged timelines from discovery to market availability, challenges in securing insurance-payer reimbursement, and regulatory hurdles about drug safety. In response to these challenges, there is a growing consideration for drug repositioning (DR), wherein drugs already approved by the FDA and known to target pathways disrupted in epilepsy are explored as potential candidates. This strategic approach aims to bypass the obstacles associated with traditional drug development and expedite the availability of effective therapies for individuals living with epilepsy [110]. Some of the drugs that are repurposed for the treatment of epilepsy are mentioned below:

7.5.6.1 Nifedipine

Nifedipine, widely prescribed as an antihypertensive medication, is primarily recognized for its role as an L-type calcium channel blocker. Its ability to regulate ion movement prompts speculation about its potential as an antiseizure and anti-epileptic agent. Extensive research in rodent seizure models has positioned nifedipine as a potential therapy for human seizures, particularly as an adjunct treatment for drug-resistant epilepsy. In this context, the researchers have explored the use of nifedipine, suggesting its viability in managing seizures. Notably, a randomized control trial conducted in 1992 assessed nifedipine as an add-on agent for refractory epilepsy [111]. Although this trial involved 22 patients, it failed to detect any significant effects with a single dose of nifedipine. However, considering the findings of *Leo Brueggeman* indicating antiseizure activity in zebrafish and corroborating results in rodents by other researchers, alongside the limitations of the small sample size and the severity of epilepsy in the 1992 trial, there is a compelling case for reconsidering nifedipine in clinical trials as a potential seizure suppressant for humans [110, 112].

7.5.6.2 Dimethyl Fumarate (DMF)

Dimethyl fumarate is a Nrf2 activator with antioxidant and anti-inflammatory characteristics that is extensively used to treat multiple sclerosis (MS). Activation of the Nrf2 pathway, which is involved in cellular defense, offers a possible technique for lowering oxidative stress and treating epilepsy. In this context, the impact of dimethyl fumarate (DMF) on epilepsy was investigated in rats. It was observed that DMF increased Nrf2 activity, lessened neuronal cell death, and reduced seizure frequency, even when treatment started after epilepsy diagnosis. Also, the behavioral deficits were reversed, and DMF showed lasting effects on symptomatic seizures post-treatment. These results highlight DMF's potential as a therapeutic avenue for preventing epileptogenesis and modifying epilepsy through Nrf2 activation, suggesting its promising role in epilepsy management [113].

7.6 Challenges

7.6.1 Validation of Different Methods

Although computational drug repurposing has made significant advancements, these methods still require validation. The primary goal of drug repurposing is to find the suitable therapeutic candidate with the highest potential of treating patients. On the other hand, computational approaches anticipate a large number of candidates. Therefore, validation techniques are required to advance this procedure toward an actual clinical therapy. One of the basic validation processes involves conducting in vitro and in vivo investigations to experimentally validate the discovered hits [114, 115].

7.6.2 Combination of Drug

Drug combination has garnered a lot of attention lately as an alternative to single agent therapy in terms of therapeutic regimen. When two drugs are taken together, their combined effect can be higher than the sum of their individual effects, yet there is no increased risk of adverse effects. This phenomenon is known as synergy. Thus, it is imperative to develop new strategies and techniques that predict the synergy of compound [116].

7.7 Summary

Owing to an increase in the life expectancy, it is predicted that the widespread prevalence of neurological diseases, particularly AD, which are now common, would rise dramatically. Therefore, in order to address this global health catastrophe, new and effective treatment approaches are desperately needed. Drug repurposing has lately acquired popularity due to its ability to reduce costs and time by prioritizing a list of drug candidates for future experimental testing. Because drug repurposing avoids the pharmacokinetic and toxicological testing involved in conventional drug discovery approaches, it can potentially expedite the process of bringing a new drug to the market [117, 118].

Acknowledgment The authors would like to acknowledge National Institute of Pharmaceutical Education and Research Guwahati, Assam, India, and the Department of Pharmaceuticals, Ministry of Chemicals and Fertilizers, Government of India, for their support in curating this present work.

References

1. Matilla-Dueñas A, et al. Rare neurodegenerative diseases: clinical and genetic update. Adv Exp Med Biol. 2017;1031:443–96.
2. Pankevich DE, et al. Improving and accelerating drug development for nervous system disorders. Neuron. 2014;84(3):546–53.
3. Lee HM, Kim Y. Drug repurposing is a new opportunity for developing drugs against neuropsychiatric disorders. Schizophr Res Treatment. 2016;2016:6378137.
4. Fava M. The promise and challenges of drug repurposing in psychiatry. World Psychiatry. 2018;17(1):28–9.
5. Xue H, et al. Review of drug repositioning approaches and resources. Int J Biol Sci. 2018;14(10):1232–44.
6. Shipp S. Structure and function of the cerebral cortex. Curr Biol. 2007;17(12):R443–9.
7. Thameem Dheen S, Kaur C, Ling E-A. Microglial activation and its implications in the brain diseases. Curr Med Chem. 2007;14(11):1189–97.
8. World Health Organization. Neurological disorders: public health challenges. Geneva: World Health Organization; 2006.
9. Naseem M, Parvez S. Role of melatonin in traumatic brain injury and spinal cord injury. Sci World J. 2014;2014:586270.
10. Fan L, et al. Associations of circulating saturated long-chain fatty acids with risk of mild cognitive impairment and Alzheimer's disease in the Alzheimer's Disease Neuroimaging Initiative (ADNI) cohort. EBioMedicine. 2023;97:104818.
11. Hoenig MC, et al. Educational level and its association with dopamine transporter loss in patients with Parkinson's disease. Parkinsonism Relat Disord. 2023;115:105844.
12. Li X, et al. Global, regional, and national burden of Alzheimer's disease and other dementias, 1990–2019. Front Aging Neurosci. 2022;14:937486.

13. Stiles J, Jernigan TL. The basics of brain development. Neuropsychol Rev. 2010;20(4):327–48.
14. Van den Heuvel MP, Sporns O. Network hubs in the human brain. Trends Cogn Sci. 2013;17(12):683–96.
15. Kovacs GG. Molecular pathology of neurodegenerative diseases: principles and practice. J Clin Pathol. 2019;72(11):725–35.
16. Scott DA, et al. A pathologic cascade leading to synaptic dysfunction in α-synuclein-induced neurodegeneration. J Neurosci. 2010;30(24):8083–95.
17. Przedborski S, Vila M, Jackson-Lewis V. Series introduction: neurodegeneration: what is it and where are we? J Clin Invest. 2003;111(1):3–10.
18. Hague S, Klaffke S, Bandmann O. Neurodegenerative disorders: Parkinson's disease and Huntington's disease. J Neurol Neurosurg Psychiatry. 2005;76(8):1058–63.
19. Harding BN, et al. Spectrum of neuropathophysiology in spinal muscular atrophy type I. J Neuropathol Exp Neurol. 2015;74(1):15–24.
20. Pickett EK, et al. Amyloid beta and tau cooperate to cause reversible behavioral and transcriptional deficits in a model of Alzheimer's disease. Cell Reports. 2019;29(11):3592–3604.e5.
21. Spires-Jones TL, Hyman BT. The intersection of amyloid beta and tau at synapses in Alzheimer's disease. Neuron. 2014;82(4):756–71.
22. Akhondzadeh S, Noroozian M. Alzheimer's disease: pathophysiology and pharmacotherapy. IDrugs. 2002;5(11):1062–9.
23. Kouli A, Torsney KM, Kuan W-L. Parkinson's disease: etiology, neuropathology, and pathogenesis. In: Parkinson's disease: pathogenesis and clinical aspects. Brisbane: Codon; 2018. p. 3–26.
24. Priyadarshi A, et al. Environmental risk factors and Parkinson's disease: a metaanalysis. Environ Res. 2001;86(2):122–7.
25. Emamzadeh FN, Surguchov A. Parkinson's disease: biomarkers, treatment, and risk factors. Front Neurosci. 2018;12:612.
26. Bartels AL, Leenders KL. Parkinson's disease: the syndrome, the pathogenesis and pathophysiology. Cortex. 2009;45(8):915–21.
27. Chaudhuri KR, Schapira AH. Non-motor symptoms of Parkinson's disease: dopaminergic pathophysiology and treatment. Lancet Neurol. 2009;8(5):464–74.
28. Moore DJ, et al. Molecular pathophysiology of Parkinson's disease. Annu Rev Neurosci. 2005;28:57–87.
29. Spires-Jones TL, Attems J, Thal DR. Interactions of pathological proteins in neurodegenerative diseases. Acta Neuropathol. 2017;134:187–205.
30. Henderson M, et al. Work and common psychiatric disorders. J R Soc Med. 2011;104(5):198–207.
31. Mullin AP, et al. Neurodevelopmental disorders: mechanisms and boundary definitions from genomes, interactomes and proteomes. Transl Psychiatry. 2013;3(12):e329.
32. Elorza Ridaura I, Sorrentino S, Moroni L. Parallels between the developing vascular and neural systems: signaling pathways and future perspectives for regenerative medicine. Adv Sci (Weinh). 2021;8(23):e2101837.
33. Hampel H, et al. Revisiting the cholinergic hypothesis in Alzheimer's disease: emerging evidence from translational and clinical research. J Prev Alzheimers Dis. 2019;6:2–15.
34. Adlimoghaddam A, et al. A review of clinical treatment considerations of donepezil in severe Alzheimer's disease. CNS Neurosci Ther. 2018;24(10):876–88.
35. Deardorff WJ, Feen E, Grossberg GT. The use of cholinesterase inhibitors across all stages of Alzheimer's disease. Drugs Aging. 2015;32:537–47.
36. Dunn B, Stein P, Cavazzoni P. Approval of aducanumab for Alzheimer disease—the FDA's perspective. JAMA Intern Med. 2021;181(10):1276–8.
37. Walsh S, et al. Aducanumab for Alzheimer's disease? BMJ. 2021;374:n1682.
38. Johnson JW, Kotermanski SE. Mechanism of action of memantine. Curr Opin Pharmacol. 2006;6(1):61–7.
39. Tariot PN, et al. Memantine treatment in patients with moderate to severe Alzheimer disease already receiving donepezil: a randomized controlled trial. JAMA. 2004;291(3):317–24.

40. Taylor C, et al. Mechanisms of action of antidepressants: from neurotransmitter systems to signaling pathways. Cell Signal. 2005;17(5):549–57.
41. Sansone RA, Sansone LA. Serotonin norepinephrine reuptake inhibitors: a pharmacological comparison. Innov Clin Neurosci. 2014;11(3–4):37–42.
42. Yasiry Z, Shorvon SD. How phenobarbital revolutionized epilepsy therapy: the story of phenobarbital therapy in epilepsy in the last 100 years. Epilepsia. 2012;53(Suppl. 8):26–39.
43. Löscher W, et al. New avenues for anti-epileptic drug discovery and development. Nat Rev Drug Discov. 2013;12(10):757–76.
44. Ghosh S, et al. Pharmacological and therapeutic approaches in the treatment of epilepsy. Biomedicine. 2021;9(5):470.
45. Lee H-M, Kim Y. Drug repurposing is a new opportunity for developing drugs against neuropsychiatric disorders. Schizophr Res Treat. 2016;2016:6378137.
46. Ankur G, et al. Trends in molecular aspects and therapeutic applications of drug repurposing for infectious diseases. In: Shailendra KS, editor. Drug repurposing. Rijeka: IntechOpen; 2021.
47. Pandita V, et al. System and network biology-based computational approaches for drug repositioning. In: Computational approaches for novel therapeutic and diagnostic designing to mitigate SARS-CoV2 infection. London: Elsevier; 2022. p. 267–90.
48. Wang X, et al. Network-based drug repurposing for the treatment of COVID-19 patients in different clinical stages. Heliyon. 2023;9(3):e14059.
49. Hua Y, et al. Drug repositioning: Progress and challenges in drug discovery for various diseases. Eur J Med Chem. 2022;234:114239.
50. Zhang C, et al. Identification of biomarkers and drug repurposing candidates based on an immune-, inflammation- and membranous glomerulonephritis-associated triplets network for membranous glomerulonephritis. BMC Med Genet. 2020;13(1):5.
51. Das S, et al. Systematic indication extension for drugs using patient stratification insights generated by combinatorial analytics. Patterns (N Y). 2022;3(6):100496.
52. Steve G, Sayoni D, Krystyna T. AI enabled precision medicine: patient stratification, drug repurposing and combination therapies. In: John WC, Belle T, editors. Artificial intelligence in oncology drug discovery and development. IntechOpen: Rijeka; 2020.
53. Beachy SH, et al. Drug repurposing and repositioning: workshop summary. Washington, DC: National Academies Press; 2014.
54. Elder D, Tindall S. The many advantages of repurposing existing drugs. Eur Pharm Rev. 2020;25(3):34–7.
55. Begley CG, et al. Drug repurposing: misconceptions, challenges, and opportunities for academic researchers. Sci Transl Med. 2021;13(612):eabd5524.
56. Katsnelson A, De Strooper B, Zoghbi HY. Neurodegeneration: from cellular concepts to clinical applications. Sci Transl Med. 2016;8(364):364ps18.
57. McColgan P, Tabrizi SJ. Huntington's disease: a clinical review. Eur J Neurol. 2018;25(1):24–34.
58. Corbett A, et al. Drug repositioning for Alzheimer's disease. Nat Rev Drug Discov. 2012;11(11):833–46.
59. Kim TW. Drug repositioning approaches for the discovery of new therapeutics for Alzheimer's disease. Neurotherapeutics. 2015;12(1):132–42.
60. Hodos RA, et al. In silico methods for drug repurposing and pharmacology. Wiley Interdiscip Rev Syst Biol Med. 2016;8(3):186–210.
61. Kaitin KI. Deconstructing the drug development process: the new face of innovation. Clin Pharmacol Ther. 2010;87(3):356–61.
62. Baker NC, et al. A bibliometric review of drug repurposing. Drug Discov Today. 2018;23(3):661–72.
63. Li L, et al. Identification of type 2 diabetes subgroups through topological analysis of patient similarity. Sci Transl Med. 2015;7(311):311ra174.
64. Murata M, Horiuchi E, Kanazawa I. Zonisamide has beneficial effects on Parkinson's disease patients. Neurosci Res. 2001;41(4):397–9.

65. Nishimura Y, et al. Overcoming obstacles to drug repositioning in Japan. Front Pharmacol. 2017;8:729.
66. Sirota M, et al. Discovery and preclinical validation of drug indications using compendia of public gene expression data. Sci Transl Med. 2011;3(96):96ra77.
67. Li J, et al. A survey of current trends in computational drug repositioning. Brief Bioinform. 2016;17(1):2–12.
68. Brum AM, et al. Connectivity map-based discovery of parbendazole reveals targetable human osteogenic pathway. Proc Natl Acad Sci U S A. 2015;112(41):12711–6.
69. van Noort V, et al. Novel drug candidates for the treatment of metastatic colorectal cancer through global inverse gene-expression profiling. Cancer Res. 2014;74(20):5690–9.
70. Dudley JT, et al. Computational repositioning of the anticonvulsant topiramate for inflammatory bowel disease. Sci Transl Med. 2011;3(96):96ra76.
71. Ferrero E, Agarwal P. Connecting genetics and gene expression data for target prioritisation and drug repositioning. BioData Min. 2018;11:7.
72. Bekris LM, et al. Genetics of Alzheimer disease. J Geriatr Psychiatry Neurol. 2010;23(4):213–27.
73. Moreno-Grau S, et al. Exploring APOE genotype effects on Alzheimer's disease risk and amyloid β burden in individuals with subjective cognitive decline: the Fundacio ACE Healthy Brain Initiative (FACEHBI) study baseline results. Alzheimers Dement. 2018;14(5):634–43.
74. Jack CR Jr, et al. NIA-AA research framework: toward a biological definition of Alzheimer's disease. Alzheimers Dement. 2018;14(4):535–62.
75. Lemm JA, et al. Identification of hepatitis C virus NS5A inhibitors. J Virol. 2010;84(1):482–91.
76. Naryshkin NA, et al. Motor neuron disease. SMN2 splicing modifiers improve motor function and longevity in mice with spinal muscular atrophy. Science. 2014;345(6197):688–93.
77. Palacino J, et al. SMN2 splice modulators enhance U1-pre-mRNA association and rescue SMA mice. Nat Chem Biol. 2015;11(7):511–7.
78. Anon. 2023 Alzheimer's disease facts and figures. Alzheimers Dement. 2023;19(4):1598–695.
79. Grabowska ME, et al. Drug repurposing for Alzheimer's disease from 2012–2022-a 10-year literature review. Front Pharmacol. 2023;14:1257700.
80. Bourdakou MM, et al. Drug repurposing on Alzheimer's disease through modulation of NRF2 neighborhood. Redox Biol. 2023;67:102881.
81. Rozin SI. Case series using montelukast in patients with memory loss and dementia. Open Neurol J. 2017;11:7.
82. Kumar A, et al. Montelukast potentiates the protective effect of rofecoxib against kainic acid-induced cognitive dysfunction in rats. Pharmacol Biochem Behav. 2012;103(1):43–52.
83. Hu M, et al. Montelukast targeting the cysteinyl leukotriene receptor 1 ameliorates Aβ1-42-induced memory impairment and neuroinflammatory and apoptotic responses in mice. Neuropharmacology. 2014;79:707–14.
84. Di Meco A, et al. Zileuton restores memory impairments and reverses amyloid and tau pathology in aged Alzheimer's disease mice. Neurobiol Aging. 2014;35(11):2458–64.
85. Sheng M, et al. Sildenafil improves vascular and metabolic function in patients with Alzheimer's disease. J Alzheimers Dis. 2017;60(4):1351–64.
86. Durães F, Pinto M, Sousa E. Old drugs as new treatments for neurodegenerative diseases. Pharmaceuticals. 2018;11(2):44.
87. Okereke OI, Meadows M-E. More evidence of an inverse association between cancer and Alzheimer disease. JAMA Netw Open. 2019;2(6):–e196167.
88. Singh RK. Recent trends in the management of Alzheimer's disease: current therapeutic options and drug repurposing approaches. Curr Neuropharmacol. 2020;18(9):868–82.
89. Ha J, et al. Pioglitazone use and reduced risk of dementia in patients with diabetes mellitus with a history of ischemic stroke. Neurology. 2023;100(17):e1799–811.
90. Van Eldik LJ, et al. The roles of inflammation and immune mechanisms in Alzheimer's disease. Alzheimers Dementia. 2016;2(2):99–109.
91. Fletcher EJR, et al. Drug repurposing strategies of relevance for Parkinson's disease. Pharmacol Res Perspect. 2021;9(4):e00841.

92. Vijiaratnam N, et al. Exenatide once weekly over 2 years as a potential disease-modifying treatment for Parkinson's disease: protocol for a multicentre, randomised, double blind, parallel group, placebo controlled, phase 3 trial: the 'Exenatide-PD3' study. BMJ Open. 2021;11(5):e047993.
93. Tong H, et al. Simvastatin inhibits activation of NADPH oxidase/p38 MAPK pathway and enhances expression of antioxidant protein in Parkinson disease models. Front Mol Neurosci. 2018;11:165.
94. Bayram E, et al. Progressive supranuclear palsy and statin use. Mov Disord. 2020;35(7):1253–7.
95. Carroll CB, et al. Simvastatin as a neuroprotective treatment for Parkinson's disease (PD STAT): protocol for a double-blind, randomised, placebo-controlled futility study. BMJ Open. 2019;9(10):e029740.
96. Fletcher EJ, et al. Targeted repositioning identifies drugs that increase fibroblast growth factor 20 production and protect against 6-hydroxydopamine-induced nigral cell loss in rats. Sci Rep. 2019;9(1):8336.
97. Mittal S, et al. β2-Adrenoreceptor is a regulator of the α-synuclein gene driving risk of Parkinson's disease. Science. 2017;357(6354):891–8.
98. Adam MP, et al. GeneReviews®. Seattle: University of Washington, Seattle; 1993.
99. Medina A, et al. Prevalence and incidence of Huntington's disease: an updated systematic review and meta-analysis. Mov Disord. 2022;37(12):2327–35.
100. Jankovic J, Clarence-Smith K. Tetrabenazine for the treatment of chorea and other hyperkinetic movement disorders. Expert Rev Neurother. 2011;11(11):1509–23.
101. Frank S, et al. Effect of deutetrabenazine on chorea among patients with Huntington disease: a randomized clinical trial. JAMA. 2016;316(1):40–50.
102. Longinetti E, Fang F. Epidemiology of amyotrophic lateral sclerosis: an update of recent literature. Curr Opin Neurol. 2019;32(5):771–6.
103. Camu W, et al. Repeated 5-day cycles of low dose aldesleukin in amyotrophic lateral sclerosis (IMODALS): a phase 2a randomised, double-blind, placebo-controlled trial. EBioMedicine. 2020;59:102844.
104. Tang Q, et al. Central role of defective interleukin-2 production in the triggering of islet autoimmune destruction. Immunity. 2008;28(5):687–97.
105. Mead RJ, et al. Amyotrophic lateral sclerosis: a neurodegenerative disorder poised for successful therapeutic translation. Nat Rev Drug Discov. 2023;22(3):185–212.
106. Menduti G, et al. Drug screening and drug repositioning as promising therapeutic approaches for spinal muscular atrophy treatment. Front Pharmacol. 2020;11:592234.
107. Calder AN, Androphy EJ, Hodgetts KJ. Small molecules in development for the treatment of spinal muscular atrophy. J Med Chem. 2016;59(22):10067–83.
108. Spector R. Ceftriaxone transport through the blood-brain barrier. J Infect Dis. 1987;156(1):209–11.
109. Nizzardo M, et al. Beta-lactam antibiotic offers neuroprotection in a spinal muscular atrophy model by multiple mechanisms. Exp Neurol. 2011;229(2):214–25.
110. Brueggeman L, et al. Drug repositioning in epilepsy reveals novel antiseizure candidates. Ann Clin Transl Neurol. 2019;6(2):295–309.
111. Larkin JG, et al. Nifedipine for epilepsy? A double-blind, placebo-controlled trial. Epilepsia. 1992;33(2):346–52.
112. El-Azab MF, Moustafa YM. Influence of calcium channel blockers on anticonvulsant and antinociceptive activities of valproic acid in pentylenetetrazole-kindled mice. Pharmacol Rep. 2012;64(2):305–14.
113. Sandouka S, et al. Repurposing dimethyl fumarate as an antiepileptogenic and disease-modifying treatment for drug-resistant epilepsy. J Transl Med. 2023;21(1):796.
114. Khatri P, et al. A common rejection module (CRM) for acute rejection across multiple organs identifies novel therapeutics for organ transplantation. J Exp Med. 2013;210(11):2205–21.

115. Xu H, et al. Validating drug repurposing signals using electronic health records: a case study of metformin associated with reduced cancer mortality. J Am Med Inform Assoc. 2015;22(1):179–91.
116. Bansal M, et al. A community computational challenge to predict the activity of pairs of compounds. Nat Biotechnol. 2014;32(12):1213–22.
117. Ma DL, Chan DS, Leung CH. Drug repositioning by structure-based virtual screening. Chem Soc Rev. 2013;42(5):2130–41.
118. Savva K, et al. Computational drug repurposing for neurodegenerative diseases. In: Roy K, editor. In silico drug design. London: Academic Press; 2019. p. 85–118.

Reshaping the Therapeutic Landscape: Drug Repurposing for Emerging Infectious Diseases and Rare Diseases

8

Amritha G. Nambiar, Anuj A. Biswas, Ashish K. Agrawal, Brahmeshwar Mishra, and Dinesh Kumar

Abbreviations

AIP	Acute intermittent porphyria
ALS	Amyotrophic lateral sclerosis
CF	Cystic fibrosis
COVID	Coronavirus disease
DNA	Deoxyribonucleic acid
FDA	Food and Drug Administration
HD	Huntington's disease
HTS	High-throughput screening
MBL	Metallo-β-lactamases
MCR	Mobile colistin resistance
MCRPE	MCR-producing Enterobacterales
MERS	Middle East respiratory syndrome
MRSA	Methicillin-resistant *Staphylococcus aureus*
NiV	Nipah virus
PBGD	Porphobilinogen deaminase
RNA	Ribonucleic acid
SARS-CoV	Severe acute respiratory syndrome coronavirus
UV	Ultraviolet ray
WHO	World Health Organization
XP	Xeroderma pigmentosum

A. G. Nambiar · A. A. Biswas · A. K. Agrawal · B. Mishra · D. Kumar (✉)
Department of Pharmaceutical Engineering and Technology, Indian Institute of Technology (BHU), Varanasi, Uttar Pradesh, India
e-mail: amrithagnambiar.rs.phe21@itbhu.ac.in; biswas.aachintya.phe19@iitbhu.ac.in; ashish.phe@itbhu.ac.in; bmishra.phe@itbhu.ac.in; dinesh.phe@itbhu.ac.in

N. Chella et al. (eds.), *Drug Repurposing*,
https://doi.org/10.1007/978-981-97-5016-0_8

Learning Objectives

Upon studying this chapter, you will be able to accomplish the following:

- Elucidate the concept of drug repurposing, its necessity, and its advantages over the traditional method of drug development, especially for emerging infectious diseases and rare diseases.
- Articulate the methods used in drug repurposing, such as in vitro screening, gene profiling, and computational methods, which include high-throughput screening, machine learning, molecular docking, and network building.
- Recognize the potential of drug repurposing in treating emerging infectious diseases, concentrating on bacterial, fungal, and viral infections.
- Recognize the complexity involved in treating rare diseases and perceive the importance of drug repurposing in offering an alternative therapeutic option.
- Explicate various case studies that exemplify the positive outcome of drug repurposing in the treatment of emerging infectious diseases and rare diseases.

8.1 Introduction

8.1.1 Drug Repurposing

Drug repurposing is finding new therapeutic applications from drugs already approved for treating other diseases, i.e., drug repositioning or drug reprofiling. Researchers examine if drugs already licensed for one ailment can be helpful in treating other diseases or conditions rather than developing new drugs from discovery to the development stage [1]. The idea of drug repurposing relies on understanding drugs with several targets and mechanisms of action in addition to their primary target and mechanism (Fig. 8.1). This implies that a drug designed to treat one disease may benefit other diseases by acting on many molecular targets.

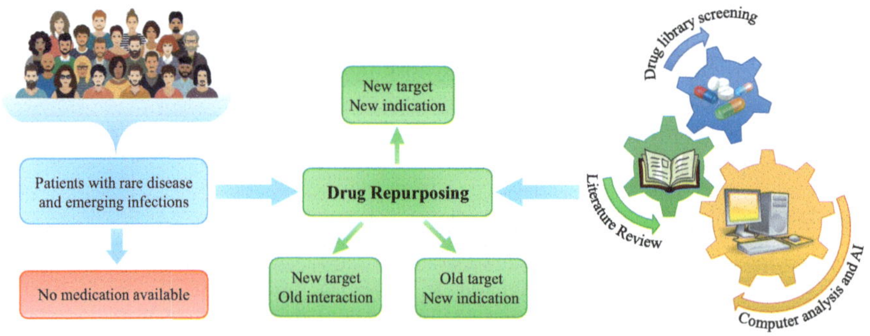

Fig. 8.1 General approaches for drug repurposing

Why is drug repurposing required?

1. Drug repurposing enables researchers and physicians to investigate alternative therapies from current drugs to increase the treatment spectrum for diseases and ailments. This may open new treatment options for patients with few or no other therapy options.
2. The process of creating a novel drug from the start is laborious and costly, which can cost billions of dollars. Drug repurposing offers an efficient and affordable route by utilizing current drugs with known pharmacokinetics, manufacturing procedures, and safety characteristics [2]. This can considerably shorten the time and cost of getting a drug to market.
3. Reusing already approved drugs lowers the risks connected to regulatory and safety problems. These safety profiles of drugs, dose recommendations, and potential adverse effects are well documented because they have already undergone rigorous research and approval procedures for their initial indications [3].
4. Repurposing enables the study of new applications without having to reevaluate safety. Drugs that have been given approval frequently have well-characterized ways of action, which means that it is known how they affect biological targets or pathways. Researchers may benefit from this information by repurposing drugs and using them to treat various ailments or diseases with comparable underlying processes. This may hasten the creation and discovery of novel therapies [4].
5. For rare diseases or ailments with few available treatments, drug repurposing presents a potential therapy for unmet medical needs. Researchers can quickly find possible candidates and accelerate their assessment for new applications by repurposing current drugs, which may be advantageous for patients with few or no approved medicines.
6. When the patent protection of a drug expires, other researchers or the pharmaceutical industry are free to study the potential of additional drug uses. Repurposing drugs can give businesses a competitive edge in the market and a means of extending the life cycle of their products.

8.1.2 Emerging Infectious Diseases

Infectious diseases that have recently emerged or have existed in the past but are expanding quickly in number or geographic scope are called emerging infectious diseases (EIDs). These diseases tend to be unique and pose a major risk to public health as humans often lack pre-existing immunity or good treatment options. According to the World Health Organization (WHO), infectious diseases kill 17 million people in a year, and nearly 30 new infectious diseases have emerged in the past two decades [5]. It is essential to know how much they have affected the world to increase an outlook on emerging infectious diseases.

1. COVID-19 has infected approximately 0.76 billion people, and mortality due to COVID-19 is more than 6.9 million worldwide [6].
2. About 30% of the population was affected by Ebola virus diseases, and death cases were about 15,266 from 1976 to 2020 [7].
3. Some lung infections like pneumonia resulted in the death of 2.5 million in 2019, in which one-third of victims were children younger than 5 years [8], and chronic obstructive pulmonary diseases caused 3.23 million deaths in 2019, according to WHO.
4. Malaria is a chronic disease caused by the Anopheles mosquito. About 247 million people were infected with malaria, and the mortality rate was about 0.61 million [9].

8.1.3 Rare Diseases

A rare disease can be defined as a health condition with low occurrence rate, impacting a small fraction of population, specifically, 1 in 2000 in Europe and 1 in 1250 in the United States. It is estimated by WHO that there are over 5000 rare diseases globally. In addition, new rare diseases are continuously being reported. Few rare diseases are more common, as approximately 350 rare diseases cover ~80% of rare disease-affected patients. The true impact of rarity estimates of illnesses in Europe and worldwide is difficult, considering that for most of these disorders, epidemiological data is not accessible. Most of the rare diseases are genetic disorders. Many genetic abnormalities adversely affect life expectancy and mental and physical health. The prominent example is inborn metabolic errors, which occur 1 in 1400 and 1 in 500 live births [10]. However, no treatment is currently available for most rare disorders.

Rare diseases are a public health issue, and there are formidable challenges to discovering and developing drugs for rare diseases. The majority of these challenges are as follows:

1. Lack of epidemiological data: The data for the number of people suffering from different rare diseases is not vast and consistent. Most rare diseases are diagnosed only at tertiary-level hospitals, making estimations difficult. Due to this, the economic cost of this disease is difficult to estimate and makes the process quite tricky. It is evident that doing systemic epidemiological research is vital to determine the population in India afflicted with rare diseases.
2. Multiple definitions and prevalence thresholds: Using multiple definitions and varied terminologies causes misunderstanding and unpredictability, ultimately affecting the research, development, and treatment of rare diseases.
3. Diagnosing rare diseases: Identifying a rare disease is one of the biggest challenges. Generally, it takes 5–7 years to diagnose the rare disease accurately. Multiple misdiagnoses are also common for a rare disease. Late diagnosis and misdiagnosis ultimately enhance the patient's suffering.

4. Challenges in the research and development related to rare diseases: The patient pool related to rare diseases is minimal and generally results in insufficient clinical explanation. Research and development are costly, and the small pool of patients discourages pharmaceutical companies from investing in rare diseases.
5. Challenges in treating rare disease: There are few medicines available to treat the rare disease. There are >5000 rare diseases, but therapies are available only for 300 rare diseases. As the number of patients is small, manufacturers do not mass produce these medicines which make the prices higher. The cost of research and development also adds up to the final prices. This significantly enhances the market prices for these medicines, making them out of reach for developing and underdeveloped countries.

Despite all the challenges mentioned above, most aspects have positive improvements. It is time for a big step forward, as evidenced by the improvements in our understanding of the mechanisms underlying many diseases and the explosion of information in genetic medicine. Governments have to enhance budgets for research, development, and manufacturing of rare diseases as incentives for the companies are very limited.

8.2 Drug Repurposing for Emerging Infectious Diseases

Advanced drug discovery and development research have enabled the treatment of most infectious diseases and saved millions of lives. However, multiple challenges remain as microbes can become resistant to the available medicines (Table 8.1). Antimicrobial resistance (AMR) is considered one of the biggest threats to human beings, which necessitates the discovery and development of novel and potent antimicrobial medicines. Drug repurposing has been demonstrated to be noteworthy in

Table 8.1 Emerging infectious diseases

Emerging bacterial infections	Emerging fungal infections	Emerging viral infections	References
Methicillin-resistant *Staphylococcus aureus* (MRSA)	Candidiasis	COVID-19	[11–13]
Carbapenem-resistant Enterobacterales (CRE)	Aspergillosis	Ebola	[14–16]
MCR-producing enterobacterales (MCRPE)	*Cryptococcus neoformans*	Zika	[17–19]
Parachlamydia acanthamoebae	Candida auris	Chikangunya virus infection	[20–22]
Waddlia chondrophila	Candida albicans	Avian influenza bird flu	[22–24]
Neoehrlichia mikurensis	Dermatophytosis	Nipah virus infection	[22, 25, 26]
Alloscardovia omnicolens	Zoonotic Sporotrichosis	Middle East respiratory syndrome	[22, 26, 27]

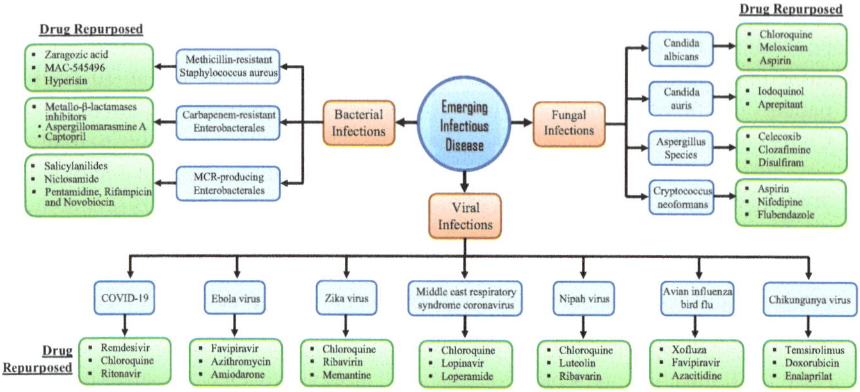

Fig. 8.2 Drug repurposing for emerging infectious diseases

treating multiple diseases with one medicine, so there is immense potential for drug repurposing to tackle antimicrobial resistance. Numerous examples exist where drug repurposing has emerged as a potential alternative to the standard treatment (Fig. 8.2).

Gallium nitrate is one of the anticancer medicines which has shown promising efficacy against *Pseudomonas aeruginosa.* Toremifene, used to treat advanced breast cancer, has also shown good antimicrobial activity and controls biofilm formation and growth. Few topoisomerase inhibitors approved for cancer treatment have also shown antibacterial properties and promising results against fluoroquinolone-resistant microorganisms. Some NSAIDs (nonsteroidal anti-inflammatory drugs) have also prevented biofilm formation and have shown good potential for being used as antimicrobial medicines. Closantel, an antihelminthic drug approved for veterinary use, has also shown good efficacy against antibiotic-resistant *Staphylococcus aureus.*

Recently, there has been a multiple-fold growth in drug-resistant invasive fungal infections. These emerging fungal infections are challenging to identify. These emerging fungal infections are also resistant to the medicines prescribed as per treatment. The underlying mechanisms at the molecular level should be investigated in these emerging fungal infections to find the possible treatment mechanism and suitable antifungal drugs. Technology-enabled drug discovery and advanced data analytics can play a massive role in discovering the ideal antifungal treatment candidate after repurposing. Fungal eye infections, particularly the resistant traits in the patients who recovered from COVID-19, have enhanced the research focus on repurposing medicines for drug-resistant fungal medication.

Drug repurposing for emerging infectious diseases can be classified into three broad categories:

(a) Drug repurposing for emerging bacterial infections.
(b) Drug repurposing for emerging fungal infections.
(c) Drug repurposing for emerging viral infections.

8.2.1 Drug Repurposing for Emerging Bacterial Infections

Uncontrolled use of antibiotics has resulted in the emergence of antibiotic resistance. Many non-antibiotic drugs show good antibacterial action and efficacy against resistant strains of bacteria (Table 8.2). Identification, verification, and then employing these drugs for drug-resistant bacteria can be an effective and sustainable strategy to control the rise of bacterial drug resistance. Repurposed medicines against Methicillin-resistant *Staphylococcus aureus* (MRSA), carbapenem-resistant enterobacterales (CRE), and MCR-producing enterobacterales (MCRPE) seem ideal examples of this.

8.2.1.1 Methicillin-Resistant *Staphylococcus aureus* (MRSA)

Staphylococcus aureus is considered the most challenging pathogen as it causes multiple infections. Many antibiotics do not show efficacy against *Staphylococcus aureus* as they develop resistance against the drug. Clinical use of semisynthetic antibiotics, which have structure similarity to methicillin, is steered to the beginning of methicillin-resistant *Staphylococcus aureus* (MRSA). The MRSA developed a penicillin-binding protein that shows poor affinity for antibiotics, which shows a similar structure to methicillin. This has resulted in high mortality rates. Clinicians use vancomycin antibiotics as a last option for the treatment of MRSA. The situation has become complicated as *Staphylococcus aureus* has also resisted vancomycin. This strain is called vancomycin-resistant *S. aureus* (VRSA). These ongoing resistant developments to the available antibiotics have justified the repurposing of already approved drugs as potential antibiotic adjuvants that offer a feasible treatment for MRSA and VRSA.

García-Fernández et al. observed that Zaragoza acid, a lipid-lowering drug, can reverse resistance to penicillin in MRSA [11]. In addition, a cell-based screen of about 45,000 diverse compounds observed that one anti-virulence molecule, MAC-545496, overturned β-lactam resistance in the MRSA strain [49]. Hypericin, an antidepressant and antiviral, has considerably enhanced the activity of β-lactam antibiotics against MRSA [50].

Farha et al. observed that wall teichoic acid (WTA) makes β-lactam antibiotics more potent against MRSA [51]. An antiplatelet medicine, ticlopidine, was observed to enhance the action of cefuroxime in multiple infection models [51]. These studies confirm that controlled WTA synthesis can provide an alternative to existing β-lactams antibiotics to block MRSA infection. Auranofin, a thioredoxin reductase (TrxR) inhibitor along with linezolid and fosfomycin, enhanced antimicrobial efficacy against MRSA [52].

8.2.1.2 Carbapenem-Resistant Enterobacterales (CREs)

Carbapenems are broad-spectrum β-lactam class antibiotics. They work by inhibiting the cell wall synthesis of microbes [53]. Carbapenems are considered the final choice medicine for treating multidrug-resistant bacterial infections [16]. Enterobacterales bacteria family includes *Klebsiella pneumoniae*, *Salmonella enterica,* and *Proteus mirabilis.* All of these show a high level of resistance to

Table 8.2 Drug repurposed for infectious diseases

Sr. No.	Name of disease	Drug name	Mechanism of action	Existing usage of the drug	References
1.	Methicillin-resistant *Staphylococcus aureus* bacterial infection	Zaragozic acid	Degrades FMMs, inhibits PBP2a oligomerization, and re-sensitizes MRSA to penicillin	Treatment of hyperlipoproteinemia and trypanosomatid infections	[28]
		MAC-545496	Inhibits GraR and reverses the β-lactam resistance in MRSA	Anti-virulence	
		Hypericin	Increases β-lactam's ability to inhibit MRSA by blocking the sarA expression	Anticancer and antidepressant	
2.	Carbapenem-resistant enterobacterales	Aspergillomarasmine A	MBL inhibitor	Treatment of hypertension	
		Captopril		Hypertension and congestive heart failure	
3.	MCR-producing enterobacterales	Salicylanilides	MCR-1 inhibition	Anti-anthelmintics	[29]
		Niclosamide			
		Pentamidine, rifampicin, and novobiocin		Antibiotics	
4.	Candida albicans	Chloroquine	Inhibition of ergosterol biosynthesis	Antimalarial	[30]
		Polymyxins (Polymyxin B)	Membrane damages on Candida albicans	Treatment of bacterial infection	
		Ribavirin	Impairs the vacuoles function of C. albicans strains	Treatment of hepatitis C	
		Auranofin	Act on reactive-oxygen-mediated cell death	Rheumatoid arthritis	
		Haloperidol	Possible action on GPCRs, which facilitate signal transmission across the cell membrane	Treatment of psychosis	
5.	Candida auris	Colistin	Membrane damages on Candida albicans	Treatment of bacterial infection	
		Sertraline	Inhibition of protein synthesis	Antidepressant	
6.	Aspergillus species	Chlorpromazine	Possible modifications of membrane	Antipsychotic	
		Disulfiram	Suppression of the multidrug transporter linked to drug resistance	Treatment of chronic alcoholism	
		Azoles (fluconazole, itraconazole)	Alter fluidity and agility of fungal membrane	Treatment of fungal infection	
7.	*Cryptococcus neoformans*	Aspirin	Stress induction via ROS- mediated damage	Anti-inflammatory	
		Echinocandins (micafungin, caspofungin, anidulafungin)	Formation of defective cell wall by inhibiting β-1,3-glucan synthase	Antifungal	
		Miltefosine	Inhibits phospholipase B1	Treatment of leishmaniasis	[31]
		Sertraline	Inhibition of protein synthesis	Antidepressant	[30]
		Disulfiram	Metal chelating suppression of the multidrug transporter linked to medication resistance	Treatment of chronic alcoholism	

8.	COVID-19	Ivermectin	Replication of SARS-CoV-2 inhibition	Antiparasitic drug	[32]
		Prolactin	Stimulation of innate and adaptive immunity	Hormone	[33]
		Azithromycin	Accelerates the production of type I and III interferons (especially interferon-β and interferon-λ), and genes involved in virus recognition such as MDA5 and RIG-I	Antibiotics	[34]
		Lopinavir and ritonavir	Inhibition of SARS-CoV main protease (3CLpro)	Treatment of HIV	[35]
		Ivermectin	Replication of SARS-CoV-2 inhibition	Antiparasitic	[32]
9.	Ebola viral infection	Favipiravir	Inhibits the influenza viral RNA-dependent RNA polymerase	Antiviral	[36]
		Clomiphene	Inhibitors of Zaire ebolavirus (EBOV) infection	Treatment of anovulatory or oligo-ovulatory infertility	[37]
		Toremifene	Inhibitors of Zaire ebolavirus (EBOV) infection	Antineoplastic hormonal agent	
		Chloroquine	Inhibitors of Zaire ebolavirus (EBOV) infection	Antimalarial	[38]
		Azithromycin	Inhibitors of Zaire ebolavirus (EBOV) infection	Antimicrobial	
10.	Zika viral infection	Chloroquine	Inhibits ZIKV infection	Antimalarial	[39]
		Sofosbuvir		Antiviral	[40]
		Daptomycin		Antibiotic	
		Azithromycin		Antimicrobial	
11.	Middle east respiratory syndrome coronavirus	Nitazoxanide	Inhibits expression of the viral N protein	Antiparasitic	[41]
		Lopinavir	MERS-CoV-induced cytopathic effect	Treatment of HIV	[35]
		Loperamide	Inhibits MERS-CoV replication	Antidiarrheal	[42]
		Chlorpromazine		Antipsychotic	
		Chloroquine		Antimalarial	
12.	Nipah virus	Favipiravir	Inhibits the Nipah virus	Antiviral	[43]
		Chloroquine	Inhibits henipaviruses	Antimalarial	[44]
		Ribavirin	Inhibit the replication of Nipah virus	Antiviral	[25]
		Luteolin	Niv-F inhibitor	Anti-inflammatory	[45]
		6-gingerol		Antioxidant	
13.	Avian influenza bird flu	Favipiravir	Induce mutagenesis in the influenza genome	Antiviral	[46]
		Baloxavir marboxil	Inhibits polymerase acidic endonuclease	Antiviral	[47]
		Azacitidine	Inhibits H5N1 virus replication	Chemotherapy drug	[48]
		Calcitriol		Treatment of low calcium level caused by kidney disease	[46]

(continued)

Table 8.2 (continued)

Sr. No.	Name of disease	Drug name	Mechanism of action	Existing usage of the drug	References
14.	Chikungunya virus	Resveratrol	Inhibits CHIKV and its replication	Grape juice (no particular use)	[20]
		Metyrapone		Diagnosis of problems related to adrenal glands	
		Enalaprilat		Treatment of hypertension	
		Lomibuvir		Treatment of chronic hepatitis C virus	
		Emetine		Treatment of amebiasis	
		Felbinac		Nonsteroidal anti-inflammatory	
		Doxorubicin		Antibiotic	
		2-fluoroadenine		Antimetabolite	
		Temsirolimus		Antineoplastic	

carbapenems antibiotics. These CREs are considered one of the biggest threats to antimicrobial resistance. They are now observed throughout the world and show high mortality rates as well.

Carbapenemase enzymes cause hydrolysis of the β-lactam ring and cause bacterial resistance to carbapenems. Combining carbapenems and metallo-β-lactamases (MBLs) inhibitors provides an effective strategy to counter-treat CREs. This combination-based repurposing offers quick development time, low cost, and better safety records. Most MBL inhibitors work either by sequestration of metals (mostly Zn2+) or by forming new covalent bonds [54].

Zn-dependent inhibition of MBLs: The active locations of Zn2+ ions are critical for MBL catalysis and blocking these Zn2+ can inhibit MBLs. Multiple metal-based chelators, including EDTA, have anti-MBL action using the same rationale. Captopril, an ACE inhibitor, has also been observed as an MBL inhibitor. It can also chelate zinc ions on its active site by employing a free thiol in captopril [55]. Other drugs that can similarly chelate metal ions and inhibit MBLs include disulfiram [56], tiopronin [57], and thiorphan [57], which are approved previously for various other diseases.

8.2.1.3 MCR-Producing Enterobacterales (MCRPE)

Colistin (polymyxin B) is used as a final retort to treat acute and chronic multidrug resistance bacterial infections triggered by resistance strains of Gram-negative bacteria. Resistance to these antibiotics was first reported in the last decade, and these reports are increasing yearly. Colistin resistance is considered a high-level threat because of its easy transmittance across different species, including humans and animals. It was first identified in 2015 and found to occur through a plasmid-intermediated mobile gene mcr-1 [19]. Multiple mcr genes including *mcr*-2, *mcr*-3, *mcr*-4, *mcr*-5, *mcr*-6, *mcr*-7, *mcr*-8 have now been identified in different species [58, 59]. This surge has alarmed the medical fraternity. Many research groups and pharmaceutical companies are struggling to find effective combinations to counter MCRPE (mobile colistin-resistant producing Enterobacterales).

Pterostilbene, one of the food additives, has been observed as an MCR-1 inhibitor [60]. Osthole, a natural product, has successfully enhanced colistin efficacy against MCR-1 [61]. Salicylanilides and niclosamide, which are FDA-approved anthelmintic medicines, have also shown *mcr-1 inhibition* [29, 62].

Gram-negative bacteria are difficult to treat as they have an extremely impervious outer membrane, which stops antibiotics from crossing this barrier. It has been found that pentamidine, which is an antiprotozoal drug, targets lipopolysaccharides in the outer membrane and helps kill gram-negative bacteria. Pentamidine combination with rifampicin and novobiocin (Gram-positive active antibiotics) shows good efficacy against MCRPE [63]. This repurposing exposed the untapped therapeutic perspective of the antiprotozoal drug pentamidine for managing MCRPEC infections. Melatonin, a food supplement, enhanced bacterial outer membrane permeability when combined with colistin and inhibited colistin resistance in mcr genes [28].

8.2.2 Drug Repurposing for Emerging Fungal Infections

Fungal pathogens cause multiple infections, for example, candidiasis, aspergillosis, or Cryptococcus. There are four classes of antifungal drugs. First includes azoles such as itraconazole and fluconazole (FLU). The second group includes amphotericin B (polyene-based). The third group includes echinocandins such as caspofungin, micafungin, and anidulafungin. 5-flucytosine comes in the fourth group of antifungal drugs. Resistance was observed against conventional antifungal medicines as well, which has resulted in a difficult situation for doctors as options to treat fungal infections were already limited [64, 65]. This shows the urgency and necessity for the development of new antifungal medicines. Repurposing other drugs for their antifungal properties can be an ideal strategy to fill this gap [3]. The mechanisms of action, cellular targets, or safety of the marketed drugs are already known. Thus, once antifungal efficacy is determined, the repurposed drug gets the advantage of accelerated regulatory approval. Multiple repurposing strategies employed for treating fungal infections, such as Candida albicans, *Candida auris*, *Aspergillus*, and *Cryptococcus* species have been discussed with examples in the following section [30].

8.2.2.1 *Candida albicans*

Candida albicans is an opportunistic fungal pathogen normally found in human gastrointestinal flora but can also make uncontrolled colonies and result in severe invasive infections. *Candida albicans* can quickly form biofilm, making them impermeable for many-antifungal medicines and enhancing the complexity of treatment [24]. Many scientific reports have proven that repurposing can effectively treat *Candida albicans* infections [66, 67].

Chloroquine [68] and quinacrine [69], which are antimalarial drugs, have shown their ability to revert resistance to azole-based antifungal medicines while treating the *Candida albicans* biofilm. Aspirin, used for reducing mild inflammation and pain, has also shown antifungal properties against *Candida* species [70]. Few other anti-inflammatory drugs, such as meloxicam, celecoxib, and etodolac, have also demonstrated promising action against resistant *Candida albicans* biofilm [71]. Disulfiram, which is prescribed for the treatment of alcohol dependence, has shown results against *Candida albicans* biofilm [72]. These examples prove the impact of repurposing to treat resistant strains of Candida *albicans*.

8.2.2.2 Candida Auris

Candida auris causes one of the most serious emerging fungal infections with a high mortality rate [73]. A few Candida auris strains are resistant to all the major classes of antifungal medicines. Echinocandins are considered the last option to treat resistant strains of C. auris. However, some strains also show resistance to echinocandins [74, 75]. In this situation, doctors prescribe cocktails of multiple antifungal medicines to treat the infection [21]. Repurposing other medicines offers a straightforward approach to treating this dangerous infection [76, 77]. Cheng et al. screened

~4000 molecules and observed six compounds showing activity against *Candida auris* [76]. Two other drugs that are repurposed for activity against *Candida auris* are iodoquinol, an anti-amebicide [77], and aprepitant, which is an anti-emetic drug [78]. These repurposed molecules offer an excellent deal for resistant strains of Candida auris when used with conventional antifungals.

8.2.2.3 Aspergillus Species

Aspergillus is a common mold observed in normal houses and neighborhoods [15]. This causes an infection known as Aspergillosis. Aspergillosis is generally observed only in people with weak immune systems, which does not affect people with good immunity. A high mortality rate of ~60% has been observed with aspergillosis in immunocompromised people [79]. The resistance has evolved in Aspergillosis because of uncontrolled use of pesticides, which shows structural similarity with azoles (the most common antifungal medicines).

Drug repurposing strategy can effectively treat *Aspergillosis* infections [80, 81]. Celecoxib, an anti-inflammatory drug, has also shown promising action against resistant *Aspergillus fumigatus* [82]. Disulfiram, which is prescribed for the treatment of alcohol dependence, has shown results against resistant aspergillosis [83]. Clofazimine, an anti-leprotic medicine, has shown promising effects against resistant strains of aspergillosis [84]. These examples prove the worth of a repurposing strategy for effectively managing resistant strains of aspergillosis.

8.2.2.4 Cryptococcus *neoformans*

Cryptococcus is a kind of fungus that is widely found in homes and neighborhoods. Cryptococcus *neoformans* can infect persons with impaired immune systems and produce cryptococcosis, possibly leading to meningitis [15]. People suffering from HIV have a weak immune system and are considered vulnerable to cryptococcosis. Management of this disease is quite peculiar as the available antifungal medicines do not show good efficacy against cryptococcosis [18]. Drug repurposing is considered as the biggest hope for managing this infection [85]. Many scientists have observed interesting results where molecules from other treatment classes have shown promising results against Cryptococcus *neoformans.* Butts et al. discovered that selective estrogen receptor modulators (SERM) which have an alkylamino functional group attached to the aromatic rings of the triphenylethylene and an aliphatic substituent at the second position of the ethylene moiety; and an electronegative moiety on the aromatic rings significantly improves the antifungal activity [85].

Aspirin and ibuprofen, which are anti-inflammatory drugs, showed significant activity against *Cryptococcus* strains [86]. Tamoxifen, an anticancer drug [18], nifedipine, an anti-hypersensitive drug [87], and flubendazole, which is a common antiparasitic medicine, have shown good efficacy against resistant strains of Cryptococcus *neoformans* [87]. This repurposed strategy suggests that repurposing can be a great tool to treat resistant strains of Cryptococcus *neoformans* where conventional antifungals are ineffective.

8.2.3 Drug Repurposing for Emerging Viral Infections

Significant advances have been observed in managing viral infections in the last few decades. Innumerable viruses can infect human cells and cause serious health problems. Finding an ideal treatment seems to be a big challenge [88]. These viruses can mutate quickly and become a massive risk to public health in a short period. Novel virus infections always keep scientists on alert, and developing new antiviral medicines is challenging. As finding and developing new drugs are costly and time-consuming, drug repurposing offers an innovative approach to managing the ever-changing viral infections. Multiple efforts exist to find broad-spectrum antiviral medicines using the drug repurposing strategy. The following section discusses emerging viral infections, their treatment strategies, a risk to the current treatments, and the importance of drug–repurposing to manage these emerging viral infections.

8.2.3.1 COVID-19 (Coronavirus Disease 2019)

The Coronavirus Disease 2019 (COVID-19) originated from a severe acute respiratory syndrome known as coronavirus 2 (SARS-CoV-2). It was initially discovered in December 2019 in Wuhan, Hubei Province, China, and rapidly propagated throughout the globe, leading to a pandemic. The total number of global deaths was about 3 million, with 1.2 million more deaths which are not reported [13]. SARS-CoV-2 belongs to the coronavirus family, encompassing viruses such as MERS-CoV and SARS-CoV. SARS-CoV-2 is a newly discovered virus that was previously unknown to humans. It stands out and has the potential to spread quickly. The procedure by which coronavirus enters the cell involves the spike glycoprotein, membrane glycoprotein, envelope protein, and nucleocapsid (N) protein. Coronavirus enters through endosomal and non-endosomal pathways [89–91]. According to WHO (World Health Organization), COVID-19 was declared a pandemic due to human-to-human spread worldwide. SARS-CoV-2 has demonstrated a high level of contagiousness, with a large percentage of infected people showing mild to severe respiratory symptoms. The disease has diverse clinical manifestations, including asymptomatic cases, making it difficult to contain.

Hydrocortisone (corticosteroids) and benzhydrocodone (opioid pro-drug) were found to be potential candidates for the management of COVID-19. Remdesivir has been observed as one of the most potent candidates to control COVID-19 [92–94]. Antimalarial drugs, like chloroquine and hydroxychloroquine, also showed promising results in managing COVID-19 complications [95].

Lopinavir, ritonavir [35], and favipiravir [96] are other antiviral medicines that showed good efficacy against COVID-19. Azithromycin is one of the most potent antibiotics and has effectively controlled pneumonia caused by COVID-19 [34]. Prolactin, a promising immunomodulator, has also shown good results against COVID-19 [33]. Mansour et al. observed that ivermectin, an antiparasitic drug, can be repurposed to manage COVID-19 complications [32].

These examples proved that it is possible to control pandemic-like situations more quickly by repurposing drugs previously authorized for use in other medical conditions. Preclinical and early clinical trials for these drugs need less time and money because they have already undergone safety testing. Drug repurposing technologies that are more precise and effective could potentially lower the risk and boost the effectiveness of pharmacological repurposing to identify more potent drugs for emerging infectious diseases.

8.2.3.2 Ebola Virus

The Ebola virus belongs to the Filoviridae family, which is highly virulent, contagious, and shows high mortality. There are frequently fatal outbreaks reported for the Ebola virus disease (EVD) in nonhuman primates and humans. The majority of Ebola virus outbreaks have been observed in Central and West African nations [14]. The symptoms of an EVD include hyperpyrexia, excruciating headache, severe muscle pain, exhaustion, vomiting, diarrhea, and, in rare cases, bleeding. These symptoms can advance quickly and cause organ failure and, ultimately, death [97]. Drug repurposing might speed up the search for possible Ebola virus therapeutics and enable their prompt availability during epidemics [98].

Drug repurposing involves testing current drugs and compounds for the Ebola virus. This extensive screening method can find drugs already approved for other conditions but also have antiviral effectiveness against the Ebola virus. Berbamine hydrochloride extracted from traditional Chinese herbs inhibits viral glycoprotein and has shown remarkable effects against the Ebola virus [99]. Favipiravir, an antiviral pro-drug, has shown promising results against the Ebola virus [36]. Azithromycin (antibiotic), chloroquine (antimalarial), and amiodarone (antiarrhythmic) have shown promising activity with low toxicity against the Ebola virus [38]. The selective estrogen receptor modulators (SERMs), clomiphene and toremifene, have demonstrated potent inhibition of the Ebola virus [37].

Repurposed drugs may directly target the Ebola virus or its life cycle. For example, certain drugs may block viruses from entering, reproducing, or releasing new viral particles. These drugs can prevent viral propagation and replication by focusing on particular phases of the life cycle of the virus. Drug repurposing is a unique approach to the treatment of various infectious diseases. Evidence from preclinical studies, clinical trials, and expert consensus within the scientific and medical community should all be considered when determining individual repurposed medications and their efficacy against the Ebola virus [100].

8.2.3.3 Zika Virus

The Zika virus is a viral infection transmitted by mosquitoes. It is named after the Zika Forest of Uganda, where it was initially discovered in 1947. Aedes mosquitoes, responsible for transmitting diseases like dengue and chikungunya, are the primary vectors for spreading the Zika virus. The Zika virus was not well known for several decades after it was first identified because it only hardly and rarely caused major pandemics in Africa and Asia [17].

Zika virus outbreaks were most noticeable from 2007 to 2016, mostly in the Pacific Island of Yap, Brazil, and other South and Central American nations [101]. The virus quickly spreads throughout the Caribbean and other regions of the Americas, eventually infecting more than 80 nations and territories. In February

2016, the World Health Organization (WHO) emphasized the need for coordinated measures to combat the virus and its potential effects on public health by classifying Zika as a worldwide concern and public health emergency [102]. In addition to mosquito bites, sexual activity, blood transfusions, and mother-to-fetus transmission during pregnancy are other ways that Zika can spread. Due to its link to neurological diseases and birth abnormalities like microcephaly, this vertical transmission is especially alarming. The Zika virus can cause mild to moderate symptoms like fever, rash, joint and muscle discomfort, headaches, and conjunctivitis [103]. It can be challenging to detect Zika infection since people who have it frequently may not show any symptoms at all.

There are many scientific reports where it has been proven that a repurposing strategy can be effectively used to treat *Zika virus* infections. Azithromycin, one of the prominent antibiotics along with daptomycin and sofosbuvir, reduces Zika viral proliferation [40]. Chloroquine, a commonly used antimalarial drug, also shows antiviral activity against Zika virus [39]. Memantine, an *N*-methyl-d-aspartate receptor blocker, reduces neuronal complications by preventing neuronal death [104]. Ribavirin, a drug used to treat hepatitis C, was repurposed and found to suppress viremia by inhibiting viral RNA-dependent RNA polymerases [105]. These examples proved the probable impact of drug repurposing in controlling Zika virus-related complications.

8.2.3.4 MERS (Middle East Respiratory Syndrome)

The first case of the viral respiratory condition, Middle East Respiratory Syndrome (MERS), was reported in Saudi Arabia in 2012. Middle East Respiratory Syndrome Coronavirus (MERS-CoV), a type of coronavirus, is the causative agent of the zoonotic disease that can be transmitted from animals to humans. Although the exact origin of the MERS-CoV virus is not entirely understood, evidence points to dromedary camels as a significant source of the infection. The virus is thought to mainly infect humans after coming into touch with infected camels, either directly or indirectly [27]. MERS has also been linked to human-to-human transmission, notably in healthcare settings and among close contacts of sick people [106]. The typical MERS symptoms include pyrexia, cough, dyspnea, and breathing problems. MERS symptoms can range from moderate to severe. In severe situations, the infection may lead to organ failure, pneumonia, and even death. MERS has a high death rate, which is thought to be over 35% [27]. MERS cannot be treated entirely with any available antiviral drug [107]. MERS-CoV vaccine research is ongoing, but as of now, there is no licensed vaccine.

Chloroquine (antimalarial), chlorpromazine (antipsychotic), loperamide (treat diarrhea), and lopinavir (antiretroviral) have shown good inhibiting propensity against MERS-CoV replication [42]. Nitazoxanide (antiparasitic) has shown a good in vitro effect against MERS via inhibiting the viral N protein and suppressing pro-inflammatory cytokines [41].

Dyall et al. screened 290 compounds against MERS and found that 27 compounds of 13 classes showed good activity against MERS [108]. Shin et al. screened 2334 drugs and found that the drug saracatinib, an inhibitor of tyrosine kinases

(SFK), is showing good effect against MERS-CoV field [109], nafamostat [110], and the ritonavir-lopinavir combination also showed good efficacy against MERS [111].

8.2.3.5 Nipah Virus Infection (NiV)

Nipah virus is also a zoonotic virus first discovered in 1999 during an outbreak in Malaysia and Singapore. Nipah virus is a Paramyxoviridae family member and is categorized as a Henipavirus and Hendra virus. Fruit bats, especially a few species of the Pteropus genus, are considered to be the natural host of the Nipah virus. These bats are prevalent in Southeast Asia. Direct contact with sick bats and close contact with infected pigs or other animals that have been exposed to bat-borne disease can result in the transmission of the virus to people.

Nipah virus transmission from person to person has also been documented, mostly in the context of close and extended contact with sick people. A variety of clinical manifestations, including asymptomatic illness, severe respiratory infection, and encephalitis (brain inflammation), can be brought on by Nipah virus infection. The Nipah virus might take a few days to many weeks to incubate. The Nipah virus can cause encephalitis, fever, headaches, muscle discomfort, and respiratory problems. Seizures, comas, and even death are possible in extreme situations. In places where the Nipah virus has spread, control measures include killing infected animals, and public health initiatives are put in place to stop the virus from spreading. Nipah virus infection does not currently have a specific antiviral treatment. Patients get supportive therapy, which includes symptom and complication management. There is no authorized vaccination or specialized antiviral medicine. Some of the FDA-approved drugs that have shown promising results for repurposing are discussed in the examples below.

Verma et al. observed that a few phytochemicals, including 6-gingerol, 4-hyroxypanduratin A, and luteolin, have shown promising results against Niv inhibition [45]. Chong et al. observed a 36% reduction in mortality without any toxicity when they used ribavirin (antiviral for hepatitis C) to treat NiV-affected patients [25]. Porotto et al. observed significant inhibition of NiV when chloroquine (an antimalarial drug) was used [44]. Lo et al. observed that remdesivir is effective against NiV [112]. Dawes et al. have demonstrated that favipiravir can be a promising medicine against NiV [43].

8.2.3.6 Avian Influenza Bird Flu

Avian influenza, sometimes known as bird flu, is a virus that mainly infects birds. It is highly contagious and is brought on by influenza-type viruses, which can affect domestic and wild avian animals, including chickens, turkeys, and ducks. While primarily affecting birds, humans can also contract specific strains of the virus, which can cause severe disease and, in some cases, death. Avian influenza viruses are categorized into two primary types based on their ability to cause severe disease in poultry: high pathogenic avian influenza (HPAI) and low pathogenic avian influenza (LPAI). While LPAI strains usually induce a moderate or undetectable illness, HPAI strains are more virulent and can result in significant fatality rates in afflicted birds. Direct contact with diseased birds, droppings, or contaminated surfaces can spread bird flu. Due to the culling of diseased or at-risk birds, trade restrictions, and lost customer confidence, outbreaks in poultry can cause significant economic losses. The potential for severe disease and mortality in humans in some bird flu strains, such as the H5N1 and H7N9 subtypes, has also sparked worry.

Infection with avian influenza, commonly known as H5N1 virus, has a significantly high mortality rate (>50%). Currently, two classes of antiviral medicines are used for the management of influenza infections: M2 ion-channel blockers, which include amantadine and rimantadine, and neuraminidase (NA) inhibitors, which include oseltamivir, zanamivir, and peramivir. M2 ion-channel blockers have been removed from the treatment option because resistance developed in the virus. Drug repositioning is an ideal and economical tactic to find a cure for the H5N1 virus because there are not many therapies for management [23].

Huang et al. tested 41 approved drugs and showed that amitriptyline HCl (an antidepressant), flavin adenine dinucleotide (an ophthalmic agent for vitamin B2 deficiency), azacitidine(anticancer), and calcitriol (an active form of vitamin D) considerably enhanced the survival against H5N1 virus [48]. Xofluza has been recently approved for anti-influenza actions [47]. Favipiravir (antiviral) and baloxavir marboxil (antiviral) are also approved to treat influenza infections [46].

8.2.3.7 Chikungunya Virus Infection

Chikungunya virus infection is caused by Aedes mosquitos, specifically Aedes aegypti and Aedes albopictus. Chikungunya symptoms often appear 3 to 7 days after being bitten by an infected insect. High fever, excruciating joint pain (arthralgia), headache, muscle pain, rash, and exhaustion are typical symptoms. Joint discomfort can be crippling, endure for weeks or months, and in some circumstances, progress into chronic pain. Currently, there is no particular antiviral medicine available to treat chikungunya infection. The primary aims of therapy are to minimize symptoms with rest, water, and pain relievers such as acetaminophen or paracetamol. NSAIDs (nonsteroidal anti-inflammatory medicines) should be avoided since they can increase bleeding risk. Hospitalization may be necessary in extreme circumstances to treat problems.

Drug repositioning can be an ideal and economical tactic to manage chikungunya virus infection, as there is no approved medicine for management. Kasabe et al. screened a library of drugs. They found out that temsirolimus (used for advanced renal cell carcinoma), 2-fluoro adenine (for lymphatic cancer), doxorubicin (breast cancer), felbinac (for arthritis), emetine (for amoebiasis), lomibuvir (for Hepatitis C), enalaprilat (for congestive heart failure), metyrapone (diagnosis of adrenal glands problems), and resveratrol (antioxidant) are showing significant activity against chikungunya [20].

Santos et al. observed that the platinum complex of ranitidine hydrochloride is highly effective against chikungunya virus [113]. Pereira et al. studied the impact of memantine hydrochloride (used for Alzheimer's) on the chikungunya virus. They observed its efficacy, which makes memantine hydrochloride a potential medicine to be repurposed for the chikungunya virus [114]. Bhakat et al. found out antiviral activity of nelfinavir against chikungunya virus [115].

8.3 Drug Repurposing for Rare Diseases

Orphan diseases, called rare diseases, are considered a major challenge for the healthcare sector. Because of its low cost and short timescale, drug repurposing for new indications is an alternate technique for discovering rare illness therapies that outperform standard drug development. These challenges make drug repurposing an ideal tool to tackle the menace of rare diseases. Drug repurposing has resulted in the identification of captivating drug candidates, some of which are in advanced stages of clinical trials or have already been authorized, with the potential to cure uncommon disorders [116]. A few examples are mentioned in Table 8.3, and a few case studies are discussed below:

8.3.1 Acute Intermittent Porphyria

Acute intermittent porphyria (AIP) is a rare genetic metabolic illness that impacts the body's ability to produce numerous enzymes and heme, a crucial component of hemoglobin. AIP is characterized by sporadic bouts of symptoms affecting the neurological system. Porphobilinogen deaminase (PBGD), commonly called the heme-producing enzyme hydroxymethylbilane synthase, is absent in AIP [118]. Due to this deficit, poisonous substances known as porphyrins and precursors, notably porphobilinogen (PBG) and aminolevulinic acid (ALA), which are harmful to the nervous system, accumulate. Hemin with glucose infusions is considered the first-line therapy in patients with AIP. Liver transplantation is the last option if the patient does not respond to the treatment. The RNA interference method was approved recently for AIP management [147]. AIP increases the neurotoxic porphyrin precursors in our body by upregulating the delta-aminolevulinic acid synthase-1 (ALAS1). The overexpression of ALAS1 (delta-aminolevulinic acid synthase 1) indicates chronic symptoms of AIP. Givosiran, an RNA interference therapy, inhibits the expression of ALAS1 (delta-aminolevulinic acid synthase 1) and controls the AIP. Hepatocytes take up givosiran via the asialoglycoprotein receptor. It then interferes with the RNA and suppresses ALAS1. Another AIP treatment involves the substitution of nonfunctional PBGD with hepatic delivery of the mRNA or protein. It was found that intravenous injection of the recombinant protein formed by connecting apolipoprotein A-I (ApoAI) to the amino terminus of human PBGD (rhApoAI-PBGD) into the blood circulation leads to easy penetration of it into hepatocytes and the blood–brain barrier. As a result, PBGD activity increases in both the liver and brain [147].

Table 8.3 Rare diseases and repurposed medicines

Sr. No.	Rare disease	Treatment	Repurposed medicines	References
1.	Acute intermittent porphyria	Hemin with glucose infusions or liver transplantation	Bicalutamide	[117, 118]
2.	Alkaptonuria	Taking a well-designed diet combination of low-protein supplements without tyrosine/phenylalanine	Nitisinone	[119, 120]
3.	Amyotrophic lateral sclerosis	Medication—riluzole, edaravone, tofersen, or sodium phenylbutyrate/taurursodiol	Bosutinib, terbutaline sulfate, amoxapine, modafinil	[121–123]
4.	Angelman syndrome	No specific therapy	Topoisomerase inhibitors, antisense oligonucleotides	[124]
5.	Ataxia telangiectasia	No treatment is available, while management includes only controlling the symptoms	Desferrioxamine	[125]
6.	Cystic fibrosis	No treatment is available, but medicines include antibiotics, anti-inflammatory, mucus thinning, etc. to relieve the symptoms	Tavaborole, 5-fluorouracil, tirapazamine, ebselen, elexacaftor, ivacaftor, and tezacaftor	[126, 127]
7.	Kawasaki disease	Aspirin infusion	Infliximab, ciclosporin	[128, 129]
8.	Huntington's disease	No treatment is available, but tetrabenazine and deuterabenazine can treat the chorea associated with the disease	Fenofibrate, glatiramer acetate, tetrabenazine, cysteamine bitartrate	[130–133]
9.	Niemann-pick disease	No treatment is available. However, bone marrow transplantation can be effective	Dipyridamole, miglustat	[134, 135]
10.	Rett syndrome	There is no cure, but for children aged two or older, USFDA has approved Trofinetide	Ketamine	[136, 137]
11.	Sickle cell anemia	Blood transfusion and medications like voxelotor	Tenofovir disoproxil fumarate, Senicapoc	[138, 139]
12.	Tay-Sachs disease	Anticonvulsant medicine can be used along with supportive care including diet, staying hydrated, and using airway clearance procedures	Pyrimethamine	[140, 141]
13.	Wilsons disease	Two chelating agents have shown therapeutic effect: Penicillamine (Cuprimine, Depen) and trientine (Syprine)	Methanobactin and zinc acetate	[142–144]

(continued)

Table 8.3 (continued)

Sr. No.	Rare disease	Treatment	Repurposed medicines	References
14.	Xeroderma pigmentosum	Sunscreens, for chemoprevention—oral isotretinoin, topical imiquimod, topical fluorouracil	Isoconazole and clemizole hydrochloride	[145, 146]

8.3.2 Alkaptonuria

An uncommon genetic metabolic condition called alkaptonuria interferes with the body's ability to break down phenylalanine and tyrosine. It is brought on by a mutation in the HGD gene, which prevents enough homogentisate 1,2-dioxygenase from being produced. The homogentisate 1,2-dioxygenase enzyme typically aids in transforming the amino acid phenylalanine and the homogentisic acid, a result of tyrosine metabolism, into other molecules that can be further metabolized or removed by the body. However, this enzyme is defective in those with alkaptonuria, which causes a buildup of homogentisic acid. The most apparent sign is dark urine, which can even turn black when exposed to air. Ochronosis, which is the darkening of connective tissues, particularly in the joints, cheeks, ears, and nose, is another prevalent sign. Ochronosis can cause arthritis and joint stiffness and may impact other organs like the heart and kidneys. While alkaptonuria typically does not result in significant health issues, it can occasionally cause musculoskeletal issues and degenerative joint deterioration. The presence of homogentisic acid in the urine can be detected by tests typically used to confirm the diagnosis of alkaptonuria.

Progression of alkaptonuria can be decreased by nitisinone. It was first created as a weed killer, which is a reversible inhibitor of 4-hydroxyphenylpyruvate dioxygenase (HPPD) [119]. This enzyme leads to homogentisic acid (HGA) formation, further aggravating alkaptonuria. Ranganath et al. observed that a 2 mg dose of nitisinone can significantly lower the homogentisate 1,2-dioxygenase [120]. Despite its benefit, nitisinone causes tyrosine accumulation by inhibiting the process of tyrosine catabolism, which can lead to corneal and dermal toxicity. Imrich et al. suggested that a diet of low-protein supplements without tyrosine/phenylalanine can control muscle mass loss while reducing nitisinone-induced hypertyrosinemia [148].

8.3.3 Amyotrophic Lateral Sclerosis

A progressive neurodegenerative condition called amyotrophic lateral sclerosis (ALS) affects the nerve cells in charge of regulating voluntary muscular activity. Motor neurons, which relay messages from the brain to the muscles, eventually deteriorate in ALS. The effect is that the muscles deteriorate and finally paralyze. Muscle weakness and twitching are early symptoms of the disease that are often misdiagnosed due to other conditions. But as time passes, the signs worsen and affect more muscle groups. Multiple parts, including the breathing, swallowing, and speaking movements, are impacted by ALS. People may experience trouble speaking correctly, eating meals, and breathing independently as the condition worsens.

Imamura et al. studied the repurposing of bosutinib for ALS and reported its safety and tolerability [121]. Paik et al. have discussed the potential of terbutaline sulfate (anti-asthmatic) for managing ALS [122]. They have found that terbutaline sulfate (TS) inhibits motor axon and neuromuscular junction degeneration without causing lethal effects in their in vivo studies. Fan et al. have found that flutriciclamide (18F-GE180), a third-generation 18-kDa translocator protein-ligand, can be an ideal diagnostic tool for studying neuron degeneration [149]. Fiscon et al. have screened many compounds and observed that amoxapine, clomipramine, mianserin, and modafinil show significant activity in ALS [123]. They have used the latest network-based algorithm, "SAveRUNNER," to draw insights from network medicine and used the drug repurposing strategy. Potential drugs are screened out to treat ALS by analyzing the relationship between disease-associated genes and various drug targets.

8.3.4 Angelman Syndrome

An uncommon neurodevelopmental condition called Angelman syndrome affects the neurological system and causes intellectual deficits and aberrant movement patterns. A genetic defect or mutation is the leading cause of Angelman syndrome. It is brought on by the loss or inactivation of a particular gene called UBE3A, found on chromosome 15 in roughly 70–80% of cases. Paternal uniparental disomy, the inheritance of two copies of chromosome 15 from the father, or other uncommon genetic pathways may cause further occurrences.

There is no specific therapy has been prescribed for the management of Angelman syndrome. Dindot et al. have recently developed an antisense oligonucleotide to reactivate the paternal *UBE3A* allele [124]. This is under clinical trial and can be an ideal treatment. Healx [150] and the Foundation for Angelman Syndrome Therapeutics (FAST) [151] are working extensively to find the repurposed molecules to treat this syndrome [152]. Scientists are also working on topoisomerase inhibitors as a potential treatment for Angelman syndrome [153]. Huang et al. discovered 12 topoisomerase I inhibitors and four topoisomerase II inhibitors that induce the reactivation of the UBE3A gene. Topoisomerase I inhibitors (camptothecin derivatives, such as silatecan (DB67), irinotecan, rubitecan, topotecan, camptothecin, belotecan (CKD602)) and indenoisoquinoline derivatives (etoposide, ICRF-193, dexrazoxane (ICRF-187), and amsacrine) [154].

8.3.5 Cystic Fibrosis

Cystic fibrosis (CF) affects two major organs: the lungs and the digestive system. The cystic fibrosis transmembrane conductance regulator (CFTR) gene, which regulates the flow of salt and fluids into and out of cells, is responsible for the disorder

[127]. In persons with cystic fibrosis, the CFTR protein is defective, resulting in thick, sticky mucus in various organs, most notably the lungs. Lung infections are more common due to the mucus blocking the airways. The digestive system is also affected because the pancreas may get clogged by thick mucus, inhibiting the synthesis of digestive enzymes and preventing nutrient absorption. The other organs that CF can affect are the liver, sinuses, and reproductive system.

There is no treatment available for the CF. However, antibiotics, anti-inflammatories, and thinning agents are prescribed to relieve the symptoms [127]. FDA has recently approved a combination of elexacaftor, ivacaftor, and tezacaftor for managing CF. This combination has improved spirometry (patient-reported respiratory symptoms) and sweat chloride concentration (a marker of CFTR activity). Rodenburg et al. screened approximately 1400 FDA-approved medicines and found that 12 can reduce fluid secretion [155]. Bonaventura et al. screened 3386 drugs and found that ebselen (anti-inflammatory and antioxidant), tirapazamine (anticancer), carmofur (anticancer), 5-fluorouracil (anticancer), and tavaborole (antifungal) can reduce chronic infections in the patients suffering from CF [126]. They found that ebselen damages the membrane cell due to augmented membrane permeability, and tirapazamine and carmofur work against planktonic cells by dispersing preformed biofilm. While 5- fluorouracil exhibits action by inhibiting thymidylate synthetase and affects pyrimidine synthesis, leading to cell death. Altogether, these four drugs act as antibiotics in treating CF-infected lungs. Puccetti et al. have successfully demonstrated the efficacy of anakinra (a non-glycosylated recombinant form of IL-1Ra) against CF. Anakinra blocks the activation of the NLRP3 inflammasome and modulates mitochondrial oxidative stress, autophagy, and proteostasis, which is impaired in cystic fibrosis and contributes to inflammation and tissue injury [156]. These examples prove that drug repurposing can be ideal for managing CF.

8.3.6 Kawasaki Disease

Kawasaki syndrome, or Kawasaki illness, is an uncommon childhood ailment that mainly affects children under the age of five. Blood vessels throughout the body, particularly the coronary arteries that deliver blood to the heart, become inflamed, causing aneurysms or weak places in the arterial walls. It increases the chance of complications such as blood clots, heart attacks, and other cardiac disorders. Although the specific cause of Kawasaki disease is unknown, it is thought to be caused by an aberrant immune response, maybe prompted by an infection or other environmental factors. The disease is neither inherited nor transmissible. Kawasaki illness symptoms appear in three phases: acute, subacute, and convalescent. Some of the drugs repurposed for Kawasaki disease include infliximab [128], daclizumab, fostamatinib, tralokinumab, mepolizumab, reslizumab [157], immunoglobulins [158], cyclosporin, cyclophosphamide, methotrexate, and corticosteroids [129]. Infliximab binds to and neutralizes tumor-necrosis factor-alpha (TNF-α), a pro-inflammatory cytokine crucial to Kawasaki disease pathogenesis [128].

8.3.7 Huntington's Disease

The progressive and inherited neurological condition known as Huntington's disease (HD) damages the brain. The most prevalent uncommon neurological condition ranges between 1/20,000 and 1/10,000 in the Caucasian population [133]. Motor abnormalities, cognitive decline, and mental problems are a trio of symptoms that define HD. A mutation in the huntingtin gene (HTT) on chromosome 4 is the root cause of HD. In particular, the basal ganglia and cerebral cortex are gradually damaged by this mutation. It led to the synthesis of a faulty form of the huntingtin protein. The main symptoms of HD are motor symptoms, cognitive symptoms, and psychiatric symptoms. Some of the drugs that are repurposed for HD were fenofibrate [159], glatiramer acetate (GA) [130], tetrabenazine [131], deutetrabenazine [160], and cysteamine bitartrate [161]. GA-reactive T cells can release brain-derived neurotrophic factor (BDNF), which can lead to an increase in the level of BDNF in the cultured peripheral blood mononuclear cells in the brain [130]. Verni et al.'s study concluded that cysteamine bitartrate was well tolerated and positively affected motor function, but further studies are required for long-term benefits [161]. Deutetrabenazine is a modified version of tetrabenazine that does not require genetic testing for CYP2D6 like tetrabenazine. Like somnolence, fatigue, akathisia, anxiety, or nausea, it has additional advantages over tetrabenazine, such as less frequent and lower daily doses, lower peak concentrations, and reduced peak concentration adverse effects [160].

8.3.8 Niemann-Pick Disease

Niemann-Pick disease (NPD) is one of the most uncommon hereditary illnesses that affect lipid metabolism, namely sphingolipids. Niemann-Pick disease comprises several subgroups, each with unique genetic bases and specific signs and symptoms [135]. The best-known subtypes of NPD are types A, B, and C. The inheritance pattern of Niemann-Pick disease is determined by its unique subtype. Because NPA and NPB are autosomal recessive disorders, a person must inherit two copies of the faulty gene, one from each parent, in order to be affected. In order to be involved, a person must have two copies of the abnormal gene passed from his parents since NPA and NPB are autosomal recessive disorders. Autosomal dominant inheritance is less prevalent than autosomal recessive inheritance in NPC [162, 163]. Dipyridamole mitigates the accumulation of fibroblast cholesterol and rescues mitochondrial deficits. The mechanism induced by dipyridamole is in response to the increased levels of extracellular adenosine due to the inhibition of ENT1. The mechanism is based on the activation of the adenosine $A_{2A}R$ subtype. The increase in adenosine levels is advantageous for Niemann-pick disease type C, thus making it suitable for repurposing [134].

8.3.9 Rett Syndrome

The rare neurodevelopmental disorder Rett syndrome primarily affects females. It is distinguished by a phase of typical development, followed by a loss of learned abilities and the emergence of several physical and cognitive deficits. Although other genetic defects may occur in some cases, mutations in the X chromosomes MECP2 gene cause Rett syndrome. Rett syndrome symptoms and indicators often appear between 6 and 18 months [137]. The illness advances through numerous stages, like early stage, regression stage, pseudo stationary stage, and late motor deterioration stage.

The anesthetic and N-methyl-D-aspartate receptor (NMDAR) antagonist ketamine repurposing is arguably the best example for treating Rett syndrome [136]. Rett syndrome leads to reduced brain-derived neurotrophic factor (BDNF), vital for neuronal survival and synaptic plasticity. Rett syndrome also causes elevated oxidative stress and respiratory dysfunction. Ketamine increases the BDNF, decreases the production of reactive oxygen species, which cause damage to neurons, and modulates the brainstem respiratory network activities [136]. Similarly, two clinical trials were reported on sarizotan (Clinical Trial Number: NCT02790034) and desipramine based on their effectiveness in preclinical experiments. However, the studies with sarizotan are yet to be published, and the studies with desipramine resulted in some side effects [164].

8.3.10 Sickle Cell Anemia

A genetic blood condition known as sickle cell anemia is caused by defective hemoglobin, a protein that carries oxygen in red blood cells. Red blood cells with this defective hemoglobin change from their typical disc shape to a stiff crescent or sickle shape. Small blood vessels may become clogged by the sickle-shaped cells, resulting in decreased blood flow, oxygen deprivation, and other issues [139]. Sickle cell anemia is inherited in an autosomal recessive way, which requires two copies of the defective gene received from both parents to manifest. Sickle cell trait is a disorder that affects people who have inherited one mutant gene and one normal gene [165]. The symptoms of sickle cell anemia include pain crisis, anemia, increased susceptibility to infections, organ damage, and delayed growth and development [166].

Khan et al. demonstrated that tenofovir disoproxil fumarate enhanced fetal hemoglobin production via γ-globin gene induction in cells and concluded that further preclinical and clinical studies need to be conducted to prove the effectiveness [138]. A few other drugs that the FDA approves for the treatment of sickle cell anemia are also repurposed, such as L-glutamine, which reduces the acute complications of the disease by protecting the red blood cells from oxidative stress and inflammation, crizanlizumab, which blocks the adhesion of sickle cells to the blood vessel walls preventing vaso-occlusion, tissue damage, and reduces the frequency of pain crises, and voxelotor, which binds to the hemoglobin molecules and stabilizes their oxygenated state inhibiting the polymerization of sickle hemoglobin, lowering the risk of anemia and improves blood flow by binding [167].

8.3.11 Wilson's Disease

Hepatolenticular degeneration, another name for Wilson's disease, is a rare genetic condition that impairs the body's capacity to metabolize copper efficiently (194). Wilson's illness causes copper to build up and become dysfunctional in several organs, notably the liver and brain. The ATP7B gene, which creates a protein that carries extra copper out of the liver and into the bile for excretion from the body, is mutated, causing copper to build up in the liver rather than be expelled [168]. Fatigue, jaundice, abdominal discomfort, and fluid buildup in the abdomen (ascites) are just a few of the symptoms that can result from the excess copper that builds up in the liver and can harm the liver. Failure of the liver may happen occasionally. However, particularly in the early stages of the disease, the liver symptoms are not always present or may be minor [169].

In 1998, zinc acetate and penicillamine (copper chelators) have been repurposed to treat Wilson's disease. Zinc acetate inhibits intestinal copper absorption, reducing the amount of copper that enters the bloodstream and accumulates in organs, including the liver and brain. It also induces metallothionein synthesis in the intestinal mucosa, which binds and sequesters copper, preventing its reabsorption [170].

8.3.12 Xeroderma Pigmentosum

An uncommon hereditary condition known as xeroderma pigmentosum (XP) is distinguished by an extraordinary sensitivity to ultraviolet (UV) rays or radiation from the sun and other sources. The dry, pigmented skin lesions defining the disorder's defining feature are referred to as "xeroderma pigmentosum" lesions [146]. The body's capacity to repair DNA damage brought on by UV radiation is impacted by XP, which increases the risk of skin cancer and other sun-related issues. The leading cause of XP is mutations in genes that play a role in the DNA repair mechanism. Mainly, mutations are discovered in genes involved in the nucleotide excision repair (NER) pathway, which is in charge of fixing UV radiation damage. When these genes are altered, the repair process is hampered or rendered ineffective, which causes a buildup of DNA damage in skin cells when exposed to sunlight [171]. People with XP gradually experience various skin abnormalities, including freckles, dryness, scaling, and pigmentation changes [172]. Through these skin abnormalities, three more types of cancer can be developed: basal cell carcinoma, squamous cell carcinoma, and melanoma.

Takaoka et al. have observed through in silico drug screening that the molecule 4E1RCat can be an ideal option against XP [173]. Panchal et al. have demonstrated the role of suppressor tRNA in controlling the XP. Suppressor tRNA could recognize the nonsense mutation in the XPA mRNA and insert a glutamine amino acid instead of a stop codon, producing a full-length and functional XPA protein in some of the XP cells [174]. Kobaisi et al. observed that isoconazole and clemizole showed impressive results against XP. Following UV exposure, isoconazole and clemizole hydrochloride were found to increase the viability and photo-resistance of XP protein C cells and promote the repair of UV-induced DNA lesions, especially 6-4 pyrimidine-pyrimidone photoproducts (6-4PP) [145]. Gabbard et al. observed that spironolactone effectively induced the xeroderma pigmentosum protein proteolytic degradation. Spironolactone can inhibit nucleotide excision repair (NER), a defective DNA repair pathway in XP. By inducing the degradation of XP protein, spironolactone can sensitize cancer cells to DNA damage and enhance the immune recognition of tumor cells [175].

8.4 Summary

Drug repurposing, also referred to as drug repositioning or reprofiling, is a strategy to identify new therapeutic uses for drugs already approved for other diseases' treatment with an understanding that drugs can have multiple targets and mechanisms of action beyond their primary ones. It is very beneficial to make a treatment for emerging infectious diseases (EIDs) and rare diseases. EIDs are considered to be unique and a significant public health threat due to the lack of pre-existing immunity or effective treatments. Similarly, rare diseases present public health concerns with substantial challenges in discovering and developing effective drugs. Drug repurposing can be achieved through a drug-oriented approach, a target-oriented approach, or a therapy/disease-oriented approach. Machine learning and artificial intelligence tools and algorithms can quickly and accurately identify promising drug candidates, expediting the process of drug repurposing. Network building is another technique that can reveal new properties at the network level and be beneficial in exploring how cellular systems produce different biological phenotypes under varying conditions. The network-building technique works by organizing various relationships among biological molecules into networks. Drug repurposing also holds the potential to combat antimicrobial resistance (AMR). AMR is regarded as one of the significant threats to human health, which necessitates the discovery and development of new and potent antimicrobial medicines.

Moreover, drug repurposing can tackle various obstacles in developing rare disease treatments. With over 5000 rare diseases identified, therapies exist for only about 300. The small patient population indicates small-scale production of medicines for rare diseases, leading to higher prices. Research and development costs increase the final prices, making these medicines unaffordable for developing and underdeveloped countries. Given these challenges, drug repurposing is emerging as an ideal solution to address issues.

References

1. Pushpakom S, Iorio F, Eyers PA, Escott KJ, Hopper S, Wells A, et al. Drug repurposing: progress, challenges and recommendations. Nat Rev Drug Discov. 2019;18(1):41–58. https://doi.org/10.1038/nrd.2018.168.
2. Low ZY, Farouk IA, Lal SK. Drug repositioning: new approaches and future prospects for life-debilitating diseases and the COVID-19 pandemic outbreak. Viruses. 2020;12(9):1058. https://doi.org/10.3390/v12091058.
3. Cha Y, Erez T, Reynolds IJ, Kumar D, Ross J, Koytiger G, et al. Drug repurposing from the perspective of pharmaceutical companies. Br J Pharmacol. 2018;175(2):168–80. https://doi.org/10.1111/bph.13798.
4. Oprea TI, Mestres J. Drug repurposing: far beyond new targets for old drugs. AAPS J. 2012;14(4):759–63. https://doi.org/10.1208/s12248-012-9390-1.
5. World Health Organization. Infectious diseases kill over 17 million people a year: WHO warns of global crisis. 1996. https://www.who.int/news/item/01-01-1996-infectious-diseases-kill-over-17-million-people-a-year-who-warns-of-global-crisis. Accessed 23 Oct 2023.
6. World Health Organization. WHO coronavirus (COVID-19) dashboard. https://covid19.who.int/. Accessed 22 Oct 2023.
7. Laupland KB, Valiquette L. Ebola virus disease. Can J Infect Dis Med Microbiol. 2014;25:527378. https://doi.org/10.1155/2014/527378.
8. Dadonaite B, Roser M. Pneumonia. 2018. https://ourworldindata.org/pneumonia. Accessed 9 Oct 2023.
9. World Health Organization. Malaria. 2023. https://www.who.int/news-room/fact-sheets/detail/malaria. Accessed 3 Oct 2023.
10. Weiner DL. Inborn errors of metabolism: Medscape. 2017. https://emedicine.medscape.com/article/804757-overview?form=fpf. Accessed 11 Nov 2023.
11. García-Fernández E, Koch G, Wagner RM, Fekete A, Stengel ST, Schneider J, et al. Membrane microdomain disassembly inhibits MRSA antibiotic resistance. Cell. 2017;171(6):1354–67. https://doi.org/10.1016/j.cell.2017.10.012.
12. Kim JH, Cheng LW, Land KM. Advances in antifungal development: discovery of new drugs and drug repurposing. Pharmaceuticals. 2022;15(7):787. https://doi.org/10.3390/ph15070787.
13. World Health Organization. The true death toll of COVID-19: estimating global excess mortality. 2021. https://www.who.int/data/stories/the-true-death-toll-of-covid-19-estimating-global-excess-mortality. Accessed 8 Nov 2023.
14. To KKW, Chan JFW, Tsang AKL, Cheng VCC, Yuen K-Y. Ebola virus disease: a highly fatal infectious disease reemerging in West Africa. Microbes Infect. 2015;17(2):84–97. https://doi.org/10.1016/j.micinf.2014.11.007.
15. Centers for Disease Control and Prevention. Aspergillosis. 2022. https://www.cdc.gov/fungal/diseases/aspergillosis/index.html. Accessed 28 Oct 2023.
16. Papp-Wallace Krisztina M, Endimiani A, Taracila Magdalena A, Bonomo RA. Carbapenems: past, present, and future. Antimicrob Agents Chemother. 2011;55(11):4943–60. https://doi.org/10.1128/aac.00296-11.
17. World Health Organization. Zika virus. 2022. https://www.who.int/news-room/fact-sheets/detail/zika-virus. Accessed 20 Nov 2023.
18. May RC, Stone NRH, Wiesner DL, Bicanic T, Nielsen K. Cryptococcus: from environmental saprophyte to global pathogen. Nat Rev Microbiol. 2016;14(2):106–17. https://doi.org/10.1038/nrmicro.2015.6.
19. Liu Y-Y, Wang Y, Walsh TR, Yi L-X, Zhang R, Spencer J, et al. Emergence of plasmid-mediated colistin resistance mechanism MCR-1 in animals and human beings in China: a microbiological and molecular biological study. Lancet Infect Dis. 2016;16(2):161–8. https://doi.org/10.1016/S1473-3099(15)00424-7.

20. Kasabe B, Ahire G, Patil P, Punekar M, Davuluri KS, Kakade M, et al. Drug repurposing approach against chikungunya virus: an in vitro and in silico study. Front Cell Infect Microbiol. 2023;13:1132538. https://doi.org/10.3389/fcimb.2023.1132538.
21. Centers for Disease Control and Prevention. Candida auris. 2023. https://www.cdc.gov/fungal/candida-auris/candida-auris-qanda.html. Accessed 14 Nov 2023.
22. Greub MVG. Emerging bacterial pathogens: the past and beyond. Clin Microbiol Infect. 2016;22(1):12–21. https://doi.org/10.1016/j.cmi.2015.10.010.
23. Pizzorno A, Terrier O, Nicolas de Lamballerie C, Julien T, Padey B, Traversier A, et al. Repurposing of drugs as novel influenza inhibitors from clinical gene expression infection signatures. Front Immunol. 2019;10:60. https://doi.org/10.3389/fimmu.2019.00060.
24. Nobile CJ, Johnson AD. Candida albicans biofilms and human disease. Annu Rev Microbiol. 2015;69(1):71–92. https://doi.org/10.1146/annurev-micro-091014-104330.
25. Chong H-T, Kamarulzaman A, Tan C-T, Goh K-J, Thayaparan T, Kunjapan SR, et al. Treatment of acute Nipah encephalitis with ribavirin. Ann Neurol. 2001;49(6):810–3. https://doi.org/10.1002/ana.1062.
26. Friedman DZP, Schwartz IS. Emerging fungal infections: new patients, new patterns, and new pathogens. J Fungi. 2019;5(3):67. https://doi.org/10.3390/jof5030067.
27. World Health Organization. Middle East respiratory syndrome coronavirus (MERS-CoV). 2022. https://www.who.int/news-room/fact-sheets/detail/middle-east-respiratory-syndrome-coronavirus-(mers-cov). Accessed 18 Oct 2023.
28. Liu Y, Jia Y, Yang K, Tong Z, Shi J, Li R, et al. Melatonin overcomes MCR-mediated colistin resistance in Gram-negative pathogens. Theranostics. 2020;10(23):10697–711. https://doi.org/10.7150/thno.45951.
29. Domalaon R, Okunnu O, Zhanel GG, Schweizer F. Synergistic combinations of anthelmintic salicylanilides oxyclozanide, rafoxanide, and closantel with colistin eradicates multidrug-resistant colistin-resistant Gram-negative bacilli. J Antibiot. 2019;72(8):605–16. https://doi.org/10.1038/s41429-019-0186-8.
30. Peyclit L, Yousfi H, Rolain J-M, Bittar F. Drug repurposing in medical mycology: identification of compounds as potential antifungals to overcome the emergence of multidrug-resistant fungi. Pharmaceuticals. 2021;14(5):488. https://doi.org/10.3390/ph14050488.
31. Spadari CC, Vila T, Rozental S, Ishida K. Miltefosine has a postantifungal effect and induces apoptosis in cryptococcus yeasts. Antimicrob Agents Chemother. 2018;62(8):e00312-18. https://doi.org/10.1128/AAC.00312-18.
32. Mansour SM, Shamma RN, Ahmed KA, Sabry NA, Esmat G, Mahmoud AA, et al. Safety of inhaled ivermectin as a repurposed direct drug for treatment of COVID-19: a preclinical tolerance study. Int Immunopharmacol. 2021;99:108004. https://doi.org/10.1016/j.intimp.2021.108004.
33. Sen A. Repurposing prolactin as a promising immunomodulator for the treatment of COVID-19: are common antiemetics the wonder drug to fight coronavirus? Med Hypotheses. 2020;144:110208. https://doi.org/10.1016/j.mehy.2020.110208.
34. Bleyzac N, Goutelle S, Bourguignon L, Tod M. Azithromycin for COVID-19: more than just an antimicrobial? Clin Drug Investig. 2020;40(8):683–6. https://doi.org/10.1007/s40261-020-00933-3.
35. Meini S, Pagotto A, Longo B, Vendramin I, Pecori D, Tascini C. Role of lopinavir/ritonavir in the treatment of Covid-19: a review of current evidence, guideline recommendations, and perspectives. J Clin Med. 2020;9(7):2050. https://doi.org/10.3390/jcm9072050.
36. Furuta Y, Takahashi K, Shiraki K, Sakamoto K, Smee DF, Barnard DL, et al. T-705 (favipiravir) and related compounds: novel broad-spectrum inhibitors of RNA viral infections. Antivir Res. 2009;82(3):95–102. https://doi.org/10.1016/j.antiviral.2009.02.198.
37. Johansen LM, Brannan JM, Delos SE, Shoemaker CJ, Stossel A, Lear C, et al. FDA-approved selective estrogen receptor modulators inhibit Ebola virus infection. Sci Transl Med. 2013;5(190):190ra79. https://doi.org/10.1126/scitranslmed.3005471.

38. Madrid PB, Panchal RG, Warren TK, Shurtleff AC, Endsley AN, Green CE, et al. Evaluation of Ebola virus inhibitors for drug repurposing. ACS Infect Dis. 2015;1(7):317–26. https://doi.org/10.1021/acsinfecdis.5b00030.
39. Delvecchio R, Higa LM, Pezzuto P, Valadão AL, Garcez PP, Monteiro FL, et al. Chloroquine, an endocytosis blocking agent, inhibits Zika virus infection in different cell models. Viruses. 2016;8(12):322. https://doi.org/10.3390/v8120322.
40. Retallack H, Di Lullo E, Arias C, Knopp KA, Laurie MT, Sandoval-Espinosa C, et al. Zika virus cell tropism in the developing human brain and inhibition by azithromycin. Proc Natl Acad Sci. 2016;113(50):14408–13. https://doi.org/10.1073/pnas.1618029113.
41. Rossignol J-F. Nitazoxanide, a new drug candidate for the treatment of Middle East respiratory syndrome coronavirus. J Infect Public Health. 2016;9(3):227–30. https://doi.org/10.1016/j.jiph.2016.04.001.
42. de Wilde AH, Jochmans D, Posthuma Clara C, Zevenhoven-Dobbe Jessika C, van Nieuwkoop S, Bestebroer Theo M, et al. Screening of an FDA-approved compound library identifies four small-molecule inhibitors of Middle East respiratory syndrome coronavirus replication in cell culture. Antimicrob Agents Chemother. 2014;58(8):4875–84. https://doi.org/10.1128/aac.03011-14.
43. Dawes BE, Kalveram B, Ikegami T, Juelich T, Smith JK, Zhang L, et al. Favipiravir (T-705) protects against Nipah virus infection in the hamster model. Sci Rep. 2018;8:7604. https://doi.org/10.1038/s41598-018-25780-3.
44. Porotto M, Orefice G, Yokoyama Christine C, Mungall Bruce A, Realubit R, Sganga Michael L, et al. Simulating henipavirus multicycle replication in a screening assay leads to identification of a promising candidate for therapy. J Virol. 2009;83(10):5148–55. https://doi.org/10.1128/jvi.00164-09.
45. Verma A, Mittal P, Pande MS, Trivedi N. In silico drug discovery of 4-hyroxypanduratin a, 6-Gingerol and Luteolin targeting Nipah virus. J Pharm Res Int. 2021;33(46B):477–84. https://doi.org/10.9734/jpri/2021/v33i46B32964.
46. Cheung PPH, Watson SJ, Choy K-T, Fun Sia S, Wong DDY, Poon LLM, et al. Generation and characterization of influenza A viruses with altered polymerase fidelity. Nat Commun. 2014;5(1):4794. https://doi.org/10.1038/ncomms5794.
47. Roche. Roche announces FDA approval of Xofluza (baloxavir marboxil) for influenza. 2018. https://www.roche.com/media/releases/med-cor-2018-10-24. Accessed 15 Oct 2023.
48. Huang F, Zhang C, Liu Q, Zhao Y, Zhang Y, Qin Y, et al. Identification of amitriptyline HCl, flavin adenine dinucleotide, azacitidine and calcitriol as repurposing drugs for influenza A H5N1 virus-induced lung injury. PLoS Pathog. 2020;16(3):e1008341. https://doi.org/10.1371/journal.ppat.1008341.
49. El-Halfawy OM, Czarny TL, Flannagan RS, Day J, Bozelli JC, Kuiack RC, et al. Discovery of an antivirulence compound that reverses β-lactam resistance in MRSA. Nat Chem Biol. 2020;16(2):143–9. https://doi.org/10.1038/s41589-019-0401-8.
50. Wang G, Li L, Wang X, Li X, Zhang Y, Yu J, et al. Hypericin enhances β-lactam antibiotics activity by inhibiting sarA expression in methicillin-resistant staphylococcus aureus. Acta Pharm Sin B. 2019;9(6):1174–82. https://doi.org/10.1016/j.apsb.2019.05.002.
51. Farha MA, Leung A, Sewell EW, D'Elia MA, Allison SE, Ejim L, et al. Inhibition of WTA synthesis blocks the cooperative action of PBPs and sensitizes MRSA to β-lactams. ACS Chem Biol. 2013;8(1):226–33. https://doi.org/10.1021/cb300413m.
52. Thangamani S, Mohammad H, Abushahba MFN, Sobreira TJP, Seleem MN. Repurposing auranofin for the treatment of cutaneous staphylococcal infections. Int J Antimicrob Agents. 2016;47(3):195–201. https://doi.org/10.1016/j.ijantimicag.2015.12.016.
53. Moellering RC Jr, Eliopoulos GM, Sentochnik DE. The carbapenems: new broad spectrum β-lactam antibiotics. J Antimicrob Chemother. 1989;24(suppl_A):1–7. https://doi.org/10.1093/jac/24.suppl_A.1.
54. Rotondo CM, Wright GD. Inhibitors of metallo-β-lactamases. Curr Opin Microbiol. 2017;39:96–105. https://doi.org/10.1016/j.mib.2017.10.026.

55. Brem J, van Berkel SS, Zollman D, Lee Sook Y, Gileadi O, McHugh Peter J, et al. Structural basis of metallo-β-lactamase inhibition by captopril stereoisomers. Antimicrob Agents Chemother. 2015;60(1):142–50. https://doi.org/10.1128/aac.01335-15.
56. Chen C, Yang K-W, Wu L-Y, Li J-Q, Sun L-Y. Disulfiram as a potent metallo-β-lactamase inhibitor with dual functional mechanisms. Chem Commun. 2020;56(18):2755–8. https://doi.org/10.1039/C9CC09074F.
57. Thomas CA, Cheng Z, Yang K, Hellwarth E, Yurkiewicz CJ, Baxter FM, et al. Probing the mechanisms of inhibition for various inhibitors of metallo-β-lactamases VIM-2 and NDM-1. J Inorg Biochem. 2020;210:111123. https://doi.org/10.1016/j.jinorgbio.2020.111123.
58. European Medicines Agenecy. Updated advice on the use of colistin products in animals within the European Union: development of resistance and possible impact on human and animal health. 2016. https://www.ema.europa.eu/en/documents/scientific-guideline/updated-advice-use-colistin-products-animals-within-european-union-development-resistance-possible_en-0.pdf. Accessed 30 Oct 2023.
59. Shen Y, Zhang R, Schwarz S, Wu C, Shen J, Walsh TR, et al. Farm animals and aquaculture: significant reservoirs of mobile colistin resistance genes. Environ Microbiol. 2020;22(7):2469–84. https://doi.org/10.1111/1462-2920.14961.
60. Zhou Y, Liu S, Wang T, Li H, Tang S, Wang J, et al. Pterostilbene, a potential MCR-1 inhibitor that enhances the efficacy of polymyxin B. Antimicrob Agents Chemother. 2018;62(4):e02146-17. https://doi.org/10.1128/aac.02146-17.
61. Zhou Y, Wang J, Guo Y, Liu X, Liu S, Niu X, et al. Discovery of a potential MCR-1 inhibitor that reverses polymyxin activity against clinical mcr-1-positive enterobacteriaceae. J Infect. 2019;78(5):364–72. https://doi.org/10.1016/j.jinf.2019.03.004.
62. Domalaon R, De Silva PM, Kumar A, Zhanel George G, Schweizer F. The anthelmintic drug niclosamide synergizes with colistin and reverses colistin resistance in Gram-negative bacilli. Antimicrob Agents Chemother. 2019;63(4):e02574-18. https://doi.org/10.1128/aac.02574-18.
63. Stokes JM, MacNair CR, Ilyas B, French S, Côté J-P, Bouwman C, et al. Pentamidine sensitizes gram-negative pathogens to antibiotics and overcomes acquired colistin resistance. Nat Microbiol. 2017;2:17028. https://doi.org/10.1038/nmicrobiol.2017.28.
64. Rajendran R, Mowat E, Jones B, Williams C, Ramage G. Prior in vitro exposure to voriconazole confers resistance to amphotericin B in Aspergillus fumigatus biofilms. Int J Antimicrob Agents. 2015;46(3):342–5. https://doi.org/10.1016/j.ijantimicag.2015.03.006.
65. Vazquez Jose A, Arganoza Maria T, Boikov D, Yoon S, Sobel Jack D, Akins RA. Stable phenotypic resistance of Candidaspecies to amphotericin B conferred by preexposure to subinhibitory levels of azoles. J Clin Microbiol. 1998;36(9):2690–5. https://doi.org/10.1128/jcm.36.9.2690-2695.1998.
66. Kathwate GH, Shinde RB, Mohan KS. Non-antifungal drugs inhibit growth, morphogenesis and biofilm formation in Candida albicans. J Antibiot. 2021;74(5):346–53. https://doi.org/10.1038/s41429-020-00403-0.
67. Kim JH, Cheng LW, Chan KL, Tam CC, Mahoney N, Friedman M, et al. Antifungal drug repurposing. Antibiotics. 2020;9(11):812. https://doi.org/10.3390/antibiotics9110812.
68. Shinde RB, Raut JS, Chauhan NM, Karuppayil SM. Chloroquine sensitizes biofilms of Candida albicans to antifungal azoles. Braz J Infect Dis. 2013;17(4):395–400. https://doi.org/10.1016/j.bjid.2012.11.002.
69. Macreadie IG, Johnson G, Schlosser T, Macreadie PI. Growth inhibition of Candida species and Aspergillus fumigatus by statins. FEMS Microbiol Lett. 2006;262(1):9–13. https://doi.org/10.1111/j.1574-6968.2006.00370.x.
70. Stepanović S, Vuković D, Ješić M, Ranin L. Influence of acetylsalicylic acid (aspirin) on biofilm production by Candida species. J Chemother. 2004;16(2):134–8. https://doi.org/10.1179/joc.2004.16.2.134.
71. Moraes DC, Ferreira-Pereira A. Insights on the anticandidal activity of non-antifungal drugs. J Mycol Méd. 2019;29(3):253–9. https://doi.org/10.1016/j.mycmed.2019.07.004.

72. Hao W, Qiao D, Han Y, Du N, Li X, Fan Y, et al. Identification of disulfiram as a potential antifungal drug by screening small molecular libraries. J Infect Chemother. 2021;27(5):696–701. https://doi.org/10.1016/j.jiac.2020.12.012.
73. Lone SA, Ahmad A. Candida auris—the growing menace to global health. Mycoses. 2019;62(8):620–37. https://doi.org/10.1111/myc.12904.
74. Kean R, Ramage G. Combined antifungal resistance and biofilm tolerance: the Global Threat of Candida auris. mSphere. 2019;4(4):e00458-19. https://doi.org/10.1128/msphere.00458-19.
75. Lockhart SR. Candida auris and multidrug resistance: defining the new normal. Fungal Genet Biol. 2019;131:103243. https://doi.org/10.1016/j.fgb.2019.103243.
76. Cheng Y-S, Roma Jose S, Shen M, Mota Fernandes C, Tsang Patricia S, Forbes He E, et al. Identification of antifungal compounds against multidrug-resistant candida auris utilizing a high-throughput drug-repurposing screen. Antimicrob Agents Chemother. 2021;65(4):e01305-20. https://doi.org/10.1128/aac.01305-20.
77. Wall G, Chaturvedi Ashok K, Wormley Floyd L, Wiederhold Nathan P, Patterson Hoja P, Patterson Thomas F, et al. Screening a repurposing library for inhibitors of multidrug-resistant candida auris identifies ebselen as a repositionable candidate for antifungal drug development. Antimicrob Agents Chemother. 2018;62(10):e01084-18. https://doi.org/10.1128/aac.01084-18.
78. Eldesouky HE, Lanman NA, Hazbun TR, Seleem MN. Aprepitant, an antiemetic agent, interferes with metal ion homeostasis of Candida auris and displays potent synergistic interactions with azole drugs. Virulence. 2020;11(1):1466–81. https://doi.org/10.1080/21505594.2020.1838741.
79. Talbot GH, Bradley J, Edwards JE Jr, Gilbert D, Scheld M, Bartlett JG. Bad bugs need drugs: an update on the development pipeline from the Antimicrobial Availability Task Force of the Infectious Diseases Society of America. Clin Infect Dis. 2006;42(5):657–68. https://doi.org/10.1086/499819.
80. Meis JF, Chowdhary A, Rhodes JL, Fisher MC, Verweij PE. Clinical implications of globally emerging azole resistance in Aspergillus fumigatus. Philos Trans R Soc Lond B Biol Sci. 2016;371(1709):20150460. https://doi.org/10.1098/rstb.2015.0460.
81. Perlin PDS, Rautemaa-Richardson R, Alastruey-Izquierdo A. The global problem of antifungal resistance: prevalence, mechanisms, and management. Lancet Infect Dis. 2017;17(12):e383–92. https://doi.org/10.1016/S1473-3099(17)30316-X.
82. Vallières C, Singh N, Alexander C, Avery SV. Repurposing nonantifungal approved drugs for synergistic targeting of fungal pathogens. ACS Infect Dis. 2020;6(11):2950–8. https://doi.org/10.1021/acsinfecdis.0c00405.
83. Khan S, Singhal S, Mathur T, Upadhyay DJ, Rattan A. Antifungal potential of disulfiram. Nihon Ishinkin Gakkai Zasshi. 2007;48(3):109–13. https://doi.org/10.3314/jjmm.48.109.
84. Osherov N, Kontoyiannis DP. The anti-Aspergillus drug pipeline: is the glass half full or empty? Med Mycol. 2017;55(1):118–24. https://doi.org/10.1093/mmy/myw060.
85. Butts A, Martin JA, DiDone L, Bradley EK, Mutz M, Krysan DJ. Structure-activity relationships for the antifungal activity of selective estrogen receptor antagonists related to tamoxifen. PLoS One. 2015;10(5):e0125927. https://doi.org/10.1371/journal.pone.0125927.
86. Kim JH, Chan KL, Cheng LW, Tell LA, Byrne BA, Clothier K, et al. High efficiency drug repurposing design for new antifungal agents. Methods Protoc. 2019;2(2):31. https://doi.org/10.3390/mps2020031.
87. Truong M, Monahan LG, Carter DA, Charles IG. Repurposing drugs to fast-track therapeutic agents for the treatment of cryptococcosis. PeerJ. 2018;6:e4761. https://doi.org/10.7717/peerj.4761.
88. Namasivayam V, Palaniappan S, Vanangamudi M. Repurposing drugs targeting epidemic viruses. Drug Discov Today. 2022;27(7):1874–94. https://doi.org/10.1016/j.drudis.2022.04.008.
89. Coutard B, Valle C, de Lamballerie X, Canard B, Seidah NG, Decroly E. The spike glycoprotein of the new coronavirus 2019-nCoV contains a furin-like cleavage site absent in CoV of the same clade. Antivir Res. 2020;176:104742. https://doi.org/10.1016/j.antiviral.2020.104742.

90. Hoffmann M, Kleine-Weber H, Schroeder S, Krüger N, Herrler T, Erichsen S, et al. SARS-CoV-2 cell entry depends on ACE2 and TMPRSS2 and is blocked by a clinically proven protease inhibitor. Cell. 2020;181(2):271. https://doi.org/10.1016/j.cell.2020.02.052.
91. Wrapp D, Wang N, Corbett KS, Goldsmith JA, Hsieh C-L, Abiona O, et al. Cryo-EM structure of the 2019-nCoV spike in the prefusion conformation. Science. 2020;367(6483):1260–3. https://doi.org/10.1126/science.abb2507.
92. Grein J, Ohmagari N, Shin D, Diaz G, Asperges E, Castagna A, et al. Compassionate use of remdesivir for patients with severe Covid-19. N Engl J Med. 2020;382(24):2327–36. https://doi.org/10.1056/NEJMoa2007016.
93. Habler K, Brügel M, Teupser D, Liebchen U, Scharf C, Schönermarck U, et al. Simultaneous quantification of seven repurposed COVID-19 drugs remdesivir (plus metabolite GS-441524), chloroquine, hydroxychloroquine, lopinavir, ritonavir, favipiravir and azithromycin by a two-dimensional isotope dilution LC–MS/MS method in human serum. J Pharm Biomed Anal. 2021;196:113935. https://doi.org/10.1016/j.jpba.2021.113935.
94. Mirmohammadi S, Kianmehr A, Sabbaghian A, Mohebbi A, Shahbazmohammadi H, Sheykharabi M, et al. In silico drug repurposing against SARS-CoV-2 using an integrative transcriptomic profiling approach: hydrocortisone and benzhydrocodone as potential drug candidates against COVID-19. Infect Genet Evol. 2022;103:105318. https://doi.org/10.1016/j.meegid.2022.105318.
95. Pathak DSK, Salunke DAA, Thivari DP, Pandey A, Nandy DK, Harish VK, Ratna D, et al. No benefit of hydroxychloroquine in COVID-19: results of systematic review and meta-analysis of randomized controlled trials. Diabetes Metab Syndr Clin Res Rev. 2020;14(6):1673–80. https://doi.org/10.1016/j.dsx.2020.08.033.
96. Cai Q, Yang M, Liu D, Chen J, Shu D, Xia J, et al. Experimental treatment with favipiravir for COVID-19: an open-label control study. Engineering. 2020;6(10):1192–8. https://doi.org/10.1016/j.eng.2020.03.007.
97. Jacob ST, Crozier I, Fischer WA, Hewlett A, Kraft CS, Vega M-AL, et al. Ebola virus disease. Nat Rev Dis Primers. 2020;6(1):13. https://doi.org/10.1038/s41572-020-0147-3.
98. Bai JPF, Hsu C-W. Drug repurposing for Ebola virus disease: principles of consideration and the animal rule. J Pharm Sci. 2018;108(2):798–806. https://doi.org/10.1016/j.xphs.2018.09.010.
99. Yi D, Li Q, Wang H, Lv K, Ma L, Wang Y, et al. Repurposing of berbamine hydrochloride to inhibit Ebola virus by targeting viral glycoprotein. Acta Pharm Sin B. 2022;12(12):4378–89. https://doi.org/10.1016/j.apsb.2022.05.023.
100. Sun W, He S, Martínez-Romero C, Kouznetsova J, Tawa G, Xu M, et al. Synergistic drug combination effectively blocks Ebola virus infection. Antivir Res. 2017;137:165–72. https://doi.org/10.1016/j.antiviral.2016.11.017.
101. Vaziri S, Pour SH, Akrami-Mohajeri F. Zika virus as an emerging arbovirus of international public health concern. Osong Public Health Res Perspect. 2022;13(5):341–51. https://doi.org/10.24171/j.phrp.2022.0101.
102. Krauer F, Riesen M, Reveiz L, Oladapo OT, Martínez-Vega R, Porgo TV, et al. Zika virus infection as a cause of congenital brain abnormalities and Guillain–Barré syndrome: systematic review. PLoS Med. 2017;14(1):e1002203. https://doi.org/10.1371/journal.pmed.1002203.
103. Contreras D, Arumugaswami V. Zika virus infectious cell culture system and the in vitro prophylactic effect of interferons. J Vis Exp. 2016;114:e54767. https://doi.org/10.3791/54767.
104. Costa Vivian V, Del Sarto JL, Rocha Rebeca F, Silva Flavia R, Doria Juliana G, Olmo Isabella G, et al. N-Methyl-d-Aspartate (NMDA) receptor blockade prevents neuronal death induced by Zika virus infection. mBio. 2017;8(2):e00350-17. https://doi.org/10.1128/mbio.00350-17.
105. Kamiyama N, Soma R, Hidano S, Watanabe K, Umekita H, Fukuda C, et al. Ribavirin inhibits Zika virus (ZIKV) replication in vitro and suppresses viremia in ZIKV-infected STAT1-deficient mice. Antivir Res. 2017;146:1–11. https://doi.org/10.1016/j.antiviral.2017.08.007.
106. Hui PDS, Azhar PEI, Kim Y-J, Memish PZA, Oh PM-D, Zumla PSA. Middle East respiratory syndrome coronavirus: risk factors and determinants of primary, household, and

nosocomial transmission. Lancet Infect Dis. 2018;18(8):e217–27. https://doi.org/10.1016/S1473-3099(18)30127-0.
107. Kim JY, Song JY, Yoon YK, Choi S-H, Song YG, Kim S-R, et al. Middle East respiratory syndrome infection control and prevention guideline for healthcare facilities. Infect Chemother. 2015;47(4):278–302. https://doi.org/10.3947/ic.2015.47.4.278.
108. Dyall J, Coleman Christopher M, Hart Brit J, Venkataraman T, Holbrook Michael R, Kindrachuk J, et al. Repurposing of clinically developed drugs for treatment of Middle East respiratory syndrome coronavirus infection. Antimicrob Agents Chemother. 2014;58(8):4885–93. https://doi.org/10.1128/aac.03036-14.
109. Shin JS, Jung E, Kim M, Baric RS, Go YY. Saracatinib inhibits Middle East respiratory syndrome-coronavirus replication in vitro. Viruses. 2018;10(6):283.
110. Yamamoto M, Matsuyama S, Li X, Takeda M, Kawaguchi Y, Inoue J-I, et al. Identification of nafamostat as a potent inhibitor of Middle East respiratory syndrome coronavirus S protein-mediated membrane fusion using the Split-protein-based cell-cell fusion assay. Antimicrob Agents Chemother. 2016;60(11):6532–9. https://doi.org/10.1128/aac.01043-16.
111. Arabi YM, Alothman A, Balkhy HH, Al-Dawood A, AlJohani S, Al Harbi S, et al. Treatment of Middle East respiratory syndrome with a combination of lopinavir-ritonavir and interferon-β1b (MIRACLE trial): study protocol for a randomized controlled trial. Trials. 2018;19(1):81. https://doi.org/10.1186/s13063-017-2427-0.
112. Lo MK, Feldmann F, Gary JM, Jordan R, Bannister R, Cronin J, et al. Remdesivir (GS-5734) protects African green monkeys from Nipah virus challenge. Sci Transl Med. 2019;11(494):eaau9242. https://doi.org/10.1126/scitranslmed.aau9242.
113. Santos IA, Pereira AKS, Guevara-Vega M, de Paiva REF, Sabino-Silva R, Bergamini FRG, et al. Repurposing potential of rimantadine hydrochloride and development of a promising platinum(II)-rimantadine metallodrug for the treatment of Chikungunya virus infection. Acta Tropica. 2022;227:106300. https://doi.org/10.1016/j.actatropica.2021.106300.
114. Pereira AKS, Santos IA, da Silva WW, Nogueira FAR, Bergamini FRG, Jardim ACG, et al. Memantine hydrochloride: a drug to be repurposed against Chikungunya virus? Pharmacol Rep. 2021;73(3):954–61. https://doi.org/10.1007/s43440-021-00216-4.
115. Bhakat S, Delang L, Kaptein S, Neyts J, Leyssen P, Jayaprakash V. Reaching beyond HIV/HCV: nelfinavir as a potential starting point for broad-spectrum protease inhibitors against dengue and chikungunya virus. RSC Adv. 2015;5(104):85938–49. https://doi.org/10.1039/C5RA14469H.
116. Roessler HI, Knoers NVAM, van Haelst MM, van Haaften G. Drug repurposing for rare diseases. Trends Pharmacol Sci. 2021;42(4):255–67. https://doi.org/10.1016/j.tips.2021.01.003.
117. Sofie Lichtwarck Bjugn F, Storjord E, Kristensen RM, Brekke O-L. Safe usage of bicalutamide and goserelin in a male patient with acute intermittent porphyria and prostate cancer. Scand J Urol. 2019;53(2–3):171–3. https://doi.org/10.1080/21681805.2018.1563628.
118. Balwani M, Sardh E, Ventura P, Peiró PA, Rees DC, Stölzel U, et al. Phase 3 trial of RNAi therapeutic givosiran for acute intermittent porphyria. N Engl J Med. 2020;382(24):2289–301. https://doi.org/10.1056/NEJMoa1913147.
119. Scherman D, Fetro C. Drug repositioning for rare diseases: knowledge-based success stories. Therapies. 2020;75(2):161–7. https://doi.org/10.1016/j.therap.2020.02.007.
120. Ranganath LR, Khedr M, Milan AM, Davison AS, Hughes AT, Usher JL, et al. Nitisinone arrests ochronosis and decreases rate of progression of Alkaptonuria: evaluation of the effect of nitisinone in the United Kingdom National Alkaptonuria Centre. Mol Genet Metab. 2018;125(1):127–34. https://doi.org/10.1016/j.ymgme.2018.07.011.
121. Imamura K, Izumi Y, Banno H, Uozumi R, Morita S, Egawa N, et al. Induced pluripotent stem cell-based Drug Repurposing for Amyotrophic lateral sclerosis Medicine (iDReAM) study: protocol for a phase I dose escalation study of bosutinib for amyotrophic lateral sclerosis patients. BMJ Open. 2019;9(12):e033131. https://doi.org/10.1136/bmjopen-2019-033131.
122. Paik H, Chung A-Y, Park H-C, Park RW, Suk K, Kim J, et al. Repurpose terbutaline sulfate for amyotrophic lateral sclerosis using electronic medical records. Sci Rep. 2015;5(1):8580. https://doi.org/10.1038/srep08580.

123. Fiscon G, Conte F, Amadio S, Volonté C, Paci P. Drug repurposing: a network-based approach to amyotrophic lateral sclerosis. Neurotherapeutics. 2021;18(3):1678–91. https://doi.org/10.1007/s13311-021-01064-z.
124. Dindot SV, Christian S, Murphy WJ, Berent A, Panagoulias J, Schlafer A, et al. An ASO therapy for Angelman syndrome that targets an evolutionarily conserved region at the start of the UBE3A-AS transcript. Sci Transl Med. 2023;15(688):eabf4077. https://doi.org/10.1126/scitranslmed.abf4077.
125. Shackelford RE, Manuszak RP, Johnson CD, Hellrung DJ, Steele TA, Link CJ, et al. Desferrioxamine treatment increases the genomic stability of Ataxia-telangiectasia cells. DNA Repair. 2003;2(9):971–81. https://doi.org/10.1016/S1568-7864(03)00090-9.
126. Di Bonaventura G, Lupetti V, Di Giulio A, Muzzi M, Piccirilli A, Cariani L, et al. Repurposing high-throughput screening identifies unconventional drugs with antibacterial and antibiofilm activities against Pseudomonas aeruginosa under experimental conditions relevant to cystic fibrosis. Microbiol Spectr. 2023;11(4):e00352–23. https://doi.org/10.1128/spectrum.00352-23.
127. National Institutes of Health—National Heart, Lungs, and Blood Institute. Cystic fibrosis treatment. 2022. https://www.nhlbi.nih.gov/health/cystic-fibrosis/treatment. Accessed 10 Nov 2023.
128. Son MB, Gauvreau K, Burns JC, Corinaldesi E, Tremoulet AH, Watson VE, et al. Infliximab for intravenous immunoglobulin resistance in Kawasaki disease: a retrospective study. J Pediatr. 2010;158(4):644–649.e1. https://doi.org/10.1016/j.jpeds.2010.10.012.
129. Eleftheriou D, Levin M, Shingadia D, Tulloh R, Klein NJ, Brogan PA. Management of Kawasaki disease. Arch Dis Child. 2014;99(1):74. https://doi.org/10.1136/archdischild-2012-302841.
130. Corey-Bloom J, Jia H, Aikin AM, Thomas EA. Disease modifying potential of glatiramer acetate in Huntington's disease. J Huntingtons Dis. 2014;3(3):311–6. https://doi.org/10.3233/JHD-140110.
131. Jankovic J, Clarence-Smith K. Tetrabenazine for the treatment of chorea and other hyperkinetic movement disorders. Expert Rev Neurother. 2011;11(11):1509–23. https://doi.org/10.1586/ern.11.149.
132. Arbez N, Roby E, Akimov S, Eddings C, Ren M, Wang X, et al. Cysteamine protects neurons from mutant huntingtin toxicity. J Huntingtons Dis. 2019;8(2):129–43. https://doi.org/10.3233/JHD-180312.
133. Ghosh R, Tabrizi SJ. Clinical features of Huntington's disease. In: Nóbrega C, Pereira de Almeida L, editors. Polyglutamine disorders, vol. 1049. Cham: Springer International Publishing; 2018. p. 1–28. https://doi.org/10.1007/978-3-319-71779-1_1.
134. Pepponi R, De Simone R, De Nuccio C, Visentin S, Matteucci A, Bernardo A, et al. Repurposing dipyridamole in Niemann pick type C disease: a proof of concept study. Int J Mol Sci. 2022;23(7):3456. https://doi.org/10.3390/ijms23073456.
135. Pfrieger FW. The Niemann-pick type diseases—a synopsis of inborn errors in sphingolipid and cholesterol metabolism. Prog Lipid Res. 2023;90:101225. https://doi.org/10.1016/j.plipres.2023.101225.
136. Katz DM, Menniti FS, Mather RJ. N-methyl-D-aspartate receptors, ketamine, and Rett syndrome: something special on the road to treatments? Biol Psychiatry. 2016;79(9):710–2. https://doi.org/10.1016/j.biopsych.2016.03.1045.
137. Weng S-M, Bailey MES, Cobb SR. Rett syndrome: from bed to bench. Pediatr Neonatol. 2011;52(6):309–16. https://doi.org/10.1016/j.pedneo.2011.08.002.
138. Khan F, Ali H, Musharraf SG. Tenofovir disoproxil fumarate-mediated γ-globin induction is correlated with the suppression of trans-acting factors in CD34+ progenitor cells: a role in the reactivation of fetal hemoglobin. Eur J Pharmacol. 2022;927:175036. https://doi.org/10.1016/j.ejphar.2022.175036.
139. Kato GJ, Piel FB, Reid CD, Gaston MH, Ohene-Frempong K, Krishnamurti L, et al. Sickle cell disease. Nat Rev Dis Primers. 2018;4(1):18010. https://doi.org/10.1038/nrdp.2018.10.

140. Osher E, Fattal-Valevski A, Sagie L, Urshanski N, Amir-Levi Y, Katzburg S, et al. Pyrimethamine increases β-hexosaminidase A activity in patients with Late Onset Tay Sachs. Mol Genet Metab. 2011;102(3):356–63. https://doi.org/10.1016/j.ymgme.2010.11.163.
141. Solovyeva VV, Shaimardanova AA, Chulpanova DS, Kitaeva KV, Chakrabarti L, Rizvanov AA. New approaches to Tay-Sachs disease therapy. Front Physiol. 2018;9:1663. https://doi.org/10.3389/fphys.2018.01663.
142. Brewer GJ, Dick RD, Johnson VD, Brunberg JA, Kluin KJ, Fink JK. Treatment of Wilson's disease with zinc: XV long-term follow-up studies. Transl Res. 1998;132(4):264–78. https://doi.org/10.1016/S0022-2143(98)90039-7.
143. Summer KH, Lichtmannegger J, Bandow N, Choi DW, DiSpirito AA, Michalke B. The biogenic methanobactin is an effective chelator for copper in a rat model for Wilson disease. J Trace Elem Med Biol. 2011;25(1):36–41. https://doi.org/10.1016/j.jtemb.2010.12.002.
144. Członkowska A, Litwin T, Dusek P, Ferenci P, Lutsenko S, Medici V, et al. Wilson disease. Nat Rev Dis Primers. 2018;4(1):21. https://doi.org/10.1038/s41572-018-0018-3.
145. Kobaisi F, Sulpice E, Barette C, Fayyad N, Fauvarque M-O, Badran B, et al. Isoconazole and Clemizole hydrochloride partially reverse the Xeroderma Pigmentosum C phenotype. Int J Mol Sci. 2021;22(15):8156. https://doi.org/10.3390/ijms22158156.
146. Schelini MC, Chaves LFOB, Toledo MC, Rodrigues FW, de Oliveira T, Isaac DLC, et al. Xeroderma pigmentosum: ocular findings in an isolated Brazilian group with an identified genetic cluster. J Ophthalmol. 2019;2019:8. https://doi.org/10.1155/2019/4818162.
147. Longo M, Paolini E, Meroni M, Dongiovanni P. Cutting-edge therapies and novel strategies for acute intermittent porphyria: step-by-step towards the solution. Biomedicine. 2022;10(3):648. https://doi.org/10.3390/biomedicines10030648.
148. Imrich R, Zatkova A, Lukacova O, Sedlakova J, Zanova E, Vlcek M, et al. Nutritional interventions for patients with alkaptonuria: a minireview. Endocr Regul. 2023;57(1):61–7. https://doi.org/10.2478/enr-2023-0008.
149. Fan Z, Calsolaro V, Atkinson RA, Femminella GD, Waldman A, Buckley C, et al. Flutriciclamide (18F-GE180) PET: first-in-human PET study of novel third-generation in vivo marker of human translocator protein. J Nucl Med. 2016;57(11):1753. https://doi.org/10.2967/jnumed.115.169078.
150. Healx. Healx: AI-powered, patient-inspired treatments for rare diseases. https://healx.ai/. Accessed 12 Nov 2023.
151. Fast. https://cureangelman.org/. Accessed 12 Nov 2023.
152. Ray F. Healx becomes FAST partner to repurpose Medicines for Angelman. Angelman Syndrome. 2020. https://angelmansyndromenews.com/news/healx-becomes-fast-partner-to-repurpose-medicines-for-angelman/. Accessed 12 Nov 2023.
153. Markati T, Duis J, Servais L. Therapies in preclinical and clinical development for Angelman syndrome. Expert Opin Investig Drugs. 2021;30(7):709–20. https://doi.org/10.1080/13543784.2021.1939674.
154. Huang H-S, Allen JA, Mabb AM, King IF, Miriyala J, Taylor-Blake B, et al. Topoisomerase inhibitors unsilence the dormant allele of Ube3a in neurons. Nature. 2012;481(7380):185–9. https://doi.org/10.1038/nature10726.
155. Rodenburg LW, Delpiano L, Railean V, Centeio R, Pinto MC, Smits SMA, et al. Drug repurposing for cystic fibrosis: identification of drugs that induce CFTR-independent fluid secretion in nasal organoids. Int J Mol Sci. 2022;23(20):12657. https://doi.org/10.3390/ijms232012657.
156. Puccetti M, Pariano M, Stincardini C, Wojtylo P, Schoubben A, Nunzi E, et al. Pulmonary drug delivery technology enables anakinra repurposing in cystic fibrosis. J Control Release. 2023;353:1023–36. https://doi.org/10.1016/j.jconrel.2022.11.043.
157. Lourdes O-F, Elio GC, Martin K, Paul L, Francisco David C, Raquel López M, et al. Identification of new risk loci shared across systemic vasculitides points towards potential target genes for drug repurposing. Ann Rheum Dis. 2023;82(6):837. https://doi.org/10.1136/ard-2022-223697.

158. Rosa SGV, Santos WC. Clinical trials on drug repositioning for COVID-19 treatment. Rev Panam Salud Pública. 2020;44:e40. https://doi.org/10.26633/RPSP.2020.40.
159. Devadiga SJ, Bharate SS. Recent developments in the management of Huntington's disease. Bioorg Chem. 2022;120:105642. https://doi.org/10.1016/j.bioorg.2022.105642.
160. Huntington Study Group. Effect of deutetrabenazine on chorea among patients with Huntington disease: a randomized clinical trial. JAMA. 2016;316(1):40–50. https://doi.org/10.1001/jama.2016.8655.
161. Verny C, Bachoud-Lévi A-C, Durr A, Goizet C, Azulay J-P, Simonin C, et al. A randomized, double-blind, placebo-controlled trial evaluating cysteamine in Huntington's disease. Mov Disord. 2017;32(6):932–6. https://doi.org/10.1002/mds.27010.
162. Patterson MC, Platt F. Therapy of Niemann–pick disease, type C. Biochim Biophys Acta. 2004;1685(1):77–82. https://doi.org/10.1016/j.bbalip.2004.08.013.
163. Schuchman EH, Desnick RJ. Types A and B Niemann-pick disease. Mol Genet Metab. 2017;120(1):27–33. https://doi.org/10.1016/j.ymgme.2016.12.008.
164. Gogliotti RG, Niswender CM. A coordinated attack: Rett syndrome therapeutic development. Trends Pharmacol Sci. 2019;40(4):233–6. https://doi.org/10.1016/j.tips.2019.02.007.
165. National Heart, Lung, and Blood Institute. Sickle cell disease causes and risk factors. National Institue of Health. 2023. https://www.nhlbi.nih.gov/health/sickle-cell-disease/causes. Accessed 15 Oct 2023.
166. Mangla A, Ehsan M, Agarwal N, Maruvada S. Sickle cell anemia. Treasure Island: StatPearls Publishing; 2023.
167. Cannon M, Phillips H, Smith S, Williams K, Brinton L, Gregory C, et al. Large-scale drug screen identifies FDA-approved drugs for repurposing in sickle-cell disease. J Clin Med. 2020;9(7):2276. https://doi.org/10.3390/jcm9072276.
168. Dev S, Kruse RL, Hamilton JP, Lutsenko S. Wilson disease: update on pathophysiology and treatment. Front Cell Dev Biol. 2022;10:871877. https://doi.org/10.3389/fcell.2022.871877.
169. European Association for the Study of the Liver. EASL clinical practice guidelines: Wilson's disease. J Hepatol. 2012;56(3):671–85. https://doi.org/10.1016/j.jhep.2011.11.007.
170. Voli F, Valli E, Lerra L, Kimpton K, Saletta F, Giorgi FM, et al. Intratumoral copper modulates PD-L1 expression and influences tumor immune evasion. Cancer Res. 2020;80(19):4129–44. https://doi.org/10.1158/0008-5472.CAN-20-0471.
171. DiGiovanna JJ, Kraemer KH. Shining a light on xeroderma pigmentosum. J Invest Dermatol. 2012;132(3):785–96. https://doi.org/10.1038/jid.2011.426.
172. Reichrath J. Sunlight, skin cancer and vitamin D: what are the conclusions of recent findings that protection against solar ultraviolet (UV) radiation causes 25-hydroxyvitamin D deficiency in solid organ-transplant recipients, xeroderma pigmentosum, and other risk groups? J Steroid Biochem Mol Biol. 2007;103(3):664–7. https://doi.org/10.1016/j.jsbmb.2006.12.029.
173. Takaoka Y, Ohta M, Tateishi S, Sugano A, Nakano E, Miura K, et al. In silico drug repurposing by structural alteration after induced fit: discovery of a candidate agent for recovery of nucleotide excision repair in xeroderma pigmentosum group D mutant (R683W). Biomedicine. 2021;9(3):249. https://doi.org/10.3390/biomedicines9030249.
174. Panchal RG, Wang S, McDermott J, Link CJ. Partial functional correction of xeroderma pigmentosum group a cells by suppressor tRNA. Hum Gene Ther. 1999;10(13):2209–19. https://doi.org/10.1089/10430349950017194.
175. Gabbard RD, Hoopes RR, Kemp MG. Spironolactone and XPB: an old drug with a new molecular target. Biomol Ther. 2020;10(5):756. https://doi.org/10.3390/biom10050756.

Clinical Trials on Repurposed Drugs: An Overview

9

Ruchira Patil, Harshad Takate, Gaurav Shanbhag, Harshada Kiran Sonawane, Amruta Prabhakar Padakanti, and Naveen Chella

Abbreviations

ART	Antiretroviral therapy
BC	Breast cancer
BCG	Bacillus Calmette-Guérin
CAD	Coronary artery disease
CFR	Code of Federal Regulations
CHF	Congestive heart failure
CSC	Cancer stem cell
CVD	Cardiovascular diseases
HBV	Hepatitis B virus
HCC	Hepatocellular carcinoma
HR	Hazard ratio
IBD	Inflammatory bowel disorder
IND	Investigational new drug
IR	Insulin resistance
MI	Myocardial infarction
NA	Not applicable
NDA	New drug application
NMIBC	Non-muscle invasive bladder cancer
NSCLC	Non-small cell lung cancer
OC	Ovarian cancer
PET	Positron emission tomography
RA	Rheumatoid arthritis

R. Patil · H. Takate · G. Shanbhag · H. K. Sonawane · A. P. Padakanti · N. Chella (✉)
Department of Pharmaceutical Technology (Formulations), National Institute of Pharmaceutical Education and Research Guwahati, Guwahati, Assam, India
e-mail: naveen@niperguwahati.in

N. Chella et al. (eds.), *Drug Repurposing*,
https://doi.org/10.1007/978-981-97-5016-0_9

RCT	Randomized controlled trial
RPPA	Reverse phase protein array
SLE	Systemic lupus erythematosus
SVR	Sustained virological response
TNBC	Triple-negative breast cancer

Learning Objectives

- Understanding the basis of clinical trials for the repurposed drugs
- Knowledge of the drugs currently under clinical trials for various therapeutic conditions such as cardiovascular, cancers, etc.

9.1 Introduction

A clinical trial, under the Code of Federal Regulations (CFR), is a scientific investigation or study involving human subjects who are assigned in advance, based on a protocol, to receive one or more interventions (or no intervention) to assess the impact of the intervention(s) on biomedical or health-related effects [1]. As a part of the de novo drug development process, a new chemical entity must undertake preclinical investigations to evaluate its pharmacokinetics and toxicity assessment. Once the safety of the drug has been confirmed in preclinical trials, an investigational new drug (IND) application must be filed. Upon approval of the IND application, a new chemical entity proceeds to phase 0 of clinical trials, often referred to as the first in human investigations. These are exploratory assessments that include a relatively small group of 10–15 individuals who receive nontherapeutic, low doses of investigational drugs. These investigations ascertain if a drug exhibits a similar response in the human body, as demonstrated in preclinical animal research. An investigational drug moves on to phase 1 clinical trial following the successful completion of phase 0, which is intended to assess the safety of treatment or drug, establish a range of safe dosages, and identify potential adverse effects. These investigations, spanning several weeks or months, include the participation of a modest cohort comprising a small group of 20–80 individuals. Subsequently, it continues to phase 2 trials, which generally span over months and enroll a cohort of 100–300 individuals with the specific target condition being studied. These investigations confirm the safety and effectiveness of the experimental drugs. Assuming that phase 2 studies demonstrate the effectiveness, an investigational drug progresses onto phase 3 trials to evaluate its risk/benefit profile and explore its effects over an extended duration in a larger population of 1000–3000 individuals. Marketing acceptance necessitates submitting a new drug application (NDA) after confirming the drug's effectiveness in phase 3 studies. Phase IV trials, called post-marketing trials, are carried out after a drug has been approved. This provides more information on the potential risks and advantages of the medication or therapy [2].

Despite this, in the current era, there is a need for more promising treatments that cannot be accomplished by the tedious, costly, and time-consuming traditional de novo drug research and development procedure that typically takes 10 to 15 years. Consequently, drug repurposing is gaining prominence as a cost-effective and time-saving approach, providing a unique therapeutic alternative for emerging diseases [3]. Drug repurposing, often called drug repositioning, is the practice of using an approved drug or a medication that is still in development for purposes other than its original intended application [4]. The FDA's approval for the emergency use of various repurposed medications to treat COVID-19 has led to a surge of interest in drug repurposing [5]. The pathway that a new chemical entity must follow to gain FDA approval through the de novo drug development process is outlined in Section 505 (b) (1). The procedure for authorizing the repurposed medications is outlined in Section 505 (b) (2), which permits the sponsor to partially depend on safety and efficacy data from earlier studies conducted for the drug's original indication. This section may obviate the requirement for preclinical and phase 1 clinical trials. Therefore, this ultimately reduces the duration required for the drug development process [3]. Several authors also suggested that a drug that has been repurposed may bypass phase 1 and go straight into phase 2. In contrast to the assertions made by some authors, our data analysis demonstrates that significant experiments using repurposed drugs are now in phase 1 rather than advancing straight to phase 2.

The oldest examples of successful drug repurposing described in the literature include sildenafil, aspirin, thalidomide, and zidovudine [6–8]. Subsequently, drug repurposing is the cornerstone approach effectively applied to advance therapeutic strategies against a wide range of life-threatening diseases [8]. This chapter presents comprehensive information on clinical studies and their progress for repurposed drugs that target various illnesses, including malignancies, viral infections, cardiovascular ailments, and inflammatory diseases. All data has been collected and summarized from https://www.anticancerfund.org/en/redo-trials-db and https://ClinicalTrials.gov.

9.2 Cancer

Cancer is a substantial worldwide public health concern, and it is anticipated that the global prevalence of cancer will rise in the coming years. Despite extensive investments in drug development, the efficacy of existing medications remains restricted. In addition, there has been a consistent decline in the average annual number of FDA approvals during the 1990s. Due to the lack of sufficient and effective anticancer treatments, there is an increasing interest in drug repurposing. Various types of research, including preclinical investigations, clinical trials, and observational studies, have shown that compounds in therapeutic classes unrelated to cancer can effectively inhibit tumor growth [9].

Drug repositioning has been undertaken for a wide range of cancer types, such as breast cancer, bone sarcoma, urological cancer, CNS cancer, endocrine cancer, gastrointestinal tract cancer, gynecological cancer, head and neck cancer, leukemia, lung cancer, lymphoma, ocular cancer, skin cancer, soft tissue sarcoma, and AIDS-related cancer.

Active pharmacological compounds with antidiabetic, analgesic, antipyretic, anti-inflammatory, antihypertensive, anti-lipidemic, anti-epileptic, anti-psychotic, and antifungal properties are being investigated for their potential in cancer treatment.

9.2.1 Breast Cancer

Breast cancer (BC) is a heterogeneous condition that causes harm to the epithelial tissue of the breast. Worldwide, it caused almost 627,000 deaths in 2018 and is the primary cause of cancer-related mortality among women. Despite significant advancements in treatment and the FDA's approval of approximately 30 new drugs for breast cancer from 2010 to 2020, specific subtypes of breast cancer, such as triple-negative breast cancer (TNBC), still lack targeted medications. Furthermore, the presence of resistance to current treatments underscores the need for the development of novel therapeutics. Therefore, medication repurposing has become a viable strategy for breast cancer researchers seeking to expedite the development of treatments. As a result, more than 12 medicines have been licensed for treating breast cancer through this approach [10].

A total of 117 studies were undertaken to repurpose medications for the treatment of breast cancer. Pharmacological classes commonly used to treat breast cancer include antidiabetic, analgesic, antipyretic, anti-inflammatory, anti-lipidemic, antifungal, anti-bacterial, antimalarial, anti-psychotic, anti-histaminic, non-steroidal anti-inflammatory drugs, and drugs used in rheumatoid arthritis, insomnia, and hypercalcemia. Most studies were done on antidiabetic (21), anti-lipidemic (13), osteoporosis (7), and hypercalcemia (10).

The study distribution in various phases is presented in Fig. 9.1a. The countries that primarily participated in these studies are China, Egypt, the United Kingdom, Bangladesh, the United States, Turkey, Russia, Canada, Italy, India, Iran, Japan, France, Australia, and Indonesia. Among all the studies on breast cancer, 1 study was terminated, 17 studies were recruiting, 6 studies were not yet recruiting, 4 studies were completed, 9 studies were active but not recruiting, 4 studies were pending, and 4 studies with an unknown status. The termination was due to challenges encountered in patient recruitment.

The majority of antidiabetic medications are repurposed for use in treating breast cancer. Metformin, an antidiabetic drug, has been extensively studied for its impact on breast cancer. The Reverse Phase Protein Array (RPPA) is a high-throughput technology that utilizes antibodies to measure the activity of cellular proteins accurately. This study evaluated changes in protein markers using Reverse Phase Protein Array (RPPA) in tumor tissue collected from women diagnosed with operable breast cancer who were undergoing a pre-surgical metformin regimen. An open-label experiment enrolled 35 non-diabetic patients with stage 0–III breast cancer. They were administered 1500 mg of metformin orally every day prior to the surgical

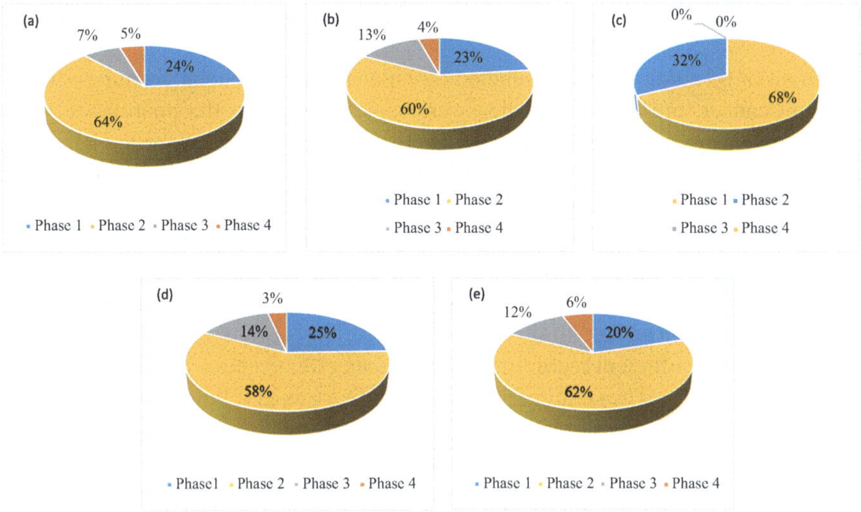

Fig. 9.1 Repurposed trials in different phases for the treatment of cancers: (**a**) breast cancer, (**b**) lung cancer, (**c**) skin cancer, (**d**) gynecological cancer, (**e**) urological cancer

removal of the tumor. Following the correction for multiple comparisons in the analysis of 32 tumors from patients treated with metformin compared to 34 untreated historical controls, it was found that 11 proteins exhibited significant differences between the two groups. These data suggest that metformin controls cellular signaling, invasion/motility, and apoptosis/cell cycle. Further investigations should assess the outcomes of metformin on a larger scale. An assessment should determine any potential links between these alterations and the clinical outcome of breast cancer (NCT00930579) [11].

9.2.2 Lung Cancer

Lung cancer is the second most common type of cancer and the primary cause of cancer-related deaths. Non-small cell lung cancer (NSCLC) is the most common subtype of lung cancer, accounting for 80% of all documented occurrences. Despite the heightened screening efforts, the median 5-year survival rate for NSCLC remained about 3%. Commonly utilized treatment strategies in the management of NSCLC encompass surgical intervention, concurrent administration of radiation and chemotherapy, and multimodal chemotherapy. NSCLC exhibits elevated levels of treatment resistance due to its heterogeneity and the adaptive capabilities of cancer stem cells (CSCs). Researchers can expedite the development of anticancer medications for patients with metastatic NSCLC and enhance survival rates by repurposing authorized existing drugs, thus avoiding the time-consuming and expensive process of creating new therapies specifically for this purpose. Hence, the use of existing drugs for new purposes can significantly transform the treatment of this destructive illness [12].

A total of 83 studies were conducted on repurposing treatments for lung cancer. For lung cancer, most of the studies were repurposed from the pharmacological class, such as antimalarial (8), anti-emetics (8), NSAID (7), antidiabetic (12), anti-inflammatory (6), and anticancer (9).

Figure 9.1b illustrates the distribution of studies in different phases. The countries participating in these studies include Egypt, the United States, the United Kingdom, France, Turkey, China, Australia, Korea, Canada, Belgium, Sweden, Netherlands, Italy, India, Japan, Germany, Mexico, and Taiwan. Among the several studies on lung cancer, one trial was terminated. The termination occurred due to factors such as insufficient recruitment, inadequate effectiveness, and an increase in adverse occurrences. The status of additional studies was as follows: recruiting (27), not yet recruiting (7), completed (1), active not recruiting (15), pending (3), ongoing (11), and other studies have an unknown state (10).

Metformin has been demonstrated to impact the metabolism of NSCLC by decreasing glucose uptake in healthy tissues. In a study, positron emission tomography (PET) scans were used to assess the effect of metformin on glucose absorption in NSCLC. One participant received a placebo, while the other 14 were randomly allocated to receive metformin. The histological analysis indicated that 6.7% of the samples exhibited poorly differentiated carcinoma, 33.3% displayed squamous cell carcinoma, and 60% showed adenocarcinoma. Contrary to the placebo group, where the metabolic disease did not change, 57% of individuals randomly selected to take metformin during the mid-treatment PET scan met the PERCIST criteria for metabolic response. Of these subjects, 75% had increasing metabolic illness, whereas 25% showed a partial metabolic response. Following 6 months, the patient who was administered a placebo had a complete metabolic response. Conversely, the group that received metformin exhibited a complete metabolic response rate of 69%, a partial metabolic response rate of 23%, and one instance of progressing metabolic illness. No severe toxicities, classified as grade ≥ 3 according to the Common Terminology Criteria for Adverse Events (CTCAE), were seen. The study's findings suggest that most persons who had metformin treatment exhibited metabolic responses on PET imaging, according to the PERCIST criteria, even with limited accumulation of data. Metformin increased metabolic activity in most malignant tumors, unlike its effect on normal physiological tissues (NCT02285855) [13].

9.2.3 Skin Cancer

Skin tumors are relatively common malignancies that affect a large number of individuals worldwide. Melanomas and nonmelanoma skin cancers are among them. Nonmelanoma skin cancers are frequently more benign clinically, although they can still be aggressive and metastatic. Melanomas rank among the most critical types of cancer. Melanomas predominantly arise in sun-exposed areas of the body. Their growth is complex, encompassing both endogenous and exogenous risk factors. In general, targeted therapy, immunotherapy, chemotherapy, and surgery are utilized to treat skin cancer, although the specific approach may vary depending on the pathological type. In recent years, there have been significant advancements in oncological treatments. However, these treatments still have disadvantages, such as limited ability to target specific cells, frequent disease recurrence, high toxicity levels, and increased expenses. In addition, obtaining clinical approval for novel medications may require an investment of 10 to 15 years and is expensive. Consequently, repurposing previously approved drugs for other ailments has emerged as a highly viable alternative [14].

Thirty-four studies are being employed to treat skin cancer using a repurposing approach. The pharmacological classes of drugs that have been repurposed for the treatment of skin cancer include antifungal, anti-inflammatory (NSAIDs), antimalarial, antidiabetic, antihypertensive, anti-epileptic, anti-inflammatory, anticancer, diuretics, immunosuppressive agents, opioid antagonists, and anti-fibrinolytic medicines. For skin cancer, antimalarial (5), antihypertensive (5), NSAID (3), antidiabetic (3), and vitamin D derivative (3) have the maximum repurposed studies.

Out of these 34 studies, the majority are in phase 1, followed by phase 2 (Fig. 9.1c). The countries participating in these studies primarily include Brazil, the United States, Belgium, France, China, Italy, the Netherlands, Australia, Japan, Canada, and Switzerland. Among all the studies on skin cancer, there are a few trials with an unknown status (5). The status of the remaining trials is as follows: recruiting (13), not yet recruiting (2), completed (2), active not recruiting (7), pending (1), suspended (2), ongoing (1).

Preclinical evidence indicates that β-adrenoceptor antagonists (β-blockers) impede tumor growth and metastasis in animal models of melanoma. The researchers proposed that administering β-blockers for comorbid conditions is linked to a decreased likelihood of developing thick (Breslow thickness > 1 mm) malignant melanoma. Two distinct patient types were discerned based on the medical records of 121 consecutive patients diagnosed with dense melanoma. Of these, 91 patients did not get any medication, whereas 30 patients (the treated subgroup) were administered β-blockers for 1 year or more. Tumor progression was noted in 3.3% of the treated group and 34.1% of the untreated group, with a median follow-up duration of 2.5 years. The Cox model analysis revealed a substantial 36% decrease in the likelihood of progression for each year of β-blocker medication, with a 95% confidence range ranging from 11% to 54% ($p = 0.002$). (Among the 24 persons in the untreated group, there were no instances of death recorded in the group). According to current knowledge, recent research indicates that exposure to β-blockers for 1 year or longer is linked, for the first time, to a decreased likelihood of acquiring thick malignant melanoma. This emphasizes the need for more extensive epidemiological investigations and randomized clinical trials [15].

9.2.4 Gynecological Cancer

Gynecological cancer encompasses any form of cancer that arises within the reproductive organs of a woman, including tumors of the cervical, endometrium, ovaries, vagina, and vulva. Antibody therapy, targeted therapy, radiation therapy, chemotherapy, and surgery are the treatments available. There is a global demand for combination therapy. However, the primary objective of the battle against cancer is consistently to discover novel treatments or targets. A comprehensive framework for new drug discovery is constructed, encompassing preclinical discoveries, in vitro and in vivo experiments, human investigations, phase I and II trials, and a phase III randomized controlled trial (RCT). The entire procedure spans over a decade and costs significantly. Hence, individuals suffering from these lethal and malignant conditions could benefit greatly from the significant cost and time savings that would result from repurposing the current drugs [16].

Seventy trials using repurposed drugs have been conducted for the treatment of gynecological cancer. Pharmacological classes like antidiabetic (24), NSAID (8), antineoplastic agents (11), anti-lipidemic agents (4), and antifungal agents (4) have the maximum repurposed studies for gynecological cancer. As shown in Fig. 9.1d, the majority of trials are in phase 2, followed by phases 1, 3, and 4. The countries participating in these studies include Canada, the United States, France, Belgium, the Netherlands, Australia, China, Norway, India, Japan, New Zealand, Iran, Korea, Italy, Sweden, Taiwan, Finland, and Egypt. Only seven gynecological cancer studies have been completed, and three are ongoing studies. All other studies are recruiting (30), suspended (3), pending (3), or unknown status (3).

Gynecological cancer primarily utilizes repurposed antidiabetic and antineoplastic medications. Vigil®, a personalized vaccine, enhances the expression of tumor neoantigens. The objective of this trial was to evaluate the safety and effectiveness of the combination of Vigil and atezolizumab in patients experiencing a recurrence of ovarian cancer, marking the first instance of such an assessment. This study is a phase 1 randomized experiment assessing the effectiveness of Vigil, a vaccine derived from a patient's tumor tissue that has been genetically engineered to include GMCSF and bi-shRNA-furin. These alterations augment immune activation and control the expression of TGFβ. Part 1 of the study examines the safety of combining atezolizumab at a dosage of 1200 mg every 21 days with Vigil at a concentration of 1 × 10e7 cells/mL. In Part 2, a randomized administration of two cycles of either atezolizumab (Atezo-1st) or Vigil (Vigil-1st) occurs, followed by a combination of both drugs. The main aim of the study was to assess the safety. The trial comprised 24 participants, with three randomly assigned to Part 1 and 21 allocated to Part 2. The decision to expand to Part 2 was predicated on patients in Part 1 effectively completing combination therapy without experiencing any adverse effects that would restrict the dosage. In Part 2, a cohort of 21 patients was evenly divided (1:1) into two groups: the Vigil-1st group, including 11 patients, and the Atezo-1st group, comprising 10 patients. When comparing Atezo-1st with Vigil-1st, the occurrence of grade 3/4 treatment-related adverse events was 17.2% and 5.1%, respectively. The Atezo-1st group had a median overall survival of 10.8 months, but the Vigil-1st group did not reach a median overall survival. The hazard ratio was 0.33. The exploratory subset analysis of patients without the BRCA mutation suggested a possible increase in the overall survival advantage. The median overall survival was not reached in the Vigil-1st group, whereas it was 5.2 months in the Atezo-1st group. The hazard ratio was 0.16, with a p-value of 0.027. The combination of Atezolizumab and Vigil-1st has shown to be a safe therapy option for individuals with BRCAwt, hence providing further support for continued research in this population [17].

9.2.5 Urological Cancer

The current treatments for bladder cancer are not universally effective, as they have specific limits in terms of their efficacy. We have identified medications that can be employed for innovative applications, particularly in managing advanced bladder cancer. This was accomplished by utilizing mathematical techniques to compare the molecular profiles (mRNAs, proteins) of bladder cancer patient tissue, previously reported, with the molecular profiles caused by drugs. This approach makes it possible to forecast numerous pharmaceuticals that can address aggressive bladder cancer. Upon further examination, it was noted that a molecular inhibitor targeting the mTOR pathway hindered the growth of various bladder cancer cell lines. Ultimately, this study presents a robust and reliable approach for repurposing pharmaceuticals by analyzing tissue molecular profiles. It identifies potential new treatments promising for aggressive bladder cancer but requires additional exploration [18].

A total of 95 studies were conducted on repurposing treatments for urological cancer. These studies investigate the use of various pharmacological classes, such as NSAIDs, vitamin D analogs, anti-lipidemic agents, antidiabetic agents, antacids, adrenergic antagonists, antiresorptive agents, antimalarial medications, antifungal medications, antidiuretic medications, alcohol antagonists, alpha-reductase inhibitors, anxiolytic drugs, anthelmintics, anti-emetics, antihypertensive drugs, anti-inflammatory drugs, cardiotonic glycosides, anti-histamines, antioxidants, anti-epileptics, antibiotics, anti-HIV drugs, immunosuppressive agents, anticancer drugs, anti-gout agents, and bronchodilators, for the treatment of urological cancer.

Figure 9.1e shows that out of these 95 studies, the maximum is in phase 2, followed by phases 1, 3, and 4. The countries participating in these studies include the United Kingdom, Canada, France, Finland, the United States, Australia, Slovakia, Netherlands, Switzerland, Egypt, Korea, China, Spain, Belgium, Italy, Sweden, Denmark, Japan, Mexico, Iran, and Norway. Of all the studies conducted, 4 have been completed, 1 has been withdrawn, and 36 are currently recruiting. The reason for termination was based on the current follow-up period of approximately 10 years. The incorporation of further data will have an insignificant effect on these findings. The status of the remaining studies is as follows: not yet recruiting (11), ongoing (13), pending (3), temporary suspension (1), and other studies have an uncertain state (2).

Antidiabetic and NSAID medications are commonly repurposed for the treatment of urological cancer. Non-muscle-invasive bladder cancer (NMIBC), the most common malignancy of the urinary system, requires continuous invasive monitoring throughout a person's life to detect any signs of the disease reoccurring. Currently, there is a lack of oral drugs that can effectively halt the progression or recurrence of the illness. Recent studies have demonstrated that metformin remains unchanged in the urine of mice. The concentration of metformin in urothelial cells is roughly 240-fold higher than in serum. When used in animal bladder cancer models, this treatment yielded favorable results. Preclinical trials have shown the possibility of oral metformin treatment in managing non-MIBC. It could potentially serve as an alternative to the current adjuvant intravesical therapy. Suppose the study's stated results are positive, they may provide a basis for conducting further phase III studies to evaluate the efficacy of metformin compared to the existing treatments for intravesical implants, such as chemotherapy and Bacillus Calmette-Guérin (BCG) (NCT03379909) [19].

9.3 Viral Diseases

9.3.1 COVID-19

The COVID-19 outbreak was officially declared a pandemic by the World Health Organization in 2019 in response to the rapid spread of the virus. The outbreak was attributed to SARS-CoV-2 or Severe Acute Respiratory Syndrome Coronavirus 2. Given the absence of targeted medication therapies or vaccinations for COVID-19, drug repurposing was considered an effective approach to combat the disease's progression and control the spread of an outbreak [20]. A wide range of drugs belonging to several pharmacological groups have shown promise for their repurposing action against COVID-19. These include pre-existing antiviral medications, anticancer, immunomodulatory agents, antimalarial, anti-inflammatory agents, anticoagulants, and monoclonal antibodies. Notably, most of the drugs repurposed for COVID-19 fall within the antiviral classification.

Since 2021 onwards, a total of 9444 clinical trials have been carried out specifically to study and address COVID-19. Most of these trials were conducted in the USA. Chloroquine, hydroxychloroquine, remdesivir, favipiravir, dexamethasone, and baricitinib were the primary drugs extensively investigated on emergency authorization for the treatment of COVID-19. Figure 9.2a illustrates the categorical classification of different phases. For the rest of the remaining studies, the study phase was found to be inapplicable. Among all of these studies undertaken, 1660 studies were found to be investigated on children aged from birth to 17 years old. Other remaining studies were carried out on adults between 18 and 64 and elderly individuals aged 65 and beyond. Among these studies, 4189 were completed, 491 were discontinued, 378 were withdrawn, and 46 were suspended. The rest of the studies were either recruiting or unknown status. Studies were terminated, suspended, and withdrawn due to various factors, mainly difficulties in enrolling participants. Other contributing reasons include inadequate financial resources,

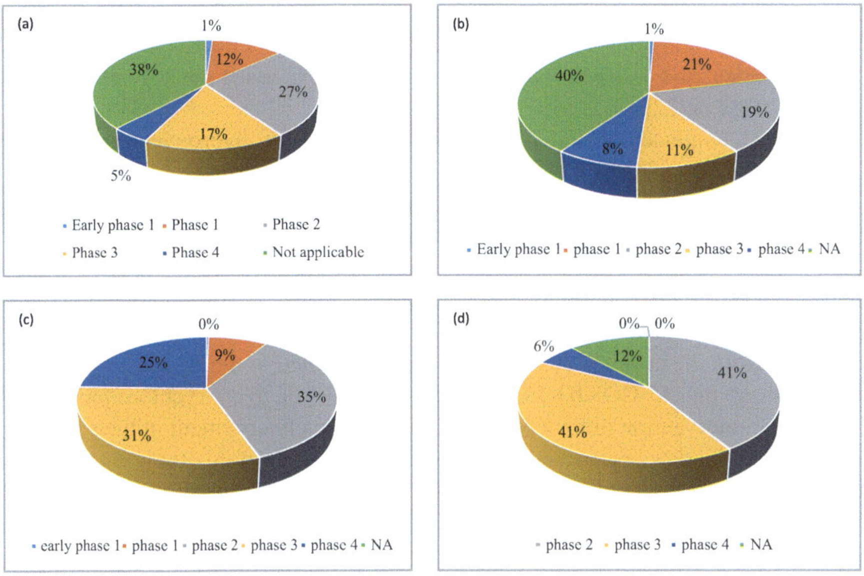

Fig. 9.2 Repurposed trials in different phases for the treatment of viral diseases: (**a**) COVID-19, (**b**) HIV, (**c**) hepatitis, (**d**) influenza

perceived unfavorable risk-to-benefit assessment, internal decision-making by the sponsor, failure to attain the primary purpose, and similar considerations.

Remdesivir, hydroxychloroquine, and favipiravir were among the medications that underwent considerable investigation for emergency approval in treating COVID-19. Among the 150 trials conducted on remdesivir, 64 have been completed. There were 292 investigations carried out on hydroxychloroquine to manage COVID-19, and 104 studies were completed.

Spinner et al. investigated the effects of remdesivir compared to standard control therapy on the clinical condition of individuals with moderate COVID-19 after 11 days. The primary aim of this study was to evaluate the clinical status of the patients on the 11th day using a 7-point ordinal scale that ranges from death to discharge from medical care. Subjects were randomly allocated to receive either a 10-day regimen of remdesivir, a 5-day regimen of remdesivir, or standard therapy. On the 11th day, those who received remdesivir for 5 days had a significantly greater chance of improving clinical status than those receiving conventional medication. Nevertheless, there was no notable disparity in the allocation of clinical state between the group receiving remdesivir for 10 days and the group receiving standard treatment (NCT04292730) [21].

Similarly, Goldman et al. conducted a study to assess the effects of remdesivir on individuals suffering from severe cases of COVID-19. The medication was administered for either a duration of 5 or 10 days. The participants were allocated randomly, with an equal number assigned to each group, to receive either a 5-day or 10-day intravenous regimen of remdesivir. The primary aim of the research was to assess the clinical status of the patients on day 14, utilizing a 7-point ordinal scale. The researchers did not observe a significant difference in effectiveness between a 5-day regimen and a 10-day regimen of remdesivir, based on the trial results (NCT04292899) [22].

Self et al. conducted a study to investigate the effect of hydroxychloroquine on the clinical status of COVID-19 patients who were hospitalized. The researchers specifically concentrated on the patients' condition after 14 days. This study aimed to evaluate the efficacy of hydroxychloroquine as a therapeutic intervention for adult patients admitted to the hospital with COVID-19. The primary outcome assessed was the clinical status of the patients 14 days post-randomization. The evaluation used a seven-point scale, with 1 representing death and 7 representing discharge from the hospital. Among individuals admitted to the hospital with respiratory illness due to COVID-19, the administration of hydroxychloroquine as a therapeutic intervention did not result in a substantial enhancement in their clinical state by the 14th day, compared to a placebo. The findings of this study do not offer justification for recommending the utilization of hydroxychloroquine for the treatment of COVID-19 in adult patients who are admitted to the hospital (NCT04332991) [23]. Barnabas et al. conducted a study investigating the efficacy of hydroxychloroquine as a prophylactic treatment following exposure to the severe acute respiratory syndrome coronavirus 2 infection. The primary outcome assessed was the incidence of newly detected SARS-CoV-2 infection, identified using PCR testing, among people who initially tested negative for SARS-CoV-2 at the start of the investigation. This study, conducted with rigorous scientific methodologies and employing randomly assigned participants, yielded no statistically significant evidence to support the effectiveness of hydroxychloroquine as a preventive strategy against SARS-CoV-2 infection in patients who had recently encountered the virus (NCT04328961) [24].

Finberg et al. undertook a study to assess the effectiveness of favipiravir as a treatment for COVID-19. The primary aim of this randomized experiment was to evaluate the duration of time required for viral clearance by day 29. The data exhibited a superior eradication of the virus in the group administered with favipiravir compared to the group undergoing standard treatment. The researchers concluded that favipiravir could be administered safely to adult COVID-19 patients admitted to the hospital (NCT04358549) [25]. However, Holubar et al. stated that the data from their trial does not provide evidence for the effectiveness of favipiravir at the usual doses in COVID-19 patients with mild symptoms. Further research is necessary to ascertain the efficacy and safety of administering higher doses of favipiravir to patients with COVID-19. This study mainly investigated the effects of favipiravir on managing COVID-19 in individuals who are not experiencing symptoms or have moderate symptoms of the disease (NCT04346628) [26].

9.3.2 HIV

Antiretroviral drugs have revolutionized HIV prevention and treatment, but patient adherence is crucial for the drugs to remain effective. Drug repurposing was therefore considered to be an effective strategy for discovering potential new HIV treatment options. The first repurposed drug for HIV, zidovudine, was originally an anticancer agent [8]. Additionally, antibiotics, antivirals, corticosteroids, and immunomodulatory agents have been repurposed for the treatment of HIV. The most researched drugs with potential for repurposing against HIV treatment are antivirals. Among various antiviral agents, anti-influenza agents, and agents effective against hepatitis viruses, cytomegalovirus has been explored for the treatment of HIV.

A total of 8311 clinical trials targeting medicines with repurposing potential against HIV have been reported. Figure 9.2b presents the phase-wise distribution of studies. Of all this research, 1937 involved children from birth to age 17, while the remaining studies focused on adults (18–64 years old) and elderly individuals (65+). The statistics showed that the maximum number of trials, 5692, were completed, 779 studies were recruiting, and 349 were active but not recruiting. Additionally, 312 studies were terminated, 250 were not recruiting yet, 191 were withdrawn, and 14 were suspended. Despite this, numerous factors led to the termination, suspension, and withdrawal of studies due to inadequate enrollment, which was a primary cause. Some additional causes include insufficient financing, potential negative risk/benefit analysis, safety issues associated with certain medicines, and a significant rate of treatment failures.

Shao et al. carried out a study to examine the effects of a fixed-dose combination of abacavir/lamivudine/zidovudine on patients co-infected with HIV and TB. The primary objectives of this study were to assess the practicability and safety of a fixed-dose combination in patients co-infected with HIV and TB, as well as to evaluate the effects of a delayed initiation approach compared to an early initiation approach for the fixed-dose combination. The main objective was to analyze the immunologic and virologic responses. The findings indicate that HIV/TB-coinfected individuals can tolerate early antiretroviral therapy (ART) well, with a minimal chance of experiencing immune reconstitution syndromes. Furthermore, initiating early ART resulted in expected advantages such as effective suppression of the virus and increased CD4 cell counts in individuals receiving the triple nucleoside combination (NCT00851630) [27].

9.3.3 Hepatitis

Chronic hepatitis infection is one of the highly prevalent viral infectious diseases globally. It is noteworthy that the hepatitis B virus is significantly more stable and around five times more contagious than the hepatitis C virus. Despite the widespread implementation of immunization programs against the human hepatitis B virus (HBV), HBV infection continues to pose a significant global public health burden. Untreated and inadequately controlled infection with HBV can result in the development of liver cirrhosis, acute and chronic hepatitis, hepatocellular carcinoma (HCC), and several associated conditions. There is a need to discover novel treatment alternatives for hepatitis due to its high prevalence rates. Due to the scarcity of novel clinical medications, drug repurposing is an emerging strategy to find new pharmacological targets [28]. Various pharmacological classes of drugs, antidiabetics, corticosteroids, statins, antituberculars, reverse transcriptase inhibitors, sedatives, antacids, antivirals, and antihistaminics, have been examined as potential candidates for repurposing against the treatment of hepatitis. Among all these classes, the medications that have received the greatest attention from researchers are reverse transcriptase inhibitors, i.e., ribavirin.

Based on the collected information, 1397 clinical investigations have been carried out by employing compounds with repurposing potential against hepatitis. Figure 9.2c depicts the distribution of the 1397 trials over several phases. The highest number of trials falls under phase 2, followed by phase 3, phase 4, phase 1, and early phase 1. The findings indicated that the maximum number of trials 1042 had been completed. A total of 112 studies were terminated: 7 were not recruiting yet, 6 were active but not recruiting, 7 studies were recruiting, and 59 studies were withdrawn. The status of the remaining 161 studies was unknown. Similarly, there were numerous reasons why studies were terminated, suspended, and withdrawn, but enrollment problems were the primary ones. Various circumstances contribute to the discontinuation of studies, such as insufficient funding, the choice of the sponsor or investigator to abandon the investigation, safety concerns related to a particular reverse transcriptase inhibitor (lamivudine), and the company's decision to conclude the trial for all five of the hydroxyzine studies.

Ribavirin, an extensively studied medication for hepatitis, has been the subject of 1080 clinical trials specifically targeting hepatitis. Among these trials, 839 have been completed, demonstrating its safety and effectiveness. Most ribavirin trials are undertaken with a combination of additional medicines. The research conducted by Manns et al. investigated the effects of ribavirin, in combination with peginterferon alfa-2b, on patients with chronic hepatitis C who have genotypes 2 and 3. The researchers concluded that a treatment protocol involving 24 weeks of peginterferon alfa-2b in conjunction with ribavirin, administered based on the patient's weight, continues to be a standard-of-care therapy (NCT00302081) [29]. Liu et al. conducted research investigating the efficacy of ribavirin and peginterferon alfa-2b in treatment-naïve patients with hepatitis C virus genotype 2 who were receiving hemodialysis. The researchers discovered that the combination of peginterferon and a low dose of ribavirin led to a more significant occurrence of sustained virological response (SVR) in comparison to the use of peginterferon alone (NCT00491244) [30].

9.3.4 Influenza

Influenza infections pose a significant and continuous worldwide public health issue because of the inherent morbidity and mortality, bacterial superinfections, and exacerbation of pre-existing illnesses. Therefore, new antiviral techniques need to be developed to counteract the emergence of newly circulating strains and strains that exhibit resistance to conventional antivirals [31]. According to the collected data, medications belonging to several pharmacological categories have demonstrated the potential to treat infections caused by influenza. Among these pharmacological classes are cholesterol modulators, antibiotics, antidiabetic, anticancer, anti-HIV, anti-hyperlipidemic, anti-inflammatory, and anti-parasitic drugs.

Based on collected data, 18 clinical trials have been conducted for influenza using repurposed medicines. The USA was the site of the majority of these trials. Figure 9.2d demonstrates that out of the 18 trials, the greatest number were under phases 2 and 3, with phase 4 coming in last. Among the repurposed trials, 8 were completed, four were terminated, one was withdrawn, two were recruiting, and one was active but not recruiting. Few studies have been discontinued due to a decrease in enrollment and the conclusion of the H1N1 pandemic. The removal of one research was attributed to an institutional issue.

Haffizulla et al. explored the impact of nitazoxanide, a medication repurposed for influenza, on adults and adolescents suffering from acute uncomplicated influenza. The study aimed to assess the efficacy and safety of nitazoxanide. Based on the data, they inferred that it was linked to a decrease in the duration of symptoms. They stated that additional trials are necessary to evaluate the effectiveness of the medicine (NCT01227421) [32].

9.4 Cardiovascular Diseases

Cardiovascular diseases (CVD) are a prominent source of concern in today's modern human adult lifestyle. According to the Global Observatory WHO Report 2019, it has been observed that the top cause of death among females and males per 100,000 population was ischemic heart disease [33]. Thus, there is a huge requirement to recognize the use of other medications that can be repurposed for cardiovascular diseases.

As per the clinical trial database, most drugs that are repurposed for CVD belong to diverse pharmacological classes such as:

1. Drugs used in rheumatoid arthritis, e.g., methotrexate.
2. Drugs used in gout, e.g., colchicine.
3. Drugs used in hyperlipidemia, e.g., atorvastatin and statin derivatives.
4. Drugs used in erectile dysfunction, e.g., sildenafil, tadalafil.
5. Drugs used in uricosurics, e.g., probenecid, etc.

The following data interprets the categories of cardiovascular diseases that come under CVDs.

9.4.1 Heart Failure

Heart failure or congestive heart failure (CHF) is a challenging health problem characterized by insufficient functioning of the myocardium, resulting in decreased blood flow throughout the body. Spironolactone, metformin, hydralazine, isosorbide, ivabradine, digoxin, losartan, and other medications from various pharmacological classes are repurposed for CHF. The effectiveness of antidiabetic drugs in treating congestive heart failure has been the subject of 195 studies, according to a comprehensive review that began in 2020. Metformin hydrochloride, dapaglifloxacin, pioglitazone, insulin, and DPP inhibitors were the drugs that were predominantly repurposed. Most trials were in phase 4, as shown in Fig. 9.3a.

Metformin hydrochloride, an anti-hyperglycemic medicine, has been investigated as a treatment option for CHF. In total, 26 trials are included. Among all trials, 14 were successfully concluded, two were voluntarily discontinued, one was terminated, six were actively seeking participants (recruiting), one was not seeking participants (not recruiting), and one was ongoing but not actively seeking participants.

Wong et al. showed that metformin treatment significantly enhanced insulin resistance and ventilatory response to exercise, as indicated by the slope of the minute ventilation rate of carbon dioxide production (VE/VCO_2 slope). The non-diabetic patients with CHF also showed significant weight loss after metformin treatment. However, the authors could not conclude their findings due to the complex mechanism of metformin and stated that some more findings are necessary before concluding the study (NCT00473876) [34].

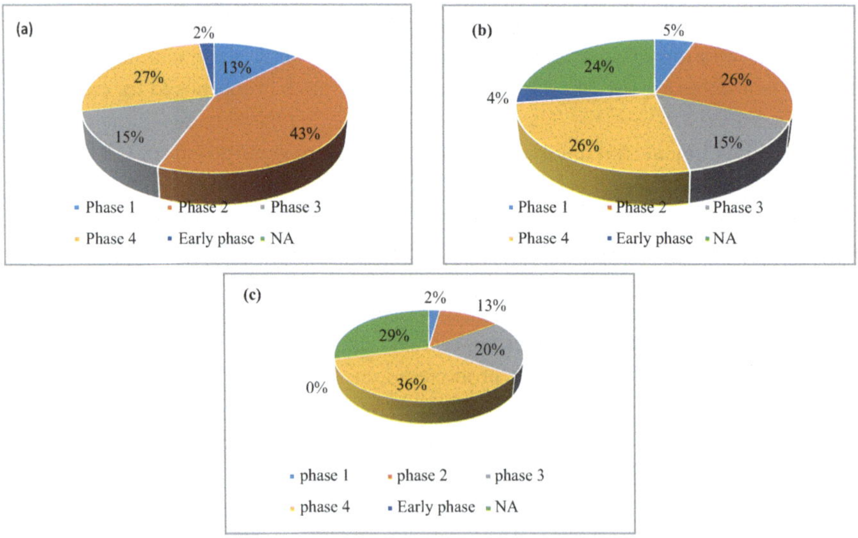

Fig. 9.3 Repurposed trials in different phases for the treatment of cardiovascular diseases: (**a**) heart failure, (**b**) hypertension, (**c**) restenosis

In a randomized, double-blind, uncontrolled study in heart failure patients (NCT00181155), intravenous infusion of allopurinol showed improvement in the myocardial energy supply via creatinine kinase that decreased myocardial oxygen consumption while maintaining the function of the heart in case of failure [35]. Another study conducted in a multicentered randomized method using 253 patients with heart failure symptoms did not show improvements in quality of life, left ventricular ejection fraction, clinical status, or exercise capacity after 24 weeks of xanthine oxidase inhibition with allopurinol (NCT00987415) [36].

An open-label, non-randomized pilot study evaluated the effect of interleukin-1 blockade on exercise capacity in patients with heart failure. The study measured the oxygen consumption rate during the exercise before and after treatment with anakinra (recombinant human interleukin-1 receptor antagonist). The findings suggested that the activation of IL-1β contributes to the decreased exercise tolerance observed in patients with systolic HF. Additionally, the study suggests blocking IL-1β as a new approach for pharmacological management (NCT01300650) [37].

9.4.2 Hypertension

Multiple pathological alterations in the pulmonary vasculature result in pulmonary arterial hypertension (PAH), an uncommon but deadly ailment. Endothelial cell dysfunction is typified by increased production of pro-proliferative, vasoconstrictive substances like endothelin and decreased release of pro-proliferative, vasodilatory molecules (prostacyclin and nitric oxide). This mismatch triggers the augmentation of tone and the growth of pulmonary artery smooth muscle cells (PASMCs) [38]. The leading cause of hypertension nowadays is a chaotic lifestyle, imbalanced diet, unusual routine, obesity, etc. The most commonly used pharmaceutical classes include aldosterone antagonists, antidiabetic medications, anti-gout drugs, and steroid hormones. The drugs that are repurposed for hypertension belong to several pharmacological classes, which include drugs such as rituximab, paclitaxel, chloroquine, etc.

Since the twentieth century, the use of repurposed drugs concerning hypertension has been increasing vigorously. It's one of the lifestyle diseases associated with slow killing capacity. In such a way, the following total trials conducted for hypertension are 168. Figure 9.3b illustrates that out of the 168 studies, the highest number of trials fall into phase 4, followed by phase 3, phase 2, phase 1, and early phase 1. Additionally, 38 trials do not fall into any single-phase category. In the following trials, 69 trials were completed, 12 were terminated, 11 were withdrawn, 7 were active but not recruiting, 11 were not yet recruiting, 31 trials had unknown status, and none were suspended. Of the 168 commenced trials, only 67 have been finished effectively. According to the information provided, there is currently no available medication for the treatment of hypertension.

A successful study has been conducted utilizing spironolactone for the treatment of hypertension. Diabetic individuals who also have hypertension face a significantly heightened chance of experiencing vascular damage, which in turn increases their susceptibility to developing kidney disease and cardiovascular disease. Blood samples and vascular tissue from people diagnosed with type 2 diabetes show higher levels of fibulin-1, an extracellular matrix glycoprotein. This research aimed to examine how the antihypertensive drug spironolactone affected plasma fibulin-1 levels. In a randomized, controlled, double-blind experiment spanning multiple centers, 119 patients diagnosed with resistant hypertension and type 2 diabetes were enrolled. At randomization, either a 25 mg dosage of spironolactone or a placebo that looked the same was added to the previous treatment. Baseline and 16-week follow-up measurements were taken for blood pressure (BP) and plasma fibulin-1. In total, 112 patients successfully concluded the investigation. At follow-up, the spironolactone group showed a decrease in all blood pressure measurements throughout the study. Plasma fibulin-1 levels fell significantly after spironolactone treatment ($p = 0.009$) but elevated after placebo ($p = 0.017$). Blood pressure and estimated glomerular filtration rate were found to be correlated with the baseline plasma levels of fibulin-1. Diabetic subjects who reported erectile dysfunction had higher levels of plasma fibulin-1 ($p = 0.004$) compared to participants who did not experience erectile dysfunction. Administering a low dose of spironolactone reduced plasma fibulin-1 levels in persons with type 2 diabetes who had resistant hypertension (NCT01062763) [39].

9.4.3 Restenosis

The term "in-stent restenosis" (ISR) reflects the recurrence of coronary artery obstruction or blockage at the location of a prior stent implantation. Even if the stent's first installation successfully restored blood flow to the heart, restenosis—a re-narrowing of the artery—poses a risk. Numerous causes, such as neointimal hyperplasia, inflammation, plaque formation, and stent design, can lead to in-stent restenosis. Various recycled cases are sent to the clinical database for the ailment restenosis. The database shows the wide range of medications used to treat conditions related to the arteries. Pharmacological classes employed in abnormalities of the restenosis type include immunosuppressants, mTOR inhibitors, anti-platelet medicines, anticancer, and NSAIDs. The most commonly utilized class during restenosis is essentially anticancer and mTOR inhibitors.

There are 378 trials in the entire database. Figure 9.3c illustrates that the maximum number of trials was in phase 4, followed by phase 3, phase 2, and phase 1. Furthermore, it was discovered that 103 trials were observed for which no phase was applied, and no trial was conducted under the early phase; 234 trials were completed, 23 were terminated, 5 were withdrawn, 2 were suspended, 9 were active but not recruiting, 8 were not yet recruiting, 75 had an unknown status, and 21 were recruited during the aforementioned trials.

The ORAR (Oral trial in ARgentina) III trial is a randomized study that evaluates the efficacy of oral rapamycin (OR) in combination with bare-metal stents (BMS) compared to drug-eluting stents (DES) for the treatment of new coronary lesions in patients. This study aimed to evaluate the cost-effectiveness outcome of each approach over a 3-year length of time. Furthermore, it is worth noting that BMS has been linked to a decrease in target vessel revascularization (TVR), but its long-term effectiveness compared to DES remains uncertain. The baseline features of the groups were comparable. The rate of follow-up after 3 years was 99%. The mortality rates due to cardiac events were 2% and 5% in the group undergoing open-heart surgery (OR group) and the group receiving drug-eluting stents (DES group), respectively. The disparity in mortality rates between the two groups did not exhibit statistical significance ($p = 0.44$). The overall rate of death, myocardial infarction, and stroke was 11% in the group that underwent open-heart surgery (OR group) and 20% in the group that received drug-eluting stents (DES group) ($p = 0.078$). The TVR rate was 14.5% in the group treated with open reduction (OR) and 17.6% in the group treated with drug-eluting stents (DES), yielding a p-value of 0.50. The cumulative expenses over 3 years were considerably lower in the group that utilized the operating room (OR) as opposed to the group that utilized drug-eluting stents (DES) ($p = 0.0001$). Furthermore, according to the non-inferiority test, the DES approach was not cost-effective. After 3 years of observation, there were no noticeable differences in effectiveness between the two approaches, and the use of DES did not demonstrate greater cost-efficiency compared to the combination of optimal medical therapy (OR) and bare-metal stents (BMS) (NCT00552669) [40].

9.5 Inflammatory Diseases

9.5.1 Rheumatoid Arthritis

Rheumatoid arthritis (RA) is a widespread inflammatory condition that may greatly influence a person's quality of life. RA is caused by antigen-specific responses of T cells, cytokine networks, and the lethal tumor-like activity of the rheumatoid synovium [41]. Many anti-RA-specific medications were developed to address this problem, and many already-existing medications were also repurposed. Various classes of drugs, such as antibiotics, antimalarials, chelators, anticancer, anti-sepsis, and anti-psoriatics, have shown potential for repurposing in treating RA. The order of drugs repurposed includes anticancer and antiseptic medications, followed by antimalarial drugs, antibiotics, chelators, and anti-psoriatic drugs. Methotrexate is the most often repurposed medication under this anticancer category for the treatment of RA.

Concerning RA, 3008 clinical trials have been carried out to date, of which approximately 1224 trials were conducted with repurposed medications, which include methotrexate, anakinra, etanercept, ustekinumab, hydroxychloroquine, tetracycline, minocycline, doxycycline, hydroxychloroquine, and bucillamine. Figure 9.4a depicts studies under different phases. Of these 1224 studies, 792 only used methotrexate from the anticancer category. Most trials took place in the United States, with China and India also being prominent locations. Of all the 1224 studies conducted, 78 involved children aged up to 17, 1035 involved adults under the age of 64, and the remaining 111 involved older persons beyond the age of 65. Several

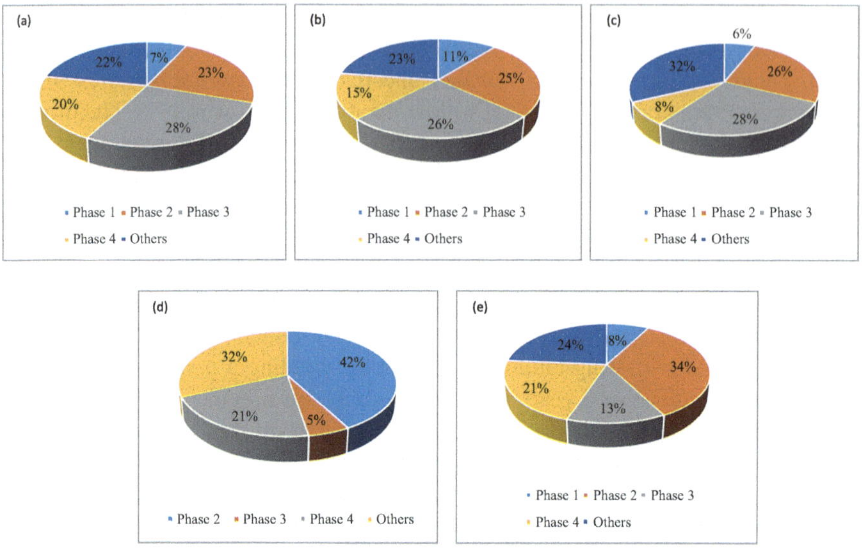

Fig. 9.4 Repurposed trials in different phases for the treatment of inflammatory diseases: (**a**) rheumatoid arthritis, (**b**) systemic lupus erythematosus, (**c**) psoriasis, (**d**) inflammatory bowel disease, (**e**) Crohn's disease

trials were prematurely terminated due to various factors, with time restrictions being the primary cause. Other factors contributing to the premature termination include a low rate of subject enrollment, challenges within the institution, and insufficient enrollment numbers. Among the 1224 repurposed clinical trials, around 690 tests yielded successful results, resulting from 857 completed trials. An investigation was successfully conducted on the effectiveness of first-line biological therapy for individuals with active rheumatoid arthritis. The study focused on combining methotrexate and Acellbia®, a biosimilar to rituximab (ALTERRA) [42].

9.5.2 Systemic Lupus Erythematosus (SLE)

Systemic lupus erythematosus (SLE) is widely recognized as a classic immune-mediated disorder because of its complex clinical and genetic characteristics [43]. Several medications employed in curing rheumatoid arthritis and chronic lymphocytic leukemia are likewise utilized in the treatment of SLE. As shown in Fig. 9.4b, most trials were under phases 2 and 3, followed by phase 4 and phase 1. The rest were either under early phase 1 or under the phases not applicable category. Most drugs that are repurposed for use in SLE fall into the category of antirheumatic medications followed by anticancer drugs which include abatacept, rituximab, and blisibimod.

Out of 887 clinical trials conducted to date, 101 used repurposed drugs. Of these 101 studies, 30 studies were using rituximab. Most of the trials took place in the

United States, with China, Egypt, France, and Japan also notable study locations. Of all the 101 studies conducted, 16 involved children aged up to 17, 68 involved adults under the age of 64, and the remaining 17 involved older persons beyond 65. Many circumstances contributed to the early termination of several studies; however, the primary cause was the sponsor's decision to end the study for commercial reasons and, in certain cases, lack of efficacy. Out of 101 repurposed clinical trials, 26 out of 48 completed trials proved successful. After several stages of effective investigation, the product was eventually launched into the market. GSK (GlaxoSmithKline) promotes Benlysta as a therapy for systemic lupus erythematosus (SLE). Benlysta contains belimumab, a drug that was originally researched for rheumatoid arthritis but has been repurposed for the treatment of systemic lupus erythematosus (SLE) [44].

9.5.3 Psoriasis

Psoriasis, a papulosquamous skin ailment, is widely acknowledged as a prominent immune-mediated disorder. Previously, it was thought to be primarily an epidermal keratinocyte problem. T cells, dendritic cells, and tumor necrosis factor α mostly influence its pathophysiology. Early-onset psoriasis (beginning before age 40) is mostly influenced by environmental triggers such as β-hemolytic streptococcal infections and HLA-Cw6 harboring [45]. Drugs from several pharmacological classes, including anti-arthritic, antineoplastic, immunomodulatory, and antidiabetics, have been repurposed for psoriasis. The majority of drugs that are repurposed for psoriasis are classified as anti-arthritic. Tofacitinib is the most often repurposed medication under this anti-arthritic category for treating psoriasis. Other drugs are ruxolitinib, baricitinib, paclitaxel, bexarotene, metformin, and azathioprine.

Out of the 2130 clinical investigations conducted so far, about 50 of them include the repurposed use; 31 of the 50 trials used tofacitinib. Figure 9.4c demonstrates that most trials were in phase 1, then phase 2 and phase 4, respectively. The remaining studies were either in the early phase 1 or categorized as inappropriate to any specific phase. The trials were mostly conducted in the United States, followed by Germany, China, India, and France. Of all the 50 studies conducted, 6 involved children aged up to 17, 43 involved adults under the age of 64, and the remaining 1 involved older persons beyond the age of 65. Since one of the studies had already achieved the long-term safety and tolerability goals, it was terminated. Among the 50 repurposed clinical trials, almost 21 out of the 29 finished investigations successfully achieved their objective. After conducting several successful trials, the producer of tofacitinib obtained a comprehensive response letter from the FDA on its supplementary new drug application (sNDA) for XELJANZ (tofacitinib citrate) to treat psoriasis. Originally intended for arthritis treatment, the drug has now been repurposed to treat psoriasis [46].

9.5.4 Inflammatory Bowel Disease

Two primary theories have arisen on the underlying cause of the pathophysiology of inflammatory bowel disorders (IBDs), which include ulcerative colitis and Crohn's disease. The first theory is that there is a significant imbalance in the mucosal immune system, leading to an excessive immune response to the commensal bacteria found in the body. The second theory is that changes in the gut microbiota composition and/or dysfunction of the epithelial barrier led to aberrant immune responses in the mucosal lining. After examining both hypotheses, it can be inferred that an atypical immune response in the mucosal lining distinguishes IBD. However, it is worth noting that microbial factors and abnormalities in epithelial cells can contribute to this response. Over the last 75–100 years, gastroenterologists and immunologists have been perplexed by inflammatory bowel diseases (IBDs) such as ulcerative colitis, Crohn's disease, and non-infectious bowel inflammations. To tackle this issue, many drugs were designed and formulated that are active against IBD [47]. There are many other drugs that were originally indicated for other treatments but were found to be active against the treatment of IBD. Major drugs that were repurposed were topiramate and cyclosporine. Cyclosporine is the most widely repurposed drug for IBD that belongs to the immunosuppressive category. Topiramate belongs to the anti-epileptic category.

Out of the 3126 clinical studies conducted thus far, 19 trials specifically focused on repurposed drugs. As seen in Fig. 9.4d, most of the trials were under phase 2, with phase 4 and phase 3 following behind. The remaining studies were either categorized as early phase 1 or fell under the category of phases not applicable. Eighteen of the 19 trials used cyclosporine, and very few studies reported the use of anti-epileptic drugs. Most trials were conducted in the United States, India, Finland, and Belgium. Of all the 19 studies conducted, 7 involved children aged up to 17, 11 involved adults under the age of 64, and the remaining 1 involved older persons beyond the age of 65. One of the studies was terminated as the Sublimity Board of Directors withdrawn the funds as the placebo studies had a response rate.

Out of the total studies of 19 repurposed clinical trials, around 9 investigations successfully attained their primary objective. Zeposia® is a commercialized product by Bristol Myers Squibb, which was first approved by USFDA for the treatment of multiple sclerosis in the year 2020, it is now repurposed in the year 2021 for the treatment of ulcerative colitis a specific type of inflammatory bowel disease.

9.5.5 Crohn's Disease

IBD primarily consists of two conditions: ulcerative colitis (UC) and Crohn's disease (CD). Relapsing inflammatory disease, Crohn's disease primarily affects the gastrointestinal tract. It often begins with fever, stomach pain, and clinical indications of diarrhea with passage of mucus or blood, or both [48]. Denosumab, vorinostat, tacrolimus, and mercaptopurine are among the very few medications that have been repurposed for the treatment of CD. Most drugs repurposed to use in CD are originally used as acute lymphoblastic leukemia (ALL), immunomodulatory agents, and antiosteoporosis drugs. The most typically repurposed drug under this ALL for CD treatment is mercaptopurine.

Out of the 1637 clinical studies conducted so far, 38 trials specifically examined the utilization of drugs for repurposed use. Figure 9.4e depicts that most trials were in phase 2, with phase 4, phase 3, and phase 1 in descending order. The remaining studies were either in the early phase 1 or fell into the category of phases not applicable. Mercaptopurine was utilized in 32 out of the 38 trials. Israel conducted most of the testing, followed by the United States, United Kingdom, Canada, and France. Of all the 38 studies conducted, 5 involved children aged up to 17, 32 involved adults under the age of 64, and the remaining 1 involved older persons beyond 65. Approximately five studies were ended because the annual accrual target was not reached, and the sponsor stopped funding the research because of organizational changes. A total of 16 clinical trials conducted with the objective of repurposing were identified as completed, with 11 resulting in positive outcomes.

9.6 Summary

The practice of repurposing drugs has gained popularity in the field of medicine. It is projected to expand further in the future, as it has the potential to accelerate the advancement of novel therapies for a range of life-threatening illnesses. Moreover, drug repurposing provides a cost-effective approach for drug discovery and diminishes the duration and resources needed for clinical trials. This literature provides extensive information on clinical studies undertaken for repurposed medications, specifically focusing on cancer, viral infections, cardiovascular disorders, and inflammatory diseases. Based on the gathered data, it is evident that most clinical trials aimed at repurposing treatments for various types of tumors primarily focus on antidiabetic drugs. Similarly, in the context of cardiovascular disorders, antidiabetics are often utilized as a repurposed class of therapy. For the treatment of new viral infections, current antiviral drugs are primarily evaluated in the event of viral infections. Amidst the pandemic, hydroxychloroquine, remdesivir, and favipiravir have been thoroughly studied as potential treatments for COVID-19 and have received emergency authorizations. Inflammatory illnesses often benefit from the repurposing of anti-arthritics and anticancer medicines. This medication repurposing highlights the importance of exploring alternative uses for existing drugs, especially in the face of new and challenging health crises. By repurposing medications originally intended for other conditions, researchers and healthcare professionals can expand treatment options and potentially improve outcomes for patients with various illnesses.

References

1. 42 CFR Part 11. Clinical trials registration and results information submission. National Archives; 2024.
2. Claiborne AB, English RA, Weisfeld V, editors. Public engagement and clinical trials: new models and disruptive technologies. Washington, DC: National Academies Press; 2011. https://doi.org/10.17226/13237.
3. Meera M, Sekar S, Mahatao R. A novel approach for drug discovery-drug repurposing. Natl J Physiol Pharm Pharmacol. 2022;12:546–51.
4. Pushpakom S, Iorio F, Eyers PA, Escott KJ, Hopper S, Wells A, Doig A, Guilliams T, Latimer J, McNamee C, Norris A, Sanseau P, Cavalla D, Pirmohamed M. Drug repurposing: progress, challenges and recommendations. Nat Rev Drug Discov. 2018;18:41–58. https://doi.org/10.1038/nrd.2018.168.
5. Krishnamurthy N, Grimshaw AA, Axson SA, Choe SH, Miller JE. Drug repurposing: a systematic review on root causes, barriers and facilitators. BMC Health Serv Res. 2022;22:970. https://doi.org/10.1186/s12913-022-08272-z.
6. Berger AC, Olson S, Johnson SG, Beachy SH, editors. Drug repurposing and repositioning. Washington, DC: National Academies Press; 2014. https://doi.org/10.17226/18731.
7. Jourdan JP, Bureau R, Rochais C, Dallemagne P. Drug repositioning: a brief overview. J Pharm Pharmacol. 2020;72:1145–51. https://doi.org/10.1111/jphp.13273.
8. Trivedi J, Mohan M, Byrareddy SN. Drug repurposing approaches to combating viral infections. J Clin Med. 2020;9:1–14. https://doi.org/10.3390/jcm9113777.
9. Sleire L, Førde-Tislevoll HE, Netland IA, Leiss L, Skeie BS, Enger PØ. Drug repurposing in cancer. Pharmacol Res. 2017;124:74–91. https://doi.org/10.1016/j.phrs.2017.07.013.
10. Zhou H, Liu H, Yu Y, Yuan X, Xiao L. Informatics on drug repurposing for breast cancer. Drug Des Devel Ther. 2023;17:1933–43. https://doi.org/10.2147/DDDT.S417563.
11. Kalinsky K, Zheng T, Hibshoosh H, Du X, Mundi P, Yang J, Refice S, Feldman SM, Taback B, Connolly E, Crew KD, Maurer MA, Hershman DL. Proteomic modulation in breast tumors after metformin exposure: results from a "window of opportunity" trial. Clin Transl Oncol. 2017;19:180–8. https://doi.org/10.1007/s12094-016-1521-1.
12. Doumat G, Daher D, Zerdan MB, Nasra N, Bahmad HF, Recine M, Poppiti R. Drug repurposing in non-small cell lung carcinoma: old solutions for new problems. Curr Oncol. 2023;30:704–19. https://doi.org/10.3390/curroncol30010055.
13. Chun SG, Liao Z, Jeter MD, Chang JY, Lin SH, Komaki RU, Guerrero TM, Mayo RC, Korah BM, Koshy SM, Heymach JV, Koong AC, Skinner HD. Metabolic responses to metformin in inoperable early-stage non–small cell lung cancer treated with stereotactic radiotherapy. Am J Clin Oncol. 2020;43:231–5. https://doi.org/10.1097/COC.0000000000000632.
14. Cortés H, Reyes-Hernández OD, Alcalá-Alcalá S, Bernal-Chávez SA, Caballero-Florán IH, González-Torres M, Sharifi-Rad J, González-Del Carmen M, Figueroa-González G, Leyva-Gómez G. Repurposing of drug candidates for treatment of skin cancer. Front Oncol. 2021;10:605714. https://doi.org/10.3389/fonc.2020.605714.
15. Dal Monte M, Calvani M, Cammalleri M, Favre C, Filippi L, Bagnoli P. β-Adrenoceptors as drug targets in melanoma: novel preclinical evidence for a role of β_3-adrenoceptors. Br J Pharmacol. 2019;176:2496–508. https://doi.org/10.1111/bph.14552.
16. Wang KH, Liu CH, Ding DC. Statins as repurposed drugs in gynecological cancer: a review. Int J Mol Sci. 2022;23:13937. https://doi.org/10.3390/ijms232213937.
17. Rocconi RP, Stevens EE, Bottsford-Miller JN, Ghamande SA, Elder J, DeMars LL, Munkarah A, Aaron P, Stanbery L, Wallraven G, Bognar E, Manley M, Horvath S, Manning L, Walter A, Galanis E, Herzog T, Monk BJ, Coleman RL, Nemunaitis J. Proof of principle study of sequential combination atezolizumab and Vigil in relapsed ovarian cancer. Cancer Gene Ther. 2022;29:369–82. https://doi.org/10.1038/s41417-021-00317-5.
18. Mokou M, Lygirou V, Angelioudaki I, Paschalidis N, Stroggilos R, Frantzi M, Latosinska A, Bamias A, Hoffmann MJ, Mischak H, Vlahou A. A novel pipeline for drug repurposing for

bladder cancer based on patients' omics signatures. Cancers (Basel). 2020;12:1–16. https://doi.org/10.3390/cancers12123519.

19. Molenaar RJ, van Hattum JW, Brummelhuis IS, Oddens JR, Savci-Heijink CD, Boevé ER, van der Meer SA, Witjes JF, Pollak MN, de Reijke TM, Wilmink JW. Study protocol of a phase II clinical trial of oral metformin for the intravesical treatment of non-muscle invasive bladder cancer. BMC Cancer. 2019;19:1133. https://doi.org/10.1186/s12885-019-6346-1.
20. Scavone C, Brusco S, Bertini M, Sportiello L, Rafaniello C, Zoccoli A, Berrino L, Racagni G, Rossi F, Capuano A. Current pharmacological treatments for COVID-19: what's next? Br J Pharmacol. 2020;177:4813–24. https://doi.org/10.1111/bph.15072.
21. Spinner CD, Gottlieb RL, Criner GJ, López JRA, Cattelan AM, Viladomiu AS, Ogbuagu O, Malhotra P, Mullane KM, Castagna A, Chai LYA, Roestenberg M, Tsang OTY, Bernasconi E, Le Turnier P, Chang S-C, SenGupta D, Hyland RH, Osinusi AO, Cao H, Blair C, Wang H, Gaggar A, Brainard DM, McPhail MJ, Bhagani S, Ahn MY, Sanyal AJ, Huhn G, Marty FM. Effect of remdesivir vs standard care on clinical status at 11 days in patients with moderate COVID-19. JAMA. 2020;324:1048. https://doi.org/10.1001/jama.2020.16349.
22. Goldman JD, Lye DCB, Hui DS, Marks KM, Bruno R, Montejano R, Spinner CD, Galli M, Ahn M-Y, Nahass RG, Chen Y-S, SenGupta D, Hyland RH, Osinusi AO, Cao H, Blair C, Wei X, Gaggar A, Brainard DM, Towner WJ, Muñoz J, Mullane KM, Marty FM, Tashima KT, Diaz G, Subramanian A. Remdesivir for 5 or 10 days in patients with severe Covid-19. N Engl J Med. 2020;383:1827–37. https://doi.org/10.1056/NEJMoa2015301.
23. Self WH, Semler MW, Leither LM, Casey JD, Angus DC, Brower RG, Chang SY, Collins SP, Eppensteiner JC, Filbin MR, Files DC, Gibbs KW, Ginde AA, Gong MN, Harrell FE, Hayden DL, Hough CL, Johnson NJ, Khan A, Lindsell CJ, Matthay MA, Moss M, Park PK, Rice TW, Robinson BRH, Schoenfeld DA, Shapiro NI, Steingrub JS, Ulysse CA, Weissman A, Yealy DM, Thompson BT, Brown SM. Effect of hydroxychloroquine on clinical status at 14 days in hospitalized patients with COVID-19. JAMA. 2020;324:2165. https://doi.org/10.1001/jama.2020.22240.
24. Barnabas RV, Brown ER, Bershteyn A, Karita HCS, Johnston C, Thorpe LE, Kottkamp A, Neuzil KM, Laufer MK, Deming M, Paasche-Orlow MK, Kissinger PJ, Luk A, Paolino K, Landovitz RJ, Hoffman R, Schaafsma TT, Krows ML, Thomas KK, Morrison S, Haugen HS, Kidoguchi L, Wener M, Greninger AL, Huang M-L, Jerome KR, Wald A, Celum C, Chu HY, Baeten JM. Hydroxychloroquine as postexposure prophylaxis to prevent severe acute respiratory syndrome coronavirus 2 infection. Ann Intern Med. 2021;174:344–52. https://doi.org/10.7326/M20-6519.
25. Finberg RW, Ashraf M, Julg B, Ayoade F, Marathe JG, Issa NC, Wang JP, Jaijakul S, Baden LR, Epstein C. US201 study: a phase 2, randomized proof-of-concept trial of favipiravir for the treatment of COVID-19. Open Forum Infect Dis. 2021;8:ofab563. https://doi.org/10.1093/ofid/ofab563.
26. Holubar M, Subramanian A, Purington N, Hedlin H, Bunning B, Walter KS, Bonilla H, Boumis A, Chen M, Clinton K, Dewhurst L, Epstein C, Jagannathan P, Kaszynski RH, Panu L, Parsonnet J, Ponder EL, Quintero O, Sefton E, Singh U, Soberanis L, Truong H, Andrews JR, Desai M, Khosla C, Maldonado Y. Favipiravir for treatment of outpatients with asymptomatic or uncomplicated coronavirus disease 2019: a double-blind, randomized, placebo-controlled, phase 2 trial. Clin Infect Dis. 2022;75:1883–92. https://doi.org/10.1093/cid/ciac312.
27. Shao HJ, Crump JA, Ramadhani HO, Uiso LO, Ole-Nguyaine S, Moon AM, Kiwera RA, Woods CW, Shao JF, Bartlett JA, Thielman NM. Early versus delayed fixed dose combination abacavir/lamivudine/zidovudine in patients with HIV and tuberculosis in Tanzania. AIDS Res Hum Retrovir. 2009;25:1277–85. https://doi.org/10.1089/aid.2009.0100.
28. Irham LM, Adikusuma W, Perwitasari DA, Dania H, Maliza R, Faridah IN, Santri IN, Phiri YVA, Chong R. The use of genomic variants to drive drug repurposing for chronic hepatitis B. Biochem Biophys Rep. 2022;31:101307. https://doi.org/10.1016/j.bbrep.2022.101307.
29. Manns M, Zeuzem S, Sood A, Lurie Y, Cornberg M, Klinker H, Buggisch P, Rössle M, Hinrichsen H, Merican I, Ilan Y, Mauss S, Abu-Mouch S, Horban A, Müller TH, Welsch C, Chen R, Faruqi R, Pedicone LD, Wedemeyer H. Reduced dose and duration of peginterferon

alfa-2b and weight-based ribavirin in patients with genotype 2 and 3 chronic hepatitis C. J Hepatol. 2011;55:554–63. https://doi.org/10.1016/j.jhep.2010.12.024.
30. Liu C-H, Liu C-J, Huang C-F, Lin J-W, Dai C-Y, Liang C-C, Huang J-F, Hung P-H, Tsai H-B, Tsai M-K, Lee C-Y, Chen S-I, Yang S-S, Su T-H, Yang H-C, Chen P-J, Chen D-S, Chuang W-L, Yu M-L, Kao J-H. Peginterferon alfa-2a with or without low-dose ribavirin for treatment-naive patients with hepatitis C virus genotype 2 receiving haemodialysis: a randomised trial. Gut. 2015;64:303–11. https://doi.org/10.1136/gutjnl-2014-307080.
31. Pizzorno A, Padey B, Terrier O, Rosa-Calatrava M. Drug repurposing approaches for the treatment of influenza viral infection: reviving old drugs to fight against a long-lived enemy. Front Immunol. 2019;10:531. https://doi.org/10.3389/fimmu.2019.00531.
32. Haffizulla J, Hartman A, Hoppers M, Resnick H, Samudrala S, Ginocchio C, Bardin M, Rossignol J-F. Effect of nitazoxanide in adults and adolescents with acute uncomplicated influenza: a double-blind, randomised, placebo-controlled, phase 2b/3 trial. Lancet Infect Dis. 2014;14:609–18. https://doi.org/10.1016/S1473-3099(14)70717-0.
33. World Health Organization. Cardiovascular diseases. Geneva: WHO; 2024.
34. Wong AKF, Symon R, AlZadjali MA, Ang DSC, Ogston S, Choy A, Petrie JR, Struthers AD, Lang CC. The effect of metformin on insulin resistance and exercise parameters in patients with heart failure. Eur J Heart Fail. 2012;14:1303–10. https://doi.org/10.1093/eurjhf/hfs106.
35. Hirsch GA, Bottomley PA, Gerstenblith G, Weiss RG. Allopurinol acutely increases adenosine triphosphate energy delivery in failing human hearts. J Am Coll Cardiol. 2012;59:802–8. https://doi.org/10.1016/j.jacc.2011.10.895.
36. Givertz MM, Anstrom KJ, Redfield MM, Deswal A, Haddad H, Butler J, Tang WHW, Dunlap ME, LeWinter MM, Mann DL, Felker GM, O'Connor CM, Goldsmith SR, Ofili EO, Saltzberg MT, Margulies KB, Cappola TP, Konstam MA, Semigran MJ, McNulty SE, Lee KL, Shah MR, Hernandez AF. Effects of xanthine oxidase inhibition in hyperuricemic heart failure patients. Circulation. 2015;131:1763–71. https://doi.org/10.1161/CIRCULATIONAHA.114.014536.
37. Van Tassell BW, Arena RA, Toldo S, Mezzaroma E, Azam T, Seropian IM, Shah K, Canada J, Voelkel NF, Dinarello CA, Abbate A. Enhanced interleukin-1 activity contributes to exercise intolerance in patients with systolic heart failure. PLoS One. 2012;7:e33438. https://doi.org/10.1371/journal.pone.0033438.
38. Prins KW, Thenappan T, Weir EK, Kalra R, Pritzker M, Archer SL. Repurposing medications for treatment of pulmonary arterial hypertension: what's old is new again. J Am Heart Assoc. 2019;8:e011343. https://doi.org/10.1161/JAHA.118.011343.
39. Oxlund CS, Cangemi C, Henriksen JE, Jacobsen IA, Gram J, Schousboe K, Tarnow L, Argraves WS, Rasmussen LM. Low-dose spironolactone reduces plasma fibulin-1 levels in patients with type 2 diabetes and resistant hypertension. J Hum Hypertens. 2015;29:28–32. https://doi.org/10.1038/jhh.2014.27.
40. Rodriguez AE, Rodriguez-Granillo AM, Antoniucci D, Mieres J, Fernandez-Pereira C, Rodriguez-Granillo GA, Santaera O, Rubilar B, Palacios IF, Serruys PW. Randomized comparison of cost-saving and effectiveness of oral rapamycin plus bare-metal stents with drug-eluting stents: three-year outcome from the randomized oral rapamycin in Argentina (ORAR) III trial. Catheter Cardiovasc Interv. 2012;80:385–94. https://doi.org/10.1002/ccd.23352.
41. Firestein GS. Evolving concepts of rheumatoid arthritis. Nature. 2003;423:356–61. https://doi.org/10.1038/nature01661.
42. ClinicalTrials.gov. Clinical trial to evaluate efficacy and safety of Acellbia® (JSC "BIOCAD") with methotrexate in first line biological therapy of patients with active rheumatoid arthritis (ALTERRA). U.S. National Library of Medicine; 2018.
43. Lam GKW, Petri M. Assessment of systemic lupus erythematosus. Clin Exp Rheumatol. 2005;23:S120–32.
44. ClinicalTrials.gov. Safety and effectiveness of BENLYSTA (belimumab) in systemic lupus erythematosus (SLE) registry (SABLE). U.S. National Library of Medicine; 2023.
45. Griffiths CE, Barker JN. Pathogenesis and clinical features of psoriasis. Lancet. 2007;370:263–71. https://doi.org/10.1016/S0140-6736(07)61128-3.

46. Pfizer. Pfizer receives complete response letter from FDA for oral XELJANZ® (tofacitinib citrate) supplemental new drug application for moderate to severe chronic plaque psoriasis. Pfizer Press Release; 2015.
47. Strober W, Fuss I, Mannon P. The fundamental basis of inflammatory bowel disease. J Clin Investig. 2007;117:514–21. https://doi.org/10.1172/JCI30587.
48. Bretto E, Ribaldone DG, Caviglia GP, Saracco GM, Bugianesi E, Frara S. Inflammatory bowel disease: emerging therapies and future treatment strategies. Biomedicine. 2023;11:2249. https://doi.org/10.3390/biomedicines11082249.

Intellectual Property Rights and Regulatory Considerations Throughout Drug Repurposing

10

Prajakta Bule, Lal Bahadur Pal, Raghunandhan Painoori, and Naveen Chella

Abbreviations

ANDA	Abbreviated new drug application
API	Active pharmaceutical ingredient
CKD	Chronic kidney disease
EMA	European medicines agency
EPC	European patent convention
EPO	European patent office
GAD	General anxiety disorder
HDAC	Histone deacetylase
IPR	Intellectual property rights
NCE	New chemical entity
NOIP	National Office of Intellectual Property
sNDA	Supplemental new drug application
TPP	Trans-pacific partnership
TRIP	Trade-related intellectual property rights
UKSC	UK Supreme Court
USPTO	United States Patent and Trademark Office

Prajakta Bule and Lal Bahadur Pal contributed equally with all other contributors.

P. Bule · L. Bahadur Pal · N. Chella (✉)
Department of Pharmaceutical Technology (Formulations), National Institute of Pharmaceutical Education and Research Guwahati, Guwahati, Assam, India
e-mail: pe2021-4014phd@niperguwahati.in; pe22phd5023@niperguwahati.in; naveen@niperguwahati.in

R. Painoori
Regulatory Affairs, Odin Pharmaceutical, LLC, Somerset, NJ, USA

N. Chella et al. (eds.), *Drug Repurposing*,
https://doi.org/10.1007/978-981-97-5016-0_10

Learning Objectives

- Understanding the potential barriers related to intellectual property rights and regulatory guidelines of drug repurposing
- Understanding the regulatory guidelines of various countries
- Understanding the case studies related to regulatory issues

10.1 Introduction

Drug discovery is a complex and risky process that requires significant financial investment and a long time. Finding a new drug requires a substantial investment of up to 1.4 billion dollars over 12–15 years [1, 2]. The process starts with identifying a target and finding a chemical or biological scaffold to act on it. Drug candidates are then synthesized, purified, and characterized for structural confirmation, and assays are developed to study their effectiveness. The candidates are tested in vitro, followed by in vivo studies for efficacy, and their safety is profiled on cell cultures and/or in animal models. Desirable and druggable properties are achieved through experimentation and optimization. A stable dosage form is formulated, and phased clinical trials in humans are conducted to conclude the process [3]. The data generated through these various stages is submitted to regulatory authorities for marketing approval. The process typically experiences a significant drop-off, with only a single molecule out of tens of thousands meeting all the criteria for eventual marketing approval [4]. Pharmaceutical industries and many scientists are now focusing on drug repurposing to cut costs and save drug discovery and development time. During repurposing, a drug candidate already approved for marketing is carefully assessed for potential alternative medical applications. This offers two significant advantages: reduction in development costs and development risk. This makes the process more efficient and cost-effective, with potential savings of up to 60%.

The COVID-19 pandemic has clarified that repurposing can be a valuable approach to drug discovery using existing therapeutic modalities to treat different disease conditions. A range of well-known drugs such as remdesivir, falcipiravir, barictinib, sotrovimab, tixagevimab co-packaged with cilgavimab and tocilizumab have received emergency use authorization for combating the SARS-CoV-2 virus from USFDA [5, 6]. Several other drugs, including hydroxychloroquine, ivermectin, doxycycline, ritonavir, and lopinavir, were also investigated for their potential in the treatment of COVID-19 infection [7]. A lot of literature is available that provides proof of the concept of drug repurposing in multiple diseases. It primarily highlights drug safety and efficacy yet fails to investigate the elaborate regulatory standards and obstacles associated with drug repurposing. However, to make the research applicable in practice, it is crucial to have a deeper understanding of regulatory terms and conditional exclusivity. The current chapter focuses on the patenting and regulatory concerns related to drug repurposing.

10.2 Drug Repurposing

Drug repositioning involves the identification of a new purpose for a drug or drug candidate that goes beyond its original intended use. Repositioning is also known as repurposing, drug reprofiling, reused medicament, therapeutic switching, second medical use, or second indication. There is a dire need in the area of rare and orphan diseases for drug repurposing due to limited financial incentives to innovators for new drug development. Health authorities are encouraging innovators by providing incentives related to marketing exclusivity. The repurposing approach would be an alternate approach to bringing existing drugs to market to address unmet medical needs. The history of medicine has been characterized by numerous instances of drugs being repurposed, even for those which have been in use for a long time. Serendipity was the only cause of most of them being discovered. Novel approaches have emerged, explicitly utilizing data analysis techniques to discover potential candidates for drug repositioning. Numerous start-ups are fully dedicated to the development of this concept. A specialized journal, *Drug Repurposing, Rescue, and Repositioning*, published by MA Liebert Inc., Boston, MA, USA, focuses exclusively on this topic. An annual international conference (*Drug Repurposing, Rescue, and Repositioning*) also discussed this concept, with the most recent 10th edition held in Washington, DC, in 2022 [8].

10.2.1 Barriers to Drug Repurposing

One important consideration when it comes to drug repurposing is the potential risk involved, which is comparable to that of developing a completely new drug. There is a chance that the repurposed drug may not demonstrate a satisfactory balance between its benefits and risks during clinical trials, which could prevent it from receiving regulatory approval for its new use. While the safety profile of drugs that have been on the market for a while is already well established, it is important to take into account how the variations in benefits in the new indication might affect the overall balance between benefits and risks. Furthermore, there are other aspects to take into account that may limit the advantages of the current comprehension of the drug's clinical characteristics. First and foremost, choosing the proper dosage is crucial to achieving the best possible therapeutic outcome for any medication, regardless of its purpose. It is important to note that in medications with multiple uses, the appropriate dosage may vary between indications. Therefore, additional research is required to determine the relationship between dosage with response in the new indication. This will ensure that trials can be conducted to support approval of the new indication.

Furthermore, with the continuous improvement of regulatory science standards, historical data, such as preclinical toxicology studies, may no longer meet the requirements for a new indication. Consequently, there may be a need to conduct repeated tests at an earlier stage. Both factors contribute to an increase in the overall cost of development, which can make the project less appealing. Ultimately, the ability to access the market with the additional indication will depend on more than regulatory evidence of safety, quality, and efficacy. It will also be subjected to a health technology evaluation considering comparative clinical effectiveness and cost-effectiveness. Many nations' regulatory systems currently include exclusivity regimes designed to incentivize corporations to invest resources in researching and developing new applications for current medications. For example, the USFDA, in collaboration with the National Center for Advancing Translational Sciences (part of the National Institute of Health), developed a database named CURE ID to understand the new uses of products and facilitate clinical trials and drug development [9]. The concern is that these regimens are unsuitable for their intended purpose and do not offer a framework for medication producers to recoup their investment, especially for pharmaceuticals for which generic versions exist.

Hence, promoting a drug repurposing approach requires balancing technological, medical, legal, and marketing goals to reduce risk, increase benefits, and achieve a profitable and effective product. However, the success of a repositioned drug relies on obtaining market exclusivity through a combination of intellectual property and regulatory exclusivity [10]. The advantages and challenges of drug repurposing are presented in Fig. 10.1.

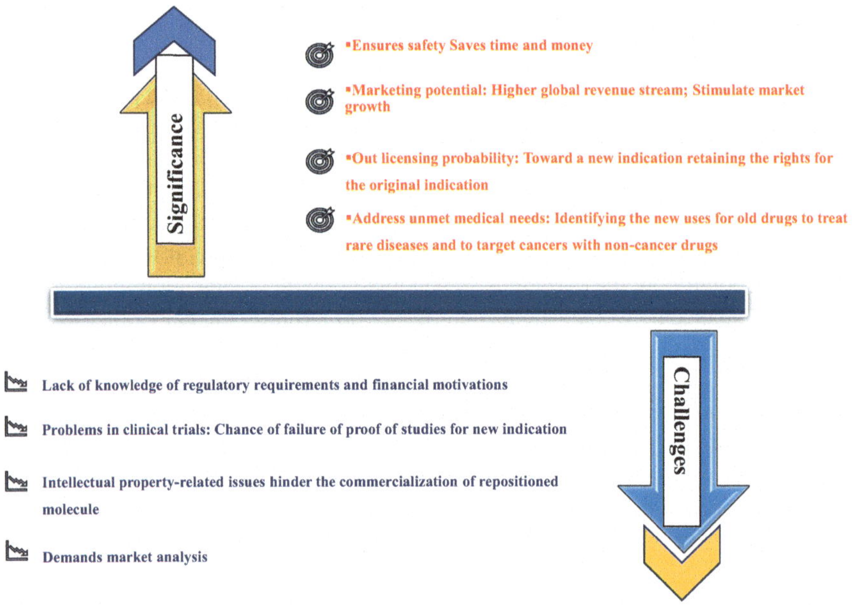

Fig. 10.1 Significance and challenges in drug repurposing

10.3 Intellectual Property Rights for Drug Repurposing

Intellectual property rights (IPR) include patents, industrial designs, trademarks, copyrights, and geographical indications. The patent is granted for an invention that meets three crucial criteria, viz. i) being new, ii) not obvious, and iii) having practical use. Any inventor can apply for intellectual property rights for their invention, allowing ample time for enhanced identification with planning, commercialization, and execution of their creativity. It significantly impacts the economy and motivates pharmaceutical companies to drive innovation. R & D costs are linked to the invention process, allowing for the potential to profit from the invention worldwide through IPR. It is a dependable method to recoup the expenses of the inventor's effort, investment, time, and money to create IPR [11]. Therefore, it fosters a positive rivalry among pharmaceutical companies and discoverers and is linked to a nation's economic growth [12]. Patents offer protection against costs associated with the inventor's hard work, financial investment, time, and resources used to generate intellectual assets; in addition, depending on the type of discovery, various health authorities give commercial exclusivities for predetermined periods. The patent applications must be submitted individually to the relevant health authorities to safeguard the inventor's work.

10.3.1 Patent Protection

Patent protection is a period when an invented drug is protected from generic drug competition. The USFDA and European Medicines Agency (EMA) initially granted patent protection for 20 years. It promotes a balance between NCE invention and generic drug competition [13]. The pharmaceutical industries protect new chemical entities by composition/formulation or delivery mechanism. The composition patent contains active pharmaceutical ingredients (API). The USFDA does not allow a successfully repurposed drug for patent protection for a new chemical entity, as the drug would not be considered a new drug then. Instead, it provides for patent protection for its alternate use, whereas in Europe, the regulation allows for patent protection for repurposed drugs.

Similarly, the protection for repurposed drugs can also be obtained by reformulation/composition of the APIs [14]. In addition, the success of such types of patents depends on the generic drugs that may be available for off-label use to accomplish the same therapeutic outcome. New generics may use "skinny labeling," which allows off-patent indications on the label and restricts the innovator from delaying the competition. A combination of new formulations with novel indications may provide attractive repositioning opportunities for previously marketed drugs. For example, a topical formulation or a dosage regimen cannot be easily reproduced with current generic products.

In Europe, a generic company can't use the original inventor's data for 8 years after getting the marketing authorization. It can use the data to support its application but can't market a product based on the available data. For example, suppose the innovator develops a distinct indication during the 8 years of exclusivity. In that case, it will be granted an additional year of exclusivity due to the new indication offering substantial advantages over established therapies.

10.3.2 Regulatory Exclusivity of Repurposed Drugs

In addition to patent protection for repurposed drugs, health authorities support innovators by offering market exclusivity. Regulatory agencies offer market exclusivity for pediatric indications and up to 7 years for orphan illnesses to promote investments in drug development for neglected diseases, as shown in Fig. 10.2 [15]. The most advantageous patent situation arises when the API in the repositioned product is solely approved for the newly patented indication and is not available elsewhere in the market. This offers protection in the market for an extended duration (6 months or more based on case) following product approval [16].

Although an excellent patent position is the most effective way to prevent competition regarding a repurposed drug product, in the absence of such patent protection, the exclusivity provided by the regulatory authority based on the repositioned use can also be a helpful approach to benefit the pharmaceutical companies [17]. In the US, regulatory exclusivity falls under two types for a drug: new chemical entity exclusivity and new formulation exclusivity.

A new chemical entity (NCE) in the United States receives exclusivity for up to 5 years under the Hatch-Waxman Act. The newly authorized drugs will benefit from this NCE exclusivity as these were never before approved or sold for the new indication. The new formulations usually are incentivized with 3 years of exclusivity. For pediatric studies, a 180-day exclusivity and the new antibiotic drugs treating infectious diseases are eligible for 5-year exclusivity. The NCE exclusivity offers the innovator 5 years, including 4 years without the risk of competition from ANDA or

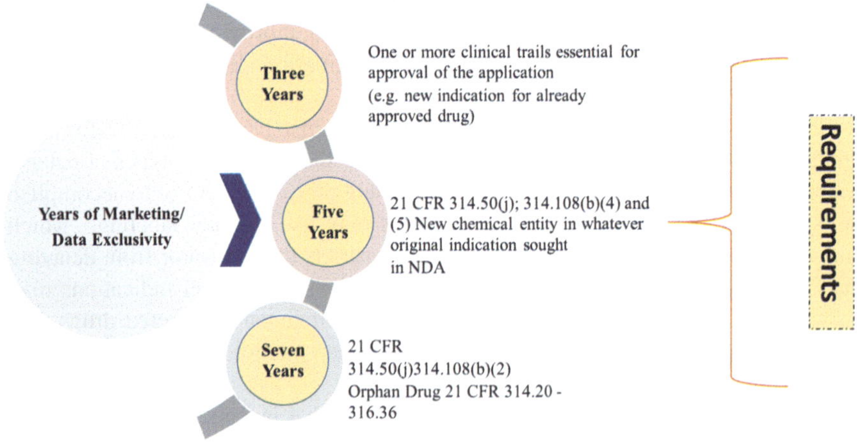

Fig. 10.2 Requirements and years of US marketing/data exclusivity

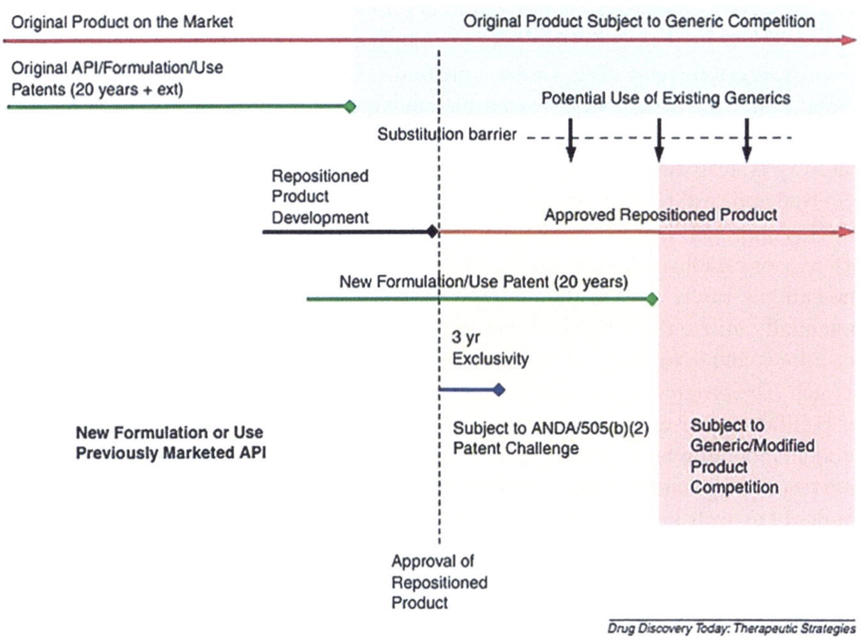

Fig. 10.3 Illustration of exclusivities for repositioned product based on previously marketed API. Reprinted with permission from Smith et al. [19], Copyright © (2011) Elsevier

505(b)(2) filings. In addition, an ANDA or 505(b)(2) applicant will be subject to an extra 30-month stay if a lawsuit is filed. In Europe, market exclusivity is granted for 10 years for orphan drugs and 8 years for data exclusivity from initial authorization that limits the generic competition by restricting the release of preclinical and clinical results [18]. Fig. 10.3 demonstrates the potential opportunities for exclusivity with a repositioned drug based on a new formulation and/or indication for an existing drug. In the United States, the new drug approvals are assigned specific classification codes that can be submitted using different regulatory pathways. Out of these types, type 1 (new molecular entity), type 6 (new indication), and supplemental new drug application (sNDA) are only used for drug repurposing submissions, whereas type 3 (new dosage form) and 4 (new combination) can be used for both repurposing and reformulation [20].

10.4 Drug Repurposing Regulatory Considerations

Pharmaceutical regulations are "the set of legal, administrative, and technical steps governments take to ensure that medicines are safe, effective, and of good quality, and that product information is correct and up to date" [4]. There are three commonly used approaches for obtaining approval for drug products in the United States, viz. i) 505(b)(1), ii) 505(j), and iii) 505(b)(2).

The 505(b)(1) approach is utilized to approve new drugs and requires extensive human and animal research for pharmacology and toxicology to prove the safety and efficacy of the new drug. Generic medicines are typically processed through the 505(j) route, which requires proving the equivalence of two drug products by conducting in vitro and/or clinical bioavailability/bioequivalence tests. The 505(b)(2) pathway is designed for a novel form or application of a drug that has previously received approval (repurposing).

This approach involves leveraging previous research to enhance the safety and efficacy of established medications. Only the necessary studies to validate the new indication's safety and/or effectiveness need to be conducted. The sponsor could potentially utilize the 505(b)(2) approach to expand upon existing chemistry, pharmacology, and toxicology research performed on the drug rather than starting from scratch. The program can utilize written works, published studies, previous findings of USFDA-approved studies from another application, approved product labels, or product monographs as sources. Establishing a scientific or technical bridge between 505 (b)(2) drugs and previously approved drugs for which the information available on the FDA website would be crucial for scientific understanding, especially in the bioavailability studies. It is essential to understand that the procedure for submitting experimental new drug applications for clinical trials of established drugs for new purposes follows a comparable process, especially from a bioequivalence perspective. This implies that existing drugs can be examined for alternative purposes without submitting chemistry and production files for the investigational new drug or seeking approval from the original owners, as long as the drug is utilized following the labeled safety instructions [21].

Regarding drugs approved under 505(b)(2) processes, patent concerns play a significant role as approval timelines might be extended due to patent or monopoly challenges. There are legal pathways like the 505(b)(2) system in countries such as Canada, Australia, and Europe. Identical to the US process, governing bodies rely on data from written literature and drug product monographs to support the new drug trials. In Europe, the European Medicines Agency (EMA) oversees a parallel approval process as outlined in Article 10 of Directive 2001/83/EC. However, Article 10 pertains to drugs that call for studies specifically designed to account for variations from reference-listed drugs. It does not establish a legal foundation for the utilization of non-proprietary studies, unlike Section 505(b)(2) of the Food, Drug, and Cosmetics Act. This particular section permits the inclusion of non-proprietary studies that have already met rigorous standards of quality and safety to substantiate various aspects of an application. Another significant distinction is that Article 10 cannot be utilized for novel molecular entities because it only applies to alterations from the reference-listed drugs. The EMA typically takes 6 months before the FDA approves new medication indications [20].

Table 10.1 The patent rules and regulatory guidelines for different countries

Country	Patentability	Statutory provision
Bangladesh	No	Notification of the Department of Patents, Designs and Trademarks (2008)
United States	Yes	Repurposing patents are under Section 101 and treatment method in the US.45 (Section 505 b(2))
Europe	Yes	EPC 2000 claims
Bhutan	Unclear	Industrial Property Act (2001)
Brunei Darussalam	Yes	Patents Order (2011), Section 14(7)
Cambodia	No	Law on Patents, Utility Model Certificates, and Industrial Designs (2006), Article 136
China	Yes	Patent Law, Article 22.2 and Guidelines for Patent Examination, Chapter 10, Part II, Section 6.2
Hong Kong SAR	Yes	Patents Ordinance (2016), Section 9(B)(5)
India	Generally not	Patent Act (1970), Section 3(d)
Indonesia	No and yes	Law of the Republic of Indonesia, No. 13, Article 4(f) (2016)
Japan	Yes	Japan Patent Office, Examination Guidelines for Patent and Utility Model (2015), Part VII, Chapter 3 "Medicinal Inventions," at. 2.2.2. (3-2-1)
Lao PDR	Unclear	Law on Intellectual Property (2017)
Korea (Democratic People's Republic of)	Unclear	Law on Inventions (1999), Article 33(3)
Korea (Republic of)	Yes	Korea Intellectual Property Office Patent Examination Guidelines (2017)
Macau SAR	Unclear	Industrial Property Code Decree-Law No 97/99/M (1999)
Malaysia	Yes	Patents Act (1983), Section 14(4)
Maldives	No	No patent law
Mongolia	Unclear	Law on Patents (1993)
Myanmar	No	Patent Law (2019)
Nepal	Unclear	Industrial Property Act (1965)
Pakistan	No	Patents Ordinance (2000)
Philippines	Yes	Revised Guidelines on the Examination of Pharmaceutical Applications Involving Known Substances (2018)
Singapore	Yes	Patents Act, Article 14(7)
Sri Lanka	Unclear	Intellectual Property Act (2003)
Taiwan (Province of China)	Generally, not	Taiwan Patent Examination Guidelines (2020)
Thailand	Yes	Department of Intellectual Property Patent Examination Guidelines (2020)
Viet Nam	Generally, not	Law on Intellectual Property (2005) and paragraphs 23.6.d and 2

10.4.1 Second Medical Use Patents

Second medical use patents pertain to safeguarding novel applications of a known pharmaceutical compound. Table 10.1 provides an overview of the statutory provisions and current status of patentability for drug repurposing in Asia and other prominent countries. In the United States, repurposing patents are under Section 101 of the US Patent Code, and further uses may be protected as a medical treatment method [10]. The European Patent Office (EPO) permitted the protection of drugs, referred to as the "Swiss style" claim, until 2007. The Swiss-style claim was no longer allowed when the European Patent Convention (EPC) was revised in 2000 and went into effect in 2007. Article 54(5) of EPC 2000 now permits second (and subsequent) medical uses to be protected under the format "substance X for use in the treatment of disease Y" if the use is novel and inventive. Another name for this form is the "EPC 2000 claims" [22]. The Philippines and South Korea prioritize the protection of second medical use patents, while Indonesia takes a different approach. In Vietnam, the protection of second medical uses is considered unavailable due to the lack of mention in national law. The 2005 IP Law also does not allow for the patenting of methods of medical treatment.

Nevertheless, due to its intentional legal ambiguity, the Vietnamese approach is significantly more intricate than previously assumed. If national law prohibits patenting second medical uses, the government may face challenges from its trading partners. With its lexical ambiguity, Vietnam's patent office (National Office of Intellectual Property—NOIP) can discreetly reject utility patents without drawing significant international scrutiny, as with **Glivec** in India [23, 24]. The infringement of the patentee's rights where the patentee rights violation resulted from incorrect prescribing or dispensing, as demonstrated by the legal cases involving pregabalin in the UK, is included as a case study (presented under 10.4.4). We have included the recent 5-year patent data from 2019 to 2024. An extensive search on the World Intellectual Property Organization's PatentScope database for international disclosure papers was performed, and some examples of global patents are presented in Table 10.2 for reference.

10.4.2 Hurdles in Second Medical Use Patenting

Many nations permit various forms of second medical use patents. Unfortunately, these inventors may fail for many kinds of reasons, such as

(i) The second (new) use lacks novelty or inventiveness according to established principles of patent law, as evidence of its existence has already been published in the scientific literature.

(ii) In many countries, such as the US and various European countries, patentees must provide a statement in their patent application demonstrating the effectiveness of the second medical use. As a result, patentees must wait until they have sufficient evidence to support the credibility of the second medical use

Table 10.2 List of recently approved patents for drug repurposing

Application number	Drug patented	Original indication	Repurposed indication	Date
WO2023182957	Varenicline	Smoking cessation	Inflammatory diseases	29/9/2023
WO/2023/154794	Promethazine	Histamine H1	Ischemic stroke	2/9/2023
WO/20222/205986	Sevelamer	Chronic kidney disease (CKD)	Hepatic carcinoma	6/8/2022
WO/2022/2138	Ramipril	Antihypertensive agents	Liver cancer	31/10/2022
IN202321043242	Niclosamide	Anthelmintic	Breast cancer	15/9/2023
IN202241072726	Panobinostat	Anticancer HDAC inhibitor to	Alzheimer's disease	13/12/2023
WO2022208546	Sigma receptor targeting ligand	Reduce cognitive deficits	GLIOMA	24/2/2023
US20220288029	Enzalutamide	Anti-androgen	COVID-19	15/9/2022
US20220273641	Floctafenine	Nonsteroidal anti-inflammatory drugs	COVID-19	01/9/2022
IN202111053983	Pimavanserin	Hallucinations and delusions associated with Parkinson's disease	Tuberculosis	28/1/2022
WO2022006232	Avoralstat, PCI-27483, SBTI	Plasma kallikrein inhibitors, coagulation factor VII inhibitors, Trypsin inhibitors	COVID-19	6/1/2022
IN202141006921	Levetiracetam	Seizures	ANTI-BACTERIAL	18/02/2021
WO/2019/220453	Methotrexate + anti-retrovirals	Anticancer	Anti-HIV	2/11/2019

before filing their application. However, there is a risk of disclosure to the public before gathering enough evidence and convincing the examiner to grant the patent.

(iii) When a generic version of a drug becomes available, developers often struggle to recover their investment in the drug's development for a new indication.

10.4.3 Challenges in Dealing with Second Medical Use Patents Due to Variations in Legal Processes Across Different Countries: A Case Study

In Vietnam, the absence of legal provisions regarding second medical uses results in their lack of protection. The 2005 IP Law also prohibits the patenting of methods of medical treatment. Nevertheless, the Vietnamese approach is far more intricate than previously assumed [25]. The various rules and regulations challenge the proper filing of repurposed drugs. To elaborate further on this subject, we have included a case study from Vietnam, where repurposed drug filing was prohibited [26]. The country's unique approach was challenged in the case of International Application No. PCT/JP2006/318675. In August of 2008, the patent application with reference number 1-2008-00901 was filed for the national phase in Vietnam. The applicant, Otsuka Pharmaceutical, has submitted a patent application for a combination drug that includes probucol and a tetrazolylalkoxy-dihydrocarbostyril derivative with superoxide suppressant effects. These medications are highly effective in preventing and treating conditions such as cerebral infarction, arteriosclerosis, diabetes, and various renal diseases. Discovering a novel medical application for established components does not make a product original and innovative since the claims did not satisfy the requirements for novelty and inventiveness. The applicant lodged an appeal against the decision in June. Over 5 years later, in November 2019, the National Office of Intellectual Property (NOIP) made a decision (No. 5698/QD-SHTT) to dismiss the appeal. This was because the composition mentioned in the claim had already been anticipated in prior art (D1: SEKIYA M. et al. American Journal of Cardiology 1998, Vol. 82, No. 2, 144–147, hereinafter referred to as D1). While the appellant acknowledged the agreement with the NOIP regarding the availability of the composition in D1, it contended that D1 failed to address the potential treatment claimed by the invention. The NOIP, referring to a specific point in Circular No. 1/2007 (amended in 2016), presented a counter-argument stating that the new treatment is not a fundamental aspect of the invention, but rather a consequence of its composition. Discovering a new medical application for established components does not make an agent original and innovative.

Otsuka Pharmaceutical provided an example emphasizing Vietnam's need to protect second medical use innovations under Trade-Related Aspects of Intellectual Property Rights (TRIPS) Article 27. This article requires governments to grant patents for inventions in all technical fields unless they violate public policy or morals (Art. 27.2) or fall under excluded subject matter (Art. 27.3). The NOIP contested the judgment, claiming that the invention was rejected because it did not fulfill the standards for uniqueness and creativity outlined in Art. 27.1. The rest of the article was ruled irrelevant since the subject matter was not protected from commercial usage or declared off-limits. Otsuka Pharmaceutical finally realized that the innovation is patentable because it has been authorized by many patent offices, including the United States Patent and Trademark Office (USPTO) and the European Patent Office (EPO). The NOIP rejected this claim, noting that Vietnam's patent law differs from other nations, particularly in terms of "the topic of new use of a recognized material" [emphasis added]. Vietnam's Decision No. 5698/QD-SHTT establishes a clear position on second medical use innovations. Despite Vietnam's participation in multiple FTAs, none except the short-lived TPP demand comparable commitments. The recently amended IP legislation does not change the requirements for second-use patents, making it impossible to get protection for these ideas [24].

10.4.4 Patentee's Rights Infringement

The challenge of differentiating between repurposed products and those of the innovator, generics, or branded generics in the market is complex and vital. The products that have been repositioned are marketed with a more concise label, where the relevant information is extracted from the original label. The skinny label pertains to the intended new use, and marketing approval is solely for this new indication.

Nevertheless, it is challenging to identify instances of patentee rights violations resulting from incorrect prescribing or dispensing, as demonstrated by the legal cases involving pregabalin in the UK [27]. In late 2018, the UK Supreme Court made a ruling on the pregabalin case, which involved a dispute over the secondary medical application of a pain medication. The patent rights for pregabalin as a drug marketed under the name Lyrica were held by the Pfizer group company, Warner-Lambert, and expired in 2013. Later, Warner-Lambert put forward and successfully secured patent rights for the utilization of pregabalin in alleviating different forms of pain, including neuropathic pain, inflammatory pain, and idiopathic pain. This marked a significant advancement in the medical application of pregabalin, which was primarily used to treat epilepsy and general anxiety disorder (GAD). Actavis started promoting pregabalin as a generic medication with a limited scope, specifically for epilepsy and GAD (not for pain), under the brand name Lecaent. Actavis was marketing Lecaent for conditions where the patent had already expired. Upon the arrival of Lecaent, Pfizer began to notice a decline in their market share, particularly in the UK, where prescriptions solely use the generic name of the medication. Pfizer attempted to educate the physicians about the potential infringement of their patent rights when prescribing Actavis' Lecaent for neuropathic pain. However, Pfizer accomplished this by sending letters to Physicians that reportedly contained "intimidating" language. Nevertheless, this ultimately proved to be unproductive. Warner sought legal action to address the infringement of their second-use patent, as they noticed a significant impact on their market share when the more affordable generic, Lecaent, entered the market.

Doctors in the UK prescribe medications using the International Non-proprietary Name (INN). As a result, chemists are unaware of the specific purpose for which the prescription is written. Consequently, they must dispense the most affordable drug, Lecaent, a generic medication. Therefore, Lecaent was also being sold for pain management. Following a series of legal proceedings, the case went to the UK Supreme Court (UKSC). In November 2018, the UKSC provided its perspective on the infringement of second-use patents by various parties, including manufacturers, physicians, chemists, and patients.

Additionally, the UKSC examined the validity of claims in Warner's second-use pregabalin patent, EP 0934061. The court ruled in favor of the Warner patent for treating inflammatory pain, deeming the claims regarding neuropathic pain invalid. The court also determined that, regardless of the validity of the claims, the manufacturer did not intend to infringe upon them. This is because the manufacturer is solely focused on producing and promoting a specialized label for treating epilepsy and GAD. The physicians, chemists, and patients were not violating the claims as they were not utilizing the drug to produce a medication, which is the basis of the Swiss claim. However, it is important to highlight another crucial aspect in this case. The court indirectly supported Pfizer's initial request to have a medication prescribed by its brand name. The court emphasized the necessity of implementing a more structured system for managing second medical use patents, highlighting the need for improved organization. Justice Arnold, one of the judges in the proceedings before the matter came to the Supreme Court, firmly believes that the most effective way to address the issue of protecting a monopoly granted by a second medical use patent while still allowing lawful generic competition for non-patented uses of the substance is to establish a clear distinction between the patented and non-patented markets. This can be achieved by ensuring that prescribers write prescriptions for the patented use using the brand name of the patent holder and for non-patented uses using the generic name of the substance. The statement mentioned above contradicts the regulations the Medical Council of India set, which require drugs to be prescribed using their generic names. In addition, should Justice Arnold's recommendations be implemented, the concern surrounding the cost and availability of repurposed and patented drugs could escalate beyond its current magnitude. This could potentially undermine the very goals of conducting research to discover a new medical application for an existing drug [28].

10.5 Summary

Drug repurposing related to IPR protection and regulatory data exclusivity guides the industry and academia in developing a new indication product in minimum time and at low cost. Although drug repurposing faces some challenges and hurdles in creating a new use formulation related to regulatory agencies, patents, and regulatory, exclusivity provides an opportunity to protect the IPR as well as to produce the new use formulations to the market at a fast pace with cost reduction in product development activities. This chapter helps practitioners understand regulatory issues for drug or drug product modifications and repositioning existing substances. It helps them guide intellectual property and product formulation research teams during repurposed drug product development activities.

References

1. Mullard A. New drugs cost US $2.6 billion to develop. Nat Rev Drug Discov. 2014;13:877.
2. Roses AD. Pharmacogenetics in drug discovery and development: a translational perspective. Nat Rev Drug Discov. 2008;7:807–17.
3. Krishnamurthy N, Grimshaw AA, Axson SA, Choe SH, Miller JE. Drug repurposing: a systematic review on root causes, barriers and facilitators. BMC Health Serv Res. 2022;22:1–17.
4. Papapetropoulos A, Szabo C. Inventing new therapies without reinventing the wheel: the power of drug repurposing. Br J Pharmacol. 2018;175:165–7.
5. Parvathaneni V, Gupta V. Utilizing drug repurposing against COVID-19–efficacy, limitations, and challenges. Life Sci. 2020;259:118275.

6. Perryman K. Agents of bioshield: the FDA, emergency use authorizations, and public trust. Ga L Rev. 2021;56:341.
7. WHO Solidarity Trial Consortium. Repurposed antiviral drugs for Covid-19—interim WHO solidarity trial results. N Engl J Med. 2021;384:497–511.
8. Arrowhead. 10th annual drug repositioning and repurposing conference. Chanhassen: Arrowhead Publishers; 2024.
9. U.S. Food and Drug Administration. Capturing clinician's experiences repurposing drugs to inform future studies in the era of COVID-19. https://www.fda.gov/media/138725/download. Accessed 26 Apr 2024.
10. Pushpakom S, Iorio F, Eyers PA, Escott KJ, Hopper S, Wells A, Doig A, Guilliams T, Latimer J, McNamee C. Drug repurposing: progress, challenges and recommendations. Nat Rev Drug Discov. 2019;18:41–58.
11. Novac N. Challenges and opportunities of drug repositioning. Trends Pharmacol Sci. 2013;34:267–72.
12. Saha CN, Bhattacharya S. Intellectual property rights: an overview and implications in pharmaceutical industry. J Adv Pharm Technol Res. 2011;2:88.
13. Rome BN, Lee CC, Kesselheim AS. Market exclusivity length for drugs with new generic or biosimilar competition, 2012–2018. Clin Pharm Ther. 2021;109:367–71.
14. Hernandez JJ, Pryszlak M, Smith L, Yanchus C, Kurji N, Shahani VM, Molinski SV. Giving drugs a second chance: overcoming regulatory and financial hurdles in repurposing approved drugs as cancer therapeutics. Front Oncol. 2017;7:273.
15. Sarpatwari A, Beall RF, Abdurrob A, He M, Kesselheim AS. Evaluating the impact of the Orphan Drug Act's seven-year market exclusivity period. Health Aff. 2018;37:732–7.
16. P. Pantziarka, P. Vandamme, G. Bouche, L. Meheus, K. Werynska, P. Atzei, A. Alencar, ON-PATENT DRUG REPURPOSING, https://www.risingtide-foundation.org/wp-content/uploads/2024/02/On-patent-Drug-Repurposing-White-Paper-02-2024. Accessed 26 Apr 2024.
17. Clements PJ. The Hatch-Waxman act and the conflict between antitrust law & (and) patent law. Idea. 2007;48:381.
18. European Medical Agency, Glossary of regulatory terms, European medicines agency. https://www.ema.europa.eu/en/homepage. Accessed 26 Apr 2024.
19. Smith RB. Repositioned drugs: integrating intellectual property and regulatory strategies. Drug Discov Today Ther Strateg. 2011;8:131–7.
20. Murteira S, Millier A, Ghezaiel Z, Lamure M. Drug reformulations and repositioning in the pharmaceutical industry and their impact on market access: regulatory implications. J Market Access Health Policy. 2014;2:22813.
21. Chen J, Zhao Z, Wang X, Huang J. Utilizing 505 (b)(2) regulatory pathway for new drug applications: an overview on the advanced formulation approach and challenges. London: IntechOpen; 2023.
22. Bilgin FY. Indirect infringement of second medical use patents. Galatasaray Üniversitesi Hukuk Fakültesi Dergisi. 2022;2:1797–816.
23. Gabble R, Kohler JC. To patent or not to patent? The case of Novartis' cancer drug Glivec in India. Glob Health. 2014;10:1–6.
24. Le VA. Second medical use patents and compensation for the delay in marketing authorisations: the curious case of Vietnam. GRUR Int. 2022;71:1048–55.
25. Paul PP. Novelty-as qualifier of patentability vis-a-vis relevancy of Section 3 (d) of Patents Act: a critical analysis of Indian approach. Indian JL Just. 2018;9:62.
26. Vecchi SE, Scown MJ. Intellectual property rights in Vietnam. UCLA Pac Basin LJ. 1992;11:67.
27. Ravi S, Jadhav S, Ghooi R. "Drug repurposing" policies and statutes: lessons from the COVID-19 pandemic: a review. Indian J Pharm Educ Res. 2023;57:301–7.
28. Supreme Court. JUDGMENT GIVEN ON case study Warner-Lambert Company LLC, Warner-Lambert Company LLC (Appellant) v Generics (UK) Ltd t/a Mylan and another 2018. 2018. https://www.supremecourt.uk/cases/docs/uksc-2016-0197-judgment.pdf. Accessed 26 Apr 2024.

Redirection to the Drug Discovery: Antidiabetic Drugs Repurposing in Cancer

11

Prajakta Bule, Tejaswini Kolipaka, Shital Ranvare, and Naveen Chella

Abbreviations

AML	Acute myeloid leukemia
AMPK	Activated protein kinase
BRC	Bromocriptine
CMap	Connectivity map
DPP-4	Dipeptidyl peptidase
DR2	dopamine receptors
ER	Estrogen receptor
GBM	Glioblastoma multiforme
Gli	Glibenclamide
GLP	Glucagon-like peptide
IGF-1	Insulin-like growth factor receptor 1
IL-6	Interleukin-6
JNK	Jun N-terminal kinase
mTOR	Mammalian targets of rapamycin
PPAR	Peroxisome proliferator-activated receptor
PSA	Prostate-specific antigen
ROS	Reactive oxygen species
SUR1	Sulfonylurea receptor
T1DM	Type 1 diabetes mellitus
T2DM	Type 2 diabetes mellitus

Prajakta Bule and Tejaswini Kolipaka contributed equally with all other contributors.

P. Bule · T. Kolipaka · S. Ranvare · N. Chella (✉)
Department of Pharmaceutical Technology (Formulations), National Institute of Pharmaceutical Education and Research Guwahati, Guwahati, Assam, India
e-mail: pe2021-4014phd@niperguwahati.in; naveen@niperguwahati.in

N. Chella et al. (eds.), *Drug Repurposing*,
https://doi.org/10.1007/978-981-97-5016-0_11

TSC2	Tumor suppressor gene
TZDs	Thiazolidinediones
VEGFR	Vascular endothelial growth factor

Learning Objectives

- The common etiology between diabetes and cancer
- Basic mechanisms responsible for the anticancer activity of antidiabetic drugs
- Clinical trial data on antidiabetic drugs repurposed for anticancer activity
- Challenges in the delivery of repurposed drugs
- Drug delivery options currently available for the effective delivery of repurposed drugs

11.1 Introduction

According to GLOBOCAN 2020, 19.3 million new instances of cancer were detected in 2020, while about 10.0 million people died from cancer. Around 28.4 million new cancer cases will be diagnosed worldwide in 2040 [1]. Drug resistance is one of the significant obstacles in cancer treatment, leading to a high mortality rate. As a result, research has emphasized discovering novel molecules. Although the scientific community has been concerned with ensuring the standard of living for cancer patients, researchers are also worried about the rising expenses associated with cancer therapy and new drug development [2]. The drug discovery process was further saddled by disease complexity, unpleasant side effects, and low response rates, leading to frequent failures [3]. Scientists in drug discovery always applaud alternate approaches to overcome these issues through drug repurposing.

Drug repositioning, drug reprofiling, drug recycling, drug re-tasking, drug redirection, and therapeutic switching are the other terms used interchangeably with drug repurposing. It is a method for discovering new applications for currently approved or investigational drugs [4]. According to the National Advancing Translational Sciences (NCATS), drug reprofiling is "Studying the medications that have already been approved to treat a disease to perceive if they are safe and effective for treating diverse conditions" [5]. The NCATS initiated a project called "Discovery of New Therapeutic Uses for Existing Molecules," in which they concentrated on recognizing new therapeutic indications and treatment possibilities for existing new chemical entities and marketed products that have off-target effects and multiple on-target responses. Drug repositioning has significant advantages over new drug discovery in terms of reduced failure rates, reduction in cost, and development time. Repurposing a drug is estimated to cost US$300 million on average, compared to $2–3 billion invested in a routine drug discovery process [6]. Figure 11.1a corresponds to drug discovery and repurposing time frames. The literature also reported a significant increase in the success rate of each phase with

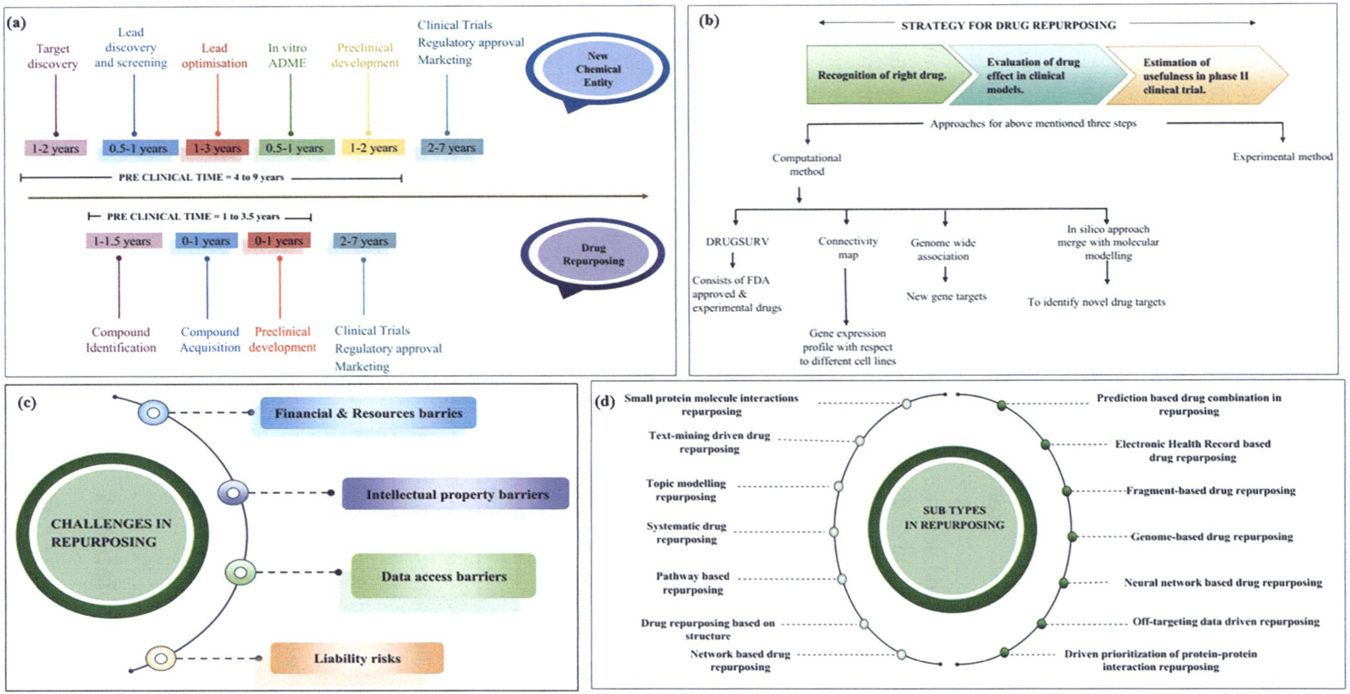

Fig. 11.1 Schematic representation of (**a**) comparison between drug discovery and drug repurposing, (**b**) Strategy for drug repurposing (**c**) challenges in repurposing, (**d**) subtypes in repurposing

drug repurposing compared to conventional drug discovery [7]. Together, these benefits have the potential to lead to a safer and faster return on investment in repurposed drug development at relatively low to intermediate costs.

Repurposing is classified based on their activity and in silico approach (Fig. 11.1bI). Activity-based drug repurposing talks about the screening of actual drugs. In contrast, repurposing a silico-based drug depends on bioinformatic tools and databases, which reveal the possible interaction between protein and drug. Each approach has advantages and disadvantages, such as faster response via in silico screening but higher chances of false hits [8]. In contrast, protein structure elucidation is not required in the activity-based method but is time-consuming [9]. Therefore, activity in silico-based drug repurposing is a substitute and interdependent approach for drug discovery. Figure 11.1bII describes the challenges in drug repurposing. Apart from this primary type, many subtypes are illustrated in Fig. 11.1c [10].

Due to the complexity and need for new drug molecules in cancer therapy, many researchers explored the option of drug repurposing in cancer. Repurposing studies on many molecules with different pharmacological activities, such as antimicrobial agents, antiepileptic agents, antihistamines, antilipidemic (statins), antidiabetics, and antiviral agents were reported in the literature [11–14]. Clinical and preclinical data suggest a biological overlap, interdependence, and common etiological attributes between cancer and diabetes [15]. Traditional risk factors such as modern lifestyle, obesity, and smoking are also commonly linked with these diseases [16, 17]. So, repurposing the antidiabetic drug in cancer is an excellent alternative to fulfill the increasing healthcare system demands in cancer. Many epidemiological and experimental investigations have found that these medications are safe and show promising anticancer properties [18–20].

To date, many researchers have explored the anticancer activity of antidiabetic agents. There is a lack of comprehensive data about antidiabetic drugs repurposed in cancer. In the current chapter, we deliberated the anticancer potential of approved and marketed antidiabetic drugs such as sulfonylureas, biguanides, phenylalanine analogues, thiazolidinediones, and inhibitors of dipeptidyl peptidase-4 (DPP-4), alpha-glucosidase, and sodium-glucose cotransporter 2 (SGLT-2). The proposed mechanism of action and various experimental approaches via distinct screening techniques (in silico, in vitro, or in vivo) were discussed to prove their anticancer activity. Clinical trial studies in different phases were also presented to provide a complete picture of antidiabetics profiling for anticancer activity. Overall, we hope this article will assist scientists and therapists in developing a more profound knowledge of drug repositioning of antidiabetics in cancer.

11.2 Overview of Diabetes

Diabetes is a chronic metabolic illness in which the body either does not make enough insulin or does not use the insulin effectually, resulting in hyperglycemia (high blood glucose levels) [21]. It is a fast-growing global problem that has an

Table 11.1 Classification of antidiabetic drugs

Class	Drug
Sulfonylureas (second generation)	Glimepiride, glipizide, gslyburide
Biguanides	Metformin
α-glucosidase inhibitors	Acarbose, miglitol
Meglitinides	Nateglinide, repaglinide
DPP-4 inhibitors	Alogliptin, saxagliptin
	Linagliptin, sitagliptin
SGLT2 inhibitors	Ertugliflozin, dapagliflozin
	Canagliflozin, empagliflozin
GLP-1 RAs agonist	Exenatide, dulaglutide
	Semaglutide, liraglutide, lixisenatide
Dopamine-2 agonist	Bromocriptine
Amylin analog	Pramlintide

accumulative impact on almost every country and economy, irrespective of age group and gender. The International Diabetes Federation (IDF) Database Atlas 2021 (Edition 10th) reports diabetes effects around 537 million people globally, with forecasts indicating that this will rise to 643 million by 2030 and 783 million by 2045 [22]. The antidiabetic drugs are listed in Table 11.1 [23].

11.3 Antidiabetic Drug Repurposing in Cancer

11.3.1 Sulfonylureas

Sulfonylureas have been the most popular medication class for treating type 2 diabetes mellitus (T2DM) since its introduction in 1950. Sulfonylureas interacts with the sulfonylurea receptor (SUR1) on the surface of β-cells and decreases blood glucose levels by directly increasing insulin secretion from β cells [24]. SUR1 is an ATP-dependent potassium channel component. The binding of a drug to SUR1 closes potassium channels, and the ensuing depolarization of the cell membrane opens voltage-dependent calcium channels. Therefore, the calcium influx from the extracellular to the intracellular compartment of the cell causes insulin release [25].

Sulfonylureas are classified into first-generation and second-generation drugs. Tolbutamide and chlorpropamide are examples of first-generation medications that are no longer prescribed. Although second-generation (glibenclamide, glipizide, gliclazide, glimepiride) medications differ in absorption, metabolism, and dosage, they are equally efficient at lowering blood glucose levels. Sulfonylureas are preferred for non-obese diabetic patients, those not responding to metformin, or where metformin is contraindicated [26]. In addition to the hyperglycemic control, some researchers observed that they also inhibit cancer progressions, as listed in Table 11.2. The detailed mechanism of anticancer activity is described in Fig. 11.2a.

Bin Yan and co-workers evaluated the apoptotic action of glibenclamide (Gli) on hepatocellular carcinoma (HCC) cells and the primary mechanism involved. In

Table 11.2 A few examples of anti-diabetic drugs showing anticancer activity with screening methodology

Drug	Cancer type	Screening method	Inhibitory cells concentration/dose	References
Sulfonylureas				
Glibenclamide	Breast cancer	In vitro cell line studies (MDA-MB-231)	25.6 ± 3.2 μM	[27]
	Gastric cancer	In vitro cell line studies (MGC-803)	5, 50, 100 μM	[28]
	Ovarian carcinoma	In vitro cell line studies (ES-2 cells)	10, 30, 100 μM	[29]
	Non-small cell lung carcinoma	In vitro cell line studies (A549, H460, H157, H1299)	A549—95.56	[30]
			H460—188.3	
			Calu-1—71.89 H157—113.3	
			H1299—143.43 mmol/L	
Gliclazide	Pancreatic cancer	In vitro cell line studies (PANC-1)	481.95 ± 0.021 μM	[31]

Biguanides				
Metformin	Pancreatic ductal adenocarcinoma	PANC-1	500 µM	[18]
		Mia PaCa2		
		Capan2		
		Xenograft mice model targeting PDX1 and BIRC5	100 mg/kg/day	
	Small-cell lung cancer and colon cancer	A549	5 mM	[32]
		HCT116		
	Multiple myeloma	RPMI8226, MM.1S, MM.1R, U266	5.95–32.18 mM	[33]
		MM cell-induced CB17/SCID mice	200 mg/kg/day	
	Hepatocellular carcinoma and breast cancer	MYCN2	110 mM	[34]
		SK-N-Be2c	155 mM	
	Pancreatic cancer	MiaPaCa-2; PANC-1; HPNE	1.5 mM	[35]
		IEC-6; MCF-10A		
		FC1242; N27		
	Neuroendocrine tumor	BON 1; GOT 1	10 mM	[36]
		NCI-H727		
	Pancreatic cancer	MiaPaCa-2 cells	1.13 mM	[37]
	Breast cancer	1. Methylnitrosourea-induced breast cancer in rats	Oral metformin 100 mg/kg/day in combination with doxorubicin	[38]
		2. 4T1 cells induced orthotopic breast cancer mouse model		
	Colorectal cancer	SW-480 HT-29	0.5 mM	[39]
	Breast cancer	MCF 7	20 mM	[40]
	HER2 + breast cancer	MCF-10, MCF-7, and BT474	15 mM	[41]
	Malignant oral cancer	2 mg/mL (AEBN)	75 mg/kg/day	[42]
		10 µg/mL (arecoline) induced female LACA Swiss Albino mice		
	Prostate cancer	MAT-LyLu	10 mM	[43]
		MAT-LyLu cells (prostate cancer induction)	250 mg/kg/day	
	Hepatocarcinogenesis	100 mg/kg 1% diethylnitrosamine (DEN) v/v in NSS once a week	200 mg/kg/day	[44]
	Ovarian cancer	SKOV3	8 mM	[45]
		OVCAR3		
		HO8910		

(continued)

Table 11.2 (continued)

Drug	Cancer type	Screening method	Inhibitory cells concentration/dose	References
Rosiglitazone	Colorectal cancer	HT-29	HT-29—5–80 μM	[46]
		HCT116		
	Myelogenous leukemia	K562 cells	70 μM	[47]
	Solid tumor	HMVEC	5 μM	[48]
		bFGF (50 ng) implanted C57BL/6 mice	50 mg/kg/day	
Troglitazone ciglitazone, and pioglitazone	Human ovarian cancer cells	OVCAR3, SKOV3, SNU-251, SNU-8, SNU-840	20–50 μM	[49]
Ciglitazone and pioglitazone	Colon cancer	HCA-7	10–50 μM	[50]
		HCT-116		
Ciglitazone	Brain tumor	Astrocytes	20 μM	[51]
Ciglitazone	Human prostate cancer	LNCaP	45 μM	[52]
	Ovarian cancer cell	A2780		[53]
		OVCAR3		
Ciglitazone and troglitazone	T cell leukemia	Jurkat T cells	30 to 100 μM	[54]
Pioglitazone	Intestinal polyp	Male C57BL/6-ApcMin/+ mice	Dietary dose of 1500 ppm for 8 weeks or 2000 ppm for 5 week	[55]
Pioglitazone	Glioma growth	Ki-67	10 μM	[56]
		C6 cells induced C57BL/6 mice	20 μM/day	
Troglitazone	Colon cancer	SW620	24 h > 16 μM	[57]
		HCT116	48 h > 13.6 μM	
			72 h > 9.1 μM	
Meglitinide analogue				
Repaglinide	High-stage neuroblastoma	SK-N-SH/FOXO3 cells	38.9 μM	[58]
Repaglinide	Hepatic, breast, and cervical cancer	MCF-7, HeLa and HepG2	89.3 ± 119.5, 87.6 ± 117.5, 109.0 ± 143.2 μg/mL	[59]

Dipeptidyl peptidase 4 (DPP-4) inhibitors				
Sitagliptin	Hepatocellular carcinoma	N-nitrosodiethylamine (NDEA)-induced HCC	20, 40, or 80 mg/kg/day	[60]
Sitagliptin	Colorectal cancer	HT-29	15 mM	[61]
Sitagliptin and vildagliptin	Colorectal cancer	HT-29	62.5 µg/mL	[62]
Vildagliptin	Lung metastases	MC 38	5.24 mM	[63]
		NCI-60 induced mice model	0.2 mg/mL/day	
Glucagon-like peptide (GLP-1) 1 receptor agonists				
Exenatide	Breast cancer	MCF-7, MDA-MB-231	0.1–10 nM	[64]
		Female athymic CAnN.Cg-Foxn1nu/CrlCrlj	300 pmol/kg	
	Endometrial cancer	Ishikawa	10 nM	[65]
		Ishikawa cells induced BALB/c mice model	24 nmol/kg/day	
	Pancreatic cancer	PaCa-2 and PANC-1	10, 100 and 1000 nmol/L	[66]
		PaCa-2 cells induced male athymic nude mice	0.2 mg/kg/day	
	Prostate cancer	LNCap	0.1–10 nmol/L	[67]
		Athymic CAnN.Cg-*Foxn1nu*/CrlCrlj mice	300 pmol/kg	
	Glioblastoma	U87 and U251	50 nM	[68]
	Ovarian carcinoma	SKOV-3 and OVCAR-3	SKOV-3	[69]
			30 µM	
			OVCAR-3	
			40 µM	
	Colon cancer	DMH (20 mg/kg/week) induced Swiss albino mice	10 or 20 mg/kg/day	[70]
	Ovarian cancer	SKOV3, OVCAR3, OVCAR4, A2780 and ES-2	1, 10, 100 nmol	[71]
		SKOV-3 cells induced-female nude mice	500 ng/day, 2 µg/day	
Exenatide with metformin	Breast cancer	MCF-7	10 nM	[72]
		MDA-MB-231 and KPL		
Liraglutide with metformin	Pancreatic cancer	MiaPaca-2	1000 nmol/L	[73]

(continued)

Table 11.2 (continued)

Drug	Cancer type	Screening method	Inhibitory cells concentration/dose	References
Liraglutide	Endometrial cancer	Ishikawa	10, 100 and 1000 nM	[74]
	Pancreatic cancer	MIA PaCa-2 PANC-1	100 nmol/L	[75]
		PaCa-2 cells induced male athymic nude mice	0.2 mg/kg/day	
	Endometrial cancer	IK and HEC-1B	IK	[76]
			0.5 mmol/L	
			HEC-1B	
			1 mmol/L	
Sodium glucose cotransporter 2 (SGLT 2)				
Canagliflozin	Lung cancer	A549	1–50 µ	[77]
		Male athymic CAnN.Cg-Foxn1nu/CrlCrlj mice	50 mg/kg/day	
Canagliflozin	Liver cancer	Huh7, HepG2	µM	[78]
		Huh7, HepG2 induced BALB/cSlc-nu/nu	100 mg/kg/day	
Canagliflozin	Breast cancer	MCF-7	3.21 µM	[79]
		MCF-7 cells induced BALB/C-nu/nu mice	100 mg/kg/day	
Canagliflozin	Anti cancer activity	SKBR3, BT-474, and MCF7	~50 µM	[80]
Canagliflozin	Hepatocellular carcinoma	MC4R-KO mice	169 mg	[81]
Tofogliflozin	Colorectal cancer	Azoxymethane (AOM)-induced colorectal pre-neoplastic diabetic and obese db/db mice	1 and 10 mg/kg/day	[82]
Dopamine-2 agonist				
Bromocriptine	Lung cancer	CCRF-CEM CEM/ADR5000	10.13 µM	[83]
		HEK293	11.78 µM	
		HEK/ABCB5	5.24 µM	
			100 µM	
Bromocriptine	Prolactinomas	MMQ	40 nM	[84]

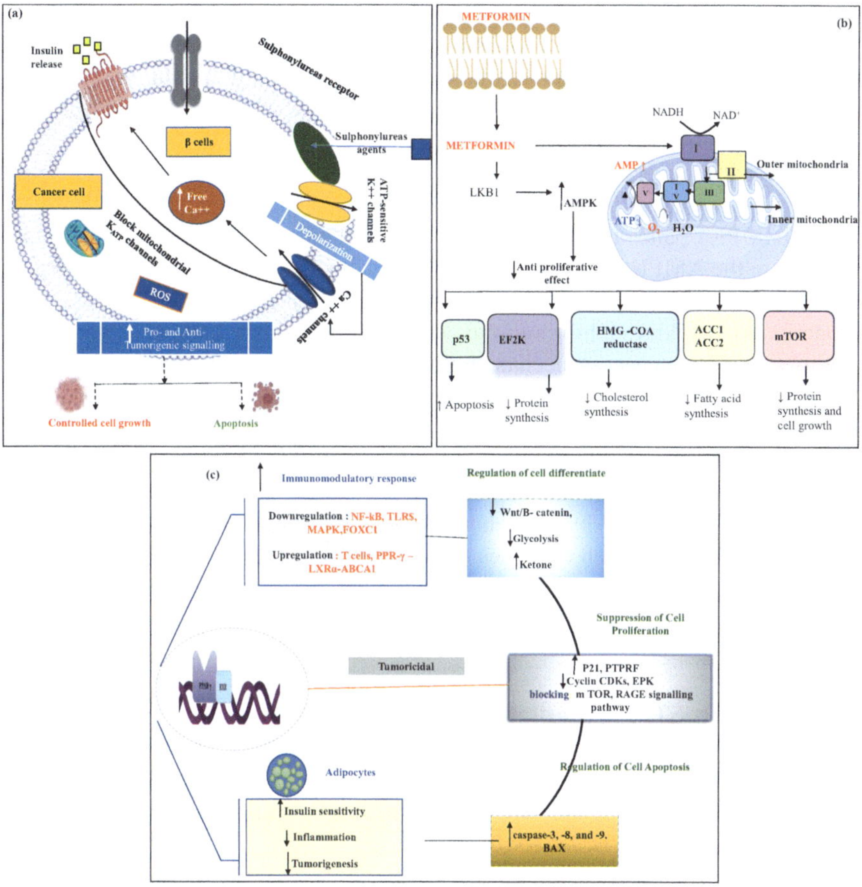

Fig. 11.2 Schematic representation of the mechanism of action of (**a**) sulfonylureas, (**b**) biguinides, (**c**) thiazolidinediones in cancer

vitro screening studies on HCC cell lines (Hepg2 and Huh7) revealed that Gli therapy inhibits cell viability by inducing apoptosis and cell cycle arrest in G2/M-phase in both cells. The primary mechanism responsible for this activity was increased production of reactive oxygen species (ROS) and JNK (Jun N-terminal kinase) pathway activation. Further, the diminished apoptotic activity of Gli in the presence of JNK pathway inhibitor SP100625 on HCC cells confirms that Gli-induced apoptosis in the HCC cell line was mediated via ROS-dependent JNK activation [85].

11.3.2 Biguanides

Biguanides improve glucose tolerance in T2DM people by reducing basal and intestinal glucose absorption and improving peripheral glucose uptake and utilization.

The primary biguanides, metformin, and phenformin were introduced in 1957 and often employed alone or with other medications [86]. Later, phenformin has been withdrawn from the market in many countries due to toxicity. Metformin has a low affinity for plasma proteins and is easily eliminated by the kidney [87]. Researchers have recently reported evidence for the potential antiproliferative effect of metformin in various cancers. The anticancer activity is either due to direct activity on cancer cells mediated through the 5′ adenosine monophosphate-activated protein kinase (AMPK) pathway or indirect action on blood glucose and insulin levels. Apart from the AMPK pathway, the other mechanisms include mammalian targets of rapamycin (mTOR) inhibition, immunomodulation, antiangiogenesis, pro-inflammatory cytokines inhibition, and cell cycle arrest through the reduction in G1 cyclins [11]. The suggested mechanism for the biguanides is summarized below:

11.3.2.1 AMPK Mediated- mTOR Inhibition

AMPK regulates cellular nutrient supply. The activation of this pathway is responsible for hypoxia and nutrient deprivation. The direct inhibition of the AMPK-mediated mTOR activity by phosphorylation of the receptor binding component decreases mTORC1, which enhances tumor progression [88]. The indirect activity in which the LKB1 pathway activates AMPK triggers the tumor suppressor gene 2 (TSC2). This results in the inactivation of mTOR and a decrease in cell growth and protein synthesis [89].

11.3.2.2 Rag GTPase Mediated-mTOR Inhibition

Rag is available in four structures—RagA, RagB, RagC, and RagD. The inactivated form of RagA ($RagA^{GDP}$) activates the $RagC^{GDP}$, which results in the activation of mTOR. Biguanides inhibit the nuclear action of RagC, leading to the inactivation of mTORC1 [90, 91].

11.3.2.3 Anti-Hyperinsulinemic Action

Clinical evidence suggests that hyperinsulinemia increases cancer risk by promoting cell growth via insulin and insulin-like growth factor receptor 1 (IGF-1). The cell growth was further supported by signal transduction of phosphoinositide3-kinase (PI3K) and mitogen-activated protein kinase (MAPK) pathways by IGF-1 activation. The decline of hyperinsulinemia by metformin causes a reduction in IGF 1R levels and exhibits antitumor activity. Further lowering blood glucose levels by metformin cuts the fuel required for cell growth, contributing to the antitumor activity [92].

Sanjana Chowdhury and a co-worker evaluated the proliferation-resistant response of metformin in 19 different cancer cell lines, including colorectal carcinoma (HCT116), prostate carcinoma (LNCaP0), non-small cell lung adenocarcinoma (HCC827), pancreas adenocarcinoma (BxPC3), hepatocellular carcinoma (HepG2), glioblastoma (T98G), and head-and-neck cancer cell lines (FaDu, SCC-16A, SCC-16B, SCC-19A, SCC-19B, SCC-74A, and SCC- 7Ab). They also studied the relationship between the antiproliferative effect and the inhibition of oxygen consumption. The authors observed a substantial variation in the proliferation-resistant response of various cell lines with a slight variation in oxygen consumption inhibition. Out of 19 cell lines studied, the prostate carcinoma (LNCaP0) cell line showed intracellularly higher metformin levels with impeccable sensitivity to the antiproliferative effect. The required concentration of metformin was much higher than its antidiabetic dose to exhibit the same response in all other cell lines. However, when tested in vivo, the antidiabetic dose achieved significant tumor accumulation and reduced hypoxic tumor fractions [93]. The activity of metformin alone or in combination against various cancers is listed in Table 11.2, and the comprehensive mechanism of action of metformin in cancer is described in Fig. 11.2b.

11.3.3 Thiazolidinediones (TZDs)

Thiazolidinediones, also called glitazones (rosiglitazone, pioglitazone, and troglitazone), are oral insulin-sensitizing medications used to treat T2DM by activating peroxisome proliferator-activated receptor γ (PPAR γ). These agents lower glucose with a minor probability of hypoglycemia. There are three forms of PPARs present in mammals: PPAR-α, PPAR-γ, and PPAR-β/δ. They are identical in biological structure but significantly differ in ligand binding affinity and tissue distribution. The α form shows two isoforms, PPARα1 and PPARα2, and these forms are highly expressed in adipose tissue and regulate glucose and lipid homeostasis. Among the two isoforms, PPARα1 is the dominant one. Activating these receptors reduces the circulation of triglycerides and free acids, fosters adipogenesis and fatty acid uptake, and improves insulin sensitivity by altering adipokine production [94]. Besides established antidiabetic activity, various studies indicated the tumor-suppressing activity of these TZDs. Most reports suggest that TZDS anticancer activity depends on the PPAR γ, as summarized in Fig. 11.2c. Antineoplastic effects are exerted through cell cycle arrest, apoptosis, and regulation of cell differentiation, as explained below.

11.3.3.1 Suppression of Cell Proliferation

The primary factor in tumor growth control is the attenuation of cell proliferation by arresting the cell cycle progression. Cyclins are the subunits of cell cycle-specific kinases, and Fta is the cell cycle process. The progression of the G1 phase is a common feature in many cancers with amplification of cyclin D1. It has been proposed that TZDs act as anticancer agents by causing G0/G1 cell cycle arrest due to reduced protein levels of cyclin D1, pRb, and Cdk4 and a rise in p21 and p27, two cyclin-dependent kinase inhibitors [95].

11.3.3.2 Regulation of Cell Apoptosis

The programmed cell death is called apoptosis, and the main characteristic of cancer is avoidance of apoptosis. TZDs provoke apoptosis by PPAR- γ activation. Apoptosis induction was assumed to be induced by intrinsic mitochondrial and extrinsic pathways [96]. PPAR- γ shows an apoptotic effect through both pathways. In the extrinsic pathway, it downregulates the expression of an antiapoptotic gene like BCL 2 gens, e.g., Bcl-2 and Bcl-XL. The intrinsic pathway increases the pro-apoptosis factor expression, e.g., BAX. It also induces reactive oxygen species (ROS) via POX, a redox enzyme in the inner mitochondrial membrane, and caspase-3, 8, and—collectively contribute to apoptosis [97].

11.3.3.3 Regulation of Cell Differentiation

TZDs are reported to inhibit cancer growth and metastasis by induction of terminal cell differentiation that reduces the stable cell phenotype. In HT-29 cancer cells, TZDs showed cell cycle arrest at the G1 phase with an apparent surge in p21Waf-1, Drg-1, and E-cadherin expression [98]. In pancreatic cancer cells, TZD treatment resulted in stimulation of p21Waf-1 and increased expression of differentiation markers, leading to reduced phenotype cell production and anticancer activity [99]. However, few other reports indicate that TZDs can exhibit antineoplastic activity independent of PPAR γ activation. S.S. Palakurthi et al. reported that TZDs' anticancer activity was due to cell proliferation inhibition caused by depletion of intracellular Ca^{+2}, stimulation of protein kinase R, and phosphorylation of eukaryotic initiation factor (eIF2α). The authors confirmed the mechanism after studying the activity in PPARγ−/− and PPARγ+/+ mouse embryonic stem cells [100]. The other pathways or receptors responsible for anticancer activity include the JNK-dependent cell death pathway, cell-dependent kinase inhibitors p27 and p21, p53- growth arrest and DNA damage (GADD45), FLICE inhibitory protein, cyclin D1, prostate-specific antigen (PSA) and androgen receptor (AR), and estrogen receptor α (ERα). Various other studies of TZDs and clinical trials based on TZDs in cancer are presented in Tables 11.2 and 11.3.

11.3.4 Meglitinide Analogs

Meglitinide analogs are a novel class of insulin secretagogues that stimulate insulin release by blocking K^+ (ATP-sensitive) channels on the beta-cell membrane by interacting with sulfonylurea receptors (SUR1/KIR 6.2) [101]. Currently, two meglitinide analogs are in the market: repaglinide and nateglinide. Meglitinide analogs are used alone or in combination with metformin, thiazolidinediones (such as pioglitazone and rosiglitazone), and long-acting insulin (NPH insulin at bedtime) [102]. The primary concern linked with these drugs is hypoglycemia. The incidence of hypoglycemia is comparable between repaglinide and sulfonylureas, although the rate of severe hypoglycemia is lower with repaglinide.

In particular, numerous in vitro and in vivo tests, observational studies, and prospective trials have demonstrated meglitinide efficacy in treating various malignancies, including liver, breast, and cervical cancers. Mechanistically, they showed their antineoplastic activity by inhibiting the Bcl-2 gene (gate controller of mitochondria-mediated apoptosis), PD-L1, FOXO3, and an increase in Beclin-1 gene expression. The inhibition of the Bcl-2 gene contributed to cell apoptosis. AntiPD1 reduced PD-L1 in cells and might increase immune cytotoxicity, and Beclin-1 was used to induce the autopsy. FOXO3 stimulates metastasis formation,

Table 11.3 Clinical trials of antidiabetic drug as an anticancer agent

Clinical trial references	Cancer type	Dosing regimen	Patients number
NCT03752099	Prostate cancer	Metformin hydrochloride	50
NCT01433913	Prostate cancer	Metformin hydrochloride	20
NCT01677897	Prostate cancer	Metformin hydrochloride	25
NCT01796028	Prostate cancer	Docetaxel + metformin	100
NCT02176161	Prostate cancer	Metformin	34
NCT01340300	Breast cancer	Metformin	
NCT00933309	Triple-negative breast cancer	Rosiglitazone + metformin hydrochloride	25
NCT01650506	Triple-negative breast cancer	Metformin + erlotinib	8
NCT01302379	Breast cancer	Metformin	333
NCT02488564	HER2 breast cancer	Metformin	49
NCT01589367	Breast cancer	Metformin HCl	208
NCT05053841	Breast cancer	Metformin	45
NCT01340300	Breast cancer	Metformin	139
NCT02153450	Pancreatic cancer	Metformin	8
NCT02005419	Pancreatic cancer	Gemcitabine + metformin	300
NCT01666730	Pancreatic cancer	Metformin hydrochloride leucovorin + fluorouracil + oxaliplatin	50 n
NCT01210911	Pancreatic cancer	Gemcitabine + erlotinib + metformin	120
NCT02048384	Pancreatic cancer	Metformin + rapamycin	22
NCT01971034	Pancreatic cancer	Paclitaxel + metformin	41
NCT01324180	Lymphoblastic leukemia	Metformin + vincristine Dexamethasone doxorubicin	14
NCT02083692	Head and neck carcinoma	Metformin	50
NCT02325401	Head and neck carcinoma	Metformin Cisplatin	20
NCT02394652	Cervix cancer	Metformin + cisplatin	16
NCT03618472	Endometrial cancer	Metformin	50
NCT02755844	Endometrial cancer	Olaparib Metformin Metronomic cyclophosphamide	35
NCT01205672	Endometrial cancer	Metformin	21
NCT01968317	Endometrial cancer	Megestrol acetate and metformin	150
NCT03183752	Benign thyroid nodules	Drug: Metformin	28
NCT03359681	Colon cancer	Drug: Metformin hydrochloride	28
NCT01725490	Colon cancer	Drug: Metformin	34
NCT00003058	Sarcoma	Troglitazone	84
NCT00780234	Lung cancer	Pioglitazone	76
NCT01655719	Thyroid cancer	Pioglitazone	15

(continued)

Table 11.3 (continued)

Clinical trial references	Cancer type	Dosing regimen	Patients number
NCT02133625	Solid tumors	Pioglitazone and carboplatin	28
NCT00099021	Head and neck cancer	Pioglitazone	21
NCT01838317	Pancreatic cancer	Pioglitazone	15
NCT00427999	Prostate cancer	Imatinib + pioglitazone + etoricoxib + dexamethasone + treosulfan	70
NCT02347813	Squamous cell skin cancer	Pioglitazone	27
NCT04883125	Myeloid leukemia	Pioglitazone	90
NCT02888964	Chronic myelogenous leukemia (CML)	Pioglitazone	90
NCT04542291	Metastatic pancreatic ductal adenocarcinoma	Dapagliflozin	15

and decreases in FOXO3 stimulate cell proliferation and migration. The overall mechanism of action is described in Fig. 11.3a.

Xiao ZX and collaborators explored the repaglinide repurposing potential in glioblastoma multiforme (GBM) through Connectivity Map (CMap). It was used to find the repaglinide gene expression signature in patients with long-time survival rates. They performed both in vitro cell line studies and in vivo animal studies. They used human GBM cells (LN229) and mouse GBM cell-induced orthotopic mouse models to assess the repaglinide efficacy in GBM. They found that repaglinide (1, 5, 10, 25, 50, 100, 200, 500, or 1000 mM) suppresses the proliferation and migration of GBM cells. Further, repaglinide was administrated at a dose of 1.04 mg/kg daily via an intraperitoneal route after GBM cell implantation, showing increased survival time in tumor-bearing mice. It accomplishes this by causing apoptosis, autophagy, and increases in the immune response. This study concluded that repaglinide is a drug that helps in GBM patients' survival rate [103]. Very few research papers are available on the meglitinide analog, as listed in Table 11.2. No clinical trial data is available for this class of drug with anticancer activity.

11.3.5 Dipeptidyl Peptidase 4 (DPP-4) Inhibitors

DPP-4 inhibitors are oral antidiabetic drugs introduced in 2006 to treat type 2 diabetes [104]. The antidiabetic action of these drugs is different from other antidiabetic drugs. They overcome hyperglycemia by inhibiting the release of glucagon from pancreatic α-cells, diminishing glucose production by the liver, and stimulating insulin secretion from β-cells in the pancreas [105]. They are used alone or with other diabetes medications such as metformin and thiazolidinediones. The safety profile of this class for T2DM is well established. DDP-4/CD 26 is the transmembrane glycoprotein receptor expressed on different cells like lymphocytes, endothelial cells, and epithelial cells. Low-molecular weight DPP-4 inhibitors like

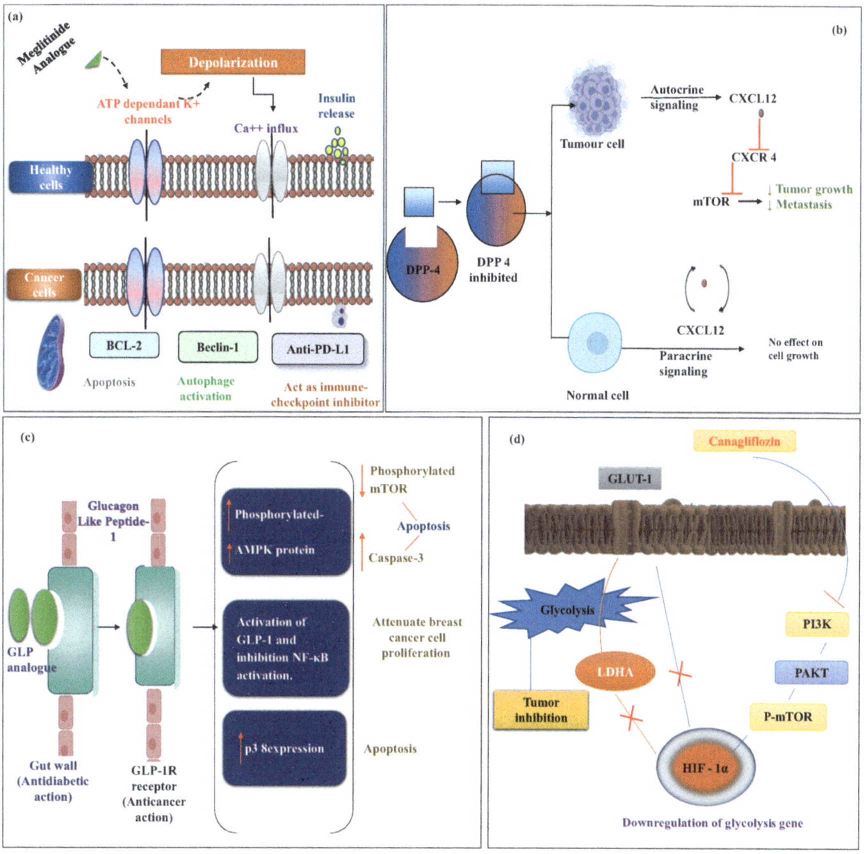

Fig. 11.3 Schematic representation of the mechanism of action of (**a**) meglitinide analogue, (**b**) DDP-4 inhibitors in cancer. (**c**) Glucagon-like peptide (GLP-1) 1 receptor agonists in cancer (**d**) sodium-glucose cotransporter 2 (SGLT)

sitagliptin, vildagliptin, saxagliptin, alogliptin, and linagliptin are used to treat T2DM [106]. Besides the antidiabetic effect, gliptins inhibit biopeptides and chemokines cleavage induced by DPP-4. However, gliptins also inhibit the DPP-4-mediated cleavage of other biopeptides and chemokines, and literature suggested that DPP-4 is overexpressed in tumorigenic conditions in serum and is considered a biomarker for human malignancy [107].

The DPP-4 has a great affinity to glucagon-like peptides 1 and 2 (GLP 1 and GLP2), which are responsible for glucose homeostasis and maintaining intestinal physiology. GLP 2 is a substrate for stromal cell-derived factor-1 (SDF-1) and insulin-like growth factor-1 (IGF-1), which plays a substantial role in the apoptosis of gastrointestinal (GI) cells. Hence, this is considered to be a promising target for intestinal neoplasm [108]. Many researchers have explored the effects of DPP-4 inhibitors, and few are listed in Table 11.2.

Yorifuji N et al. evaluated the role of DPP-4 inhibitor (sitagliptin) in colon tumorigenesis suppression using a leptin-deficient (ob/ob) C57BL/6J mice model.

According to the authors, the long-term administration of DPP-4 inhibitors such as sitagliptin significantly reduces the IL-6 expression and inhibits carcinogenesis in mice with type 2 diabetes. Mucosal DPP-4 activity, adipokine-related proteins analysis, plasma as well as mucosal GLPs, and mucosal interleukin-6 (IL-6) mRNA expression analysis were performed to the evolution of the antitumor effect of sitagliptin. The polymerase chain reaction findings reveal that oral administration of sitagliptin in the long term inhibits mucosal IL-6 expression and tumorigenesis. They also reported that this mechanism was independent of DPP-4 and GLP activity. It was concluded that DPP-4 inhibitors show benefits to diabetic patients with a risk of colon cancer [109]. The detailed mechanism of action of DPP-4 inhibitors in cancer regulation is given in Fig. 11.3b.

11.3.6 Glucagon-Like Peptide 1 (GLP-1) Receptor Agonists

GLP-1 is an incretin hormone released after oral carbohydrate/fat intake in the gut enteroendocrine cells. GLP-1 increases glucose-dependent insulin secretion and reduces glucagon release from pancreatic beta cells. It helps beta-cell proliferation, slowing gastric emptying time and appetite. The available clinical evidence indicates that GLP-1 agonists aid in the restoration of insulin secretion, leading to enhanced glycemic control and decreased body weight among individuals diagnosed with T2DM. Currently, seven GLP-1 receptor agonists are available in the market, including exenatide, lixisenatide, liraglutide, dulaglutide, semaglutide, and albiglutide. Albiglutide was discontinued from the market in 2017 due to some economic issues [110].

In addition to antidiabetic action, several researchers reported that they also show cytotoxic effects and can be repurposed for cancer. They inhibit cell proliferation by apoptosis. The detailed mechanism of action is described in Fig. 11.3c.

Liraglutide (GLP-1) has 95% structural similarity with GLP-1 hormone. Xian Lu and co-workers evaluated its cytotoxic effect in NKT cells and hepatocellular carcinoma subcutaneous model. They found that liraglutide stops the signaling between IL-6 and STAT3, which makes NKG2D ligands (MULT1, RAE1, H60, and MICA/B) more active. Further, downregulating IL-10 and TGF- (a negative NKT cell function regulator) makes liraglutide more efficient in NKT cells. It also improves antitumor immune responses by inhibiting glycolysis, which alters the tumor microenvironment [19]. It is already a well-established drug for T2DM, and the researcher's finding opens the door to liraglutide being used as a potential immunotherapeutic agent in oncology clinical trials. Numerous studies have also contributed to discovering GLP-1's anticancer activity (Table 11.2). However, no clinical trials have been conducted on this class of drugs.

11.3.7 Sodium-Glucose Cotransporter 2 (SGLT-2) Inhibitors

SGLT-2 proteins, located in the proximal convoluted tubules of the kidney, play a crucial role in the process of glucose reabsorption. SGLT-2 inhibitors demonstrate antihyperglycemic activity by inhibiting the SGLT-2 protein expression in the renal proximal convoluted tubules [111]. Besides the antidiabetic effect, increasing evidence suggests that SGLT-2 inhibitors also show an anticancer activity by inhibiting PI3K, p-AKT, p-mTOR, HIF-1α, GLUT-1, LDHA, and, Otto Warburg effect (i.e., tumor cells depend more on aerobic glycolysis, accelerating glucose uptake) (Fig. 11.3d) [112]. They are all mechanistically linked to the metabolic programming of cancer cells, interfering with glycolysis, inhibiting cancer cell proliferation, and inducing apoptosis. Consequently, targeting metabolic reprogramming may be a promising cancer treatment [113]. They directly inhibit renal, liver, and colon cancer by blocking cell glucose uptake (Table 11.2).

Duiyue Xu and co-workers investigated the anticancer effects of canagliflozin in combination with gemcitabine on pancreatic cancer in vitro and in vivo. In vitro investigation performed on the human pancreatic cancer cell lines (Capan-1 and PANC- 1) showed a dose-dependent decline in cell viability with IC50 values of about 77 and 68 μM, respectively. In the combination therapy, there was a significant improvement in the gemcitabine's ability to inhibit cell proliferation and colony formation in pancreatic cancer. In vivo studies indicated a reduction in tumor growth with minimal toxicity to the mice. Mechanistic analysis revealed that the antitumor activity was due to the repression of glycolysis mediated via the PI3K/AKT/mTOR/HIF signaling pathway by downregulating GLUT-1 and LDHA at the mRNA and protein levels. Overall, the authors concluded that the novel application of SGLT 2 inhibitors might help SGLT-2 in the drug discovery process and significantly reduce time and cost [114]. Two clinical investigations are mentioned in the literature showing the anticancer effect of CANA, but only one completed clinical trial mentioned regarding dapagliflozin is described in Table 11.3.

11.3.8 Dopamine-2 Agonists

Dopamine receptor agonists help to reduce blood glucose levels in combination with diet and exercise. Bromocriptine (BRC) is the only drug in this category that is approved by the United States Food and Drug Administration (US FDA) for treating T2DM and shows a modest significant effect on glycemic indices, cholesterol, and increasing body weight [115]. BRC is also testified to reduce the risk of the cardiovascular system in DM.

The role of the dopamine receptors (DR2 family) is unclear in cancer cells. It was reported that activation of D2 receptors shows cell apoptosis and cell proliferation inhibition. Similarly, D2 antagonists were also described to have anticancer potential via activation of the cAMP/PKA pathway [116].

BRC, an ergot derivative dopamine D2 receptor agonist, inhibits the central nervous system's serotonin turnover. A single dose of BRC in the morning resets the circadian clock [117]. Several researchers noted that DRD2 agonists show inhibitory effects in preclinical models of gastric cancer, lung cancer, and leukemia, implying that DRD2 activation could be a potential target in these cancers. They suppress tumor angiogenesis by inhibiting vascular endothelial growth factor (VEGFR)-2 assisted signaling in endothelial cells. It is also reported that dopamine analog specifically inhibits the VEGF-induced proliferation and migration of human umbilical vein endothelial cells (HUVEC) [118]. It is a commonly used treatment for prolactinoma, the most common pituitary adenoma [119, 120]. One study by E-J Seo et al. revealed a strong interaction of bromocriptine with the DNA bound NF-κB compared to the free form, indicating that bromocriptine inhibits the DNA binding with NF-κB, which may be responsible for its anticancer activity [83].

Fabio Giuseppe and co-researchers identified the activity of the BRC against myelodysplastic syndromes (MDS) and acute myeloid leukemia (AML). They used bioinformatic analysis of existing transcriptomic data to create a gene-based signature for the growth and evolution of MDS. Statistically significant cMAP, sscMap was then used to find new therapeutic compounds with this signature. They identified the genetic signature of MDS progression. This study's gene expression dataset was reported from two previously available datasets (GSE13159 and GSE15061). To improve the statistical significance and the likelihood of a successful hit, they used multiple gene signatures combined with chip definition file (CDF) microarray probe mapping. Further, in sscMap, they identified three potential antileukemic compounds, and bromocriptine was tested further in preclinical and clinical studies. The study result showed synergism of bromocriptine with cytarabine in the OCI/AML-3 cell line at all test concentrations. Several reviews and research papers support the use of D2R in chemotherapy, but more literature is needed on the BRC anticancer effect.

11.3.9 Amylin Analogue (Pramlintide)

Pramlintide is a non-natural human amylin that lowers postprandial glucagon release, decreases food intake, and slows stomach emptying [121]. Pramlintide is prescribed as an additional medication for patients with T1 or T2DM who utilize mealtime insulin therapy but cannot regulate their blood glucose to their satisfaction despite optimum insulin therapy and/or a sulfonylurea agent and/or metformin [122].

Notably, data from preclinical experiments, observational studies, and prospective trials conclusively demonstrated that pramlintide also shows antineoplastic properties along with an antidiabetic effect. On a mechanical level, the antineoplastic property of the drug is related to the p53 protein family that suppresses tumors by stimulating intracellular apoptosis or autophagy [123].

Maha S Al-Keilani and co-workers studied the antineoplastic effect of pramlintide in colorectal cancer using HCT116 and HT-29 cells. Pramlintide showed dose-dependent inhibition with IC50 values at 48.67 and 9.06 μg/mL, respectively. Their findings suggested that the enhanced apoptosis was due to the activation of p53 transcription in HT-29 cells. This recommends that pramlintide has a high potential for overcoming chemotherapeutic resistance in p53-mutated tumors. Further, they observe that this amylin analog synergizes the anticancer effect of drugs like 5-FU, OXA, and IRN in both cell lines [124].

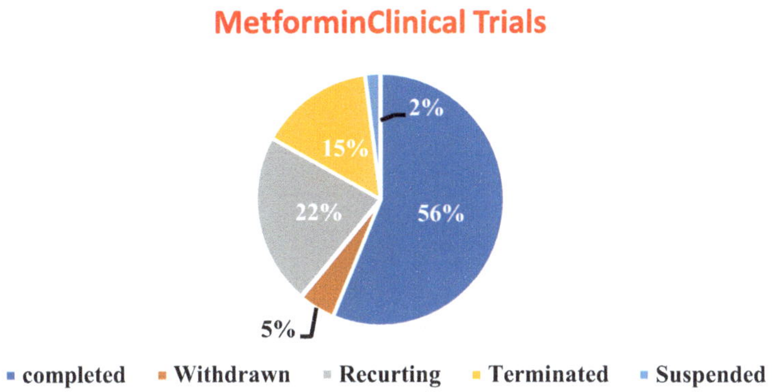

Fig. 11.4 Schematic representation of clinical trials carried out on metformin

In another study by Yang Y et al., pramlintide showed inhibition of cell growth and increased cell death in osteosarcoma cell lines at concentrations ranging from 15 to 20 μg/mL. The inhibition was found to be the result of glycolysis inhibition in tumors (osteosarcoma). The effect of the drug on tumor cells is dependent on glycolysis and cells glycolytic capacity (GC) [125]. As far as we know, there is one clinical trial of pramlintide maintained in literature, which started on February 5, 2020, for resectable cutaneous squamous cell carcinoma.

11.4 Clinical Trials

The findings from preclinical studies indicating the anticancer potential of antidiabetic drugs prompted several clinical trials. Among all antidiabetic drugs, more clinical trials were registered with biguanides at the time of writing. About 431 clinical trials have been reported as on 14-03-2024 on the clinical trail.gov.in website. The details of these are described in Fig. 11.4. Chae YK and researchers reported metformin clinical trials up to 2016 [126]. Here in the current manuscript, we have given the completed clinical trials of metformin from 2017 to till date, and for other classes, complete clinical trial data was presented in Table 11.3.

11.5 Formulation Development of the Antidiabetic Drug as an Anticancer Agent

Based on the available preclinical and clinical trial data, different scientists developed various formulations to deliver the drug efficiently to the target site. From the available literature, we get to know that metformin was explored to a greater extent. Table 11.4 presents information related to the formulations and their inhibitory concentrations.

Table 11.4 Formulation development of antidiabetic drugs for different types of cancers

Drug	Formulation	Cancer	In vitro		References
			Cell line	IC 50	
Metformin	Chitosomes	Thoracic cancer	H2052	1.6 ± 0.3 mM	[127]
			MSTO-211H 1	1.5 ± 0.4 mM	
			H2452	3.7 ± 0.7 mM	
			ORT	3.2 ± 0.2 mM	
			ROB	3.8 ± 1.1 mM	
			HAY	3.2 ± 0.16 mM	
Metformin + silibin	Nano particles	Lung cancer	A549	1.1 (12.65) mM	[128]
Metformin + silibin	Nano particles	Breast cancer	T47D	0.91 (9.61) mM	[129]
Metformin	Sono complexation	Pancreatic carcinoma	MiaPaCa-2	1.4 mM	[130]
Metformin + cisplatin	Self-assembled nanoparticles	Lung cancer	H460	35 ± 1 μM	[131]
Metformin + cisplatin	Cubosomal nanoformulation	Leukemia	697	6.5 mM	[132]
Metformin	Surface-modified cellulose nanofiber	Metastasis	B16F10	17.6 mg/mL	[133]
Metformin	Hyaluronic acid co-functionalized gold nanoparticle complex	Lung cancer	HepG2	4 μg/mL	[134]
Metformin+epirubicin	PEGylated liposomes	Sarcoma	S180	10 μM	[135]
Metformin+herceptin	Liposome	Breast cancer cells	BCS	10 mM	[136]
			MCF-7		
Metformin+ doxorubicin	Liposomes			5 and 10 μg/mL	[137]
Metformin + cisplatin	Nanocubosomes	Colorectal cancer	CT-116	7 mM	[138]
Metformin + topotecan	Nanoparticles	Tumors	MDA-MB-231 and 4T1	3.95 mM	[139]
				1.15 mM	
				3.59 mM	
				1.01 mM	

Metformin	Nanocarrier	Breast cancer	4T1.2	0.348 µg/mL	[140]
Metformin	Nanoparticles	Breast cancer	T47D	10 mM	[141]
			MDA-MB-231	28 mM	
Bromocriptine	Multi-walled carbon nanotubes	Lung cancer	MRC5 (pure drug)	94.61 µg/mL	[142]
			QU-DB (pure drug)	42.21 µg/mL	
			A549 (pure drug)	30.40 µg/mL	
			MRC5 (formulation)	209.3 µg/mL	
			QUDB (formulation)	36.08 µg/mL	
			A549 (formulation)	27.79 µg/mL	

11.6 Opinion on Anticancer Drug Repurposing in Cancer

The growing global burden of cancer increases the need for new drug development. Cancer drug discovery and development are challenging for pharmaceutical scientists with significantly lower success rates. Pharmaceutical scientists recently realized that drug repurposing is an alternative approach with a higher success rate. The time for the drug discovery using the repurposing approach may take from 3 to 12 years, depending on the data availability, with a 50–60% cost reduction. Various public health programs have been started to encourage the investigation of the repurposing potential of existing drugs.

Diabetes and cancer have biological similarities and are interlinked with each other. Traditional risk factors like modern lifestyle, obesity, smoking, etc., link these two diseases. Hence, using antidiabetics to treat cancer seems logical. Based on this, many scientists explored the anticancer activity of antidiabetic agents. Advanced in silico tools, in vitro cell line studies, and a few preclinical studies have been reported in the literature, proving the effectiveness of antidiabetics in cancer. The anticancer mechanism was established for a few drugs that may be dependent or independent of their regular pharmacological activity. It still needs to be clarified for a few of them. Based on these, a few drugs were also under clinical trials. This may help pharmaceutical companies bring molecules into the market quickly without involving much cost. This will also help the patients get effective therapy at an affordable cost.

However, there are challenges, like the anticancer activity reported was inconsistent, and the antidiabetic drugs causing cancer in some cases. This may be drug-specific. Metformin and glitazones inhibit cancer proliferation, whereas sulfonylureas have some inconsistent effects. Insulin and insulin analogues have some agonistic impact on cancer. This may be due to other factors such as comorbidities, cancer progression stage, other medication, etc.

Finally, there should be more preclinical and clinical studies to understand the effectiveness of antidiabetics in cancer. These studies should be meticulously designed to cover the aspects of specificity, disease progression, other comorbidities, etc. There are many areas of research that can be used to answer the current challenges. To conclude, a meticulous evaluation may help the healthcare industry obtain potential molecules for the next generation of anticancer therapeutics.

11.7 Summary

The pharmaceutical industry and academic researchers are increasingly focused on drug repurposing as a faster and less expensive option to extend the arsenal of licensed cancer therapies. Since cancer and diabetes share several etiological and pathological characteristics, antidiabetic medication repurposing has the potential to boost cancer therapies. Consequently, the data from preclinical research, clinical trials, and experimental studies (Tables 11.2 and 11.3) have revealed the potential of antidiabetic drugs against cancer. The mechanism behind the anticancer activity may vary from triggering cancer cell death, changing the cell behavior in tumors, or inhibiting cancer growth. This chapter provided a summary of potential drug repurposing opportunities in cancer and publications explaining the molecular mechanisms underlying their antineoplastic actions.

References

1. Sung H, Ferlay J, Siegel RL, Laversanne M, Soerjomataram I, Jemal A, Bray F. Global cancer statistics 2020: GLOBOCAN estimates of incidence and mortality worldwide for 36 cancers in 185 countries. CA Cancer J Clin. 2021;71(3):209–49.
2. Gelosa P, Castiglioni L, Camera M, Sironi L. Repurposing of drugs approved for cardiovascular diseases: opportunity or mirage? Biochem Pharmacol. 2020;177:113895.
3. Hait WN. Anticancer drug development: the grand challenges. Nat Rev Drug Discov. 2010;9(4):253–4.
4. Badria FA. Drug repurposing: hypothesis, molecular aspects and therapeutic applications. London: IntechOpen; 2020.
5. Allison M. NCATS launches drug repurposing program. Nat Biotechnol. 2012;30(7):571.
6. Pushpakom S, Iorio F, Eyers PA, Escott KJ, Hopper S, Wells A, et al. Drug repurposing: progress, challenges and recommendations. Nature reviews Drug discovery. 2019;18(1):41–58.
7. Gupta SC, Sung B, Prasad S, Webb LJ, Aggarwal BB. Cancer drug discovery by repurposing: teaching new tricks to old dogs. Trends Pharmacol Sci. 2013;34(9):508–17.
8. Kutkat O, Moatasim Y, Al-Karmalawy AA, Abulkhair HS, Gomaa MR, El-Taweel AN, Abo Shama NM, GabAllah M, Mahmoud DB, Kayali G. Robust antiviral activity of commonly prescribed antidepressants against emerging coronaviruses: in vitro and in silico drug repurposing studies. Sci Rep. 2022;12(1):1–12.
9. Gan J-h, Liu J-x, Liu Y, Chen S-w, Dai W-t, Xiao Z-X, Cao Y. DrugRep: an automatic virtual screening server for drug repurposing. Acta Pharmacol Sin. 2022;44:888–96.
10. Krishnamurthy N, Grimshaw AA, Axson SA, Choe SH, Miller JE. Drug repurposing: a systematic review on root causes, barriers and facilitators. BMC Health Serv Res. 2022;22(1):1–17.
11. Daugan M, Dufaÿ Wojcicki A, d'Hayer B, Boudy V. Metformin: an antidiabetic drug to fight cancer. Pharmacol Res. 2016;113:675–85.
12. Jiang W, Hu J-W, He X-R, Jin W-L, He X-Y. Statins: a repurposed drug to fight cancer. J Exp Clin Cancer Res. 2021;40(1):241.
13. Shah RR, Stonier PD. Repurposing old drugs in oncology: opportunities with clinical and regulatory challenges ahead. J Clin Pharm Ther. 2019;44(1):6–22.
14. Srirangam A, Milani M, Mitra R, Guo Z, Rodriguez M, Kathuria H, Fukuda S, Rizzardi A, Schmechel S, Skalnik DG, Pelus LM, Potter DA. The human immunodeficiency virus protease inhibitor ritonavir inhibits lung cancer cells, in part, by inhibition of survivin. J Thorac Oncol. 2011;6(4):661–70.
15. Schein CH. Repurposing approved drugs for cancer therapy. Br Med Bull. 2021;137(1):13–27.
16. Sidhu T, Lemetyinen H, Edge D. 'Diabetes doesn't matter as long as we're keeping traditions alive': a qualitative study exploring the knowledge and awareness of type 2 diabetes and related risk factors amongst the young Punjabi Sikh population in the UK. Ethn Health. 2022;27(4):781–99.
17. Sloan FA, Gelband H. Cancer causes and risk factors and the elements of cancer control, cancer control opportunities in low-and middle-income countries. Washington, DC: National Academies Press; 2007.
18. Liu S-H, Yu J, Creeden JF, Sutton JM, Markowiak S, Sanchez R, Nemunaitis J, Kalinoski A, Zhang J-T, Damoiseaux R. Repurposing metformin, simvastatin and digoxin as a combination for targeted therapy for pancreatic ductal adenocarcinoma. Cancer Lett. 2020;491:97–107.

19. Lu X, Xu C, Dong J, Zuo S, Zhang H, Jiang C, Wu J, Wei J. Liraglutide activates nature killer cell-mediated antitumor responses by inhibiting IL-6/STAT3 signaling in hepatocellular carcinoma. Transl Oncol. 2021;14(1):100872.
20. Sahra IB, Laurent K, Loubat A, Giorgetti-Peraldi S, Colosetti P, Auberger P, Tanti J-F, Marchand-Brustel L, Bost F. The antidiabetic drug metformin exerts an antitumoral effect in vitro and in vivo through a decrease of cyclin D1 level. Oncogene. 2008;27(25):3576–86.
21. Joseph JJ, Deedwania P, Acharya T, Aguilar D, Bhatt DL, Chyun DA, Di Palo KE, Golden SH, Sperling LS, American Heart Association Diabetes Committee of the Council on Lifestyle and Cardiometabolic Health; Council on Arteriosclerosis, Thrombosis and Vascular Biology; Council on Clinical Cardiology; and Council on Hypertension. Comprehensive management of cardiovascular risk factors for adults with type 2 diabetes: a scientific statement From the American Heart Association. Circulation. 2022;145(9):e722–59.
22. Sun H, Saeedi P, Karuranga S, Pinkepank M, Ogurtsova K, Duncan BB, Stein C, Basit A, Chan JC, Mbanya JC. IDF diabetes atlas: global, regional and country-level diabetes prevalence estimates for 2021 and projections for 2045. Diabetes Res Clin Pract. 2022;183:109119.
23. Mane PB, Antre RV, Oswal RJ. Antidiabetic drugs: an overview. Int J Pharm Chem Sci. 2012;1(1):301–6.
24. Krentz AJ, Bailey CJ. Oral antidiabetic agents. Drugs. 2005;65(3):385–411.
25. Groop LC. Sulfonylureas in NIDDM. Diabetes Care. 1992;15(6):737–54.
26. Sola D, Rossi L, Schianca GPC, Maffioli P, Bigliocca M, Mella R, Corlianò F, Fra GP, Bartoli E, Derosa G. State of the art paper sulfonylureas and their use in clinical practice. Arch Med Sci. 2015;11(4):840–8.
27. Núñez M, Medina V, Cricco G, Croci M, Cocca C, Rivera E, Bergoc R, Martín G. Glibenclamide inhibits cell growth by inducing G0/G1 arrest in the human breast cancer cell line MDA-MB-231. BMC Pharmacol Toxicol. 2013;14(1):1–13.
28. Qian X, Li J, Ding J, Wang Z, Duan L, Hu G. Glibenclamide exerts an antitumor activity through reactive oxygen species–c-jun NH (2)-terminal kinase pathway in human gastric cancer cell line MGC-803. Biochem Pharmacol. 2008;76(12):1705–15.
29. Yasukagawa T, Niwa Y, Simizu S, Umezawa K. Suppression of cellular invasion by glybenclamide through inhibited secretion of platelet-derived growth factor in ovarian clear cell carcinoma ES-2 cells. FEBS Lett. 2012;586(10):1504–9.
30. Xu K, Sun G, Li M, Chen H, Zhang Z, Qian X, Li P, Xu L, Huang W, Wang X. Glibenclamide targets sulfonylurea receptor 1 to inhibit p70S6K activity and upregulate KLF4 expression to suppress non-small cell lung carcinoma targeting SUR1 with glibenclamide in NSCLC. Mol Cancer Ther. 2019;18(11):2085–96.
31. Sliwinska A, Rogalska A, Szwed M, Kasznicki J, Jozwiak Z, Drzewoski J. Gliclazide may have an antiapoptotic effect related to its antioxidant properties in human normal and cancer cells. Mol Biol Rep. 2012;39(5):5253–67.
32. Muaddi H, Chowdhury S, Vellanki R, Zamiara P, Koritzinsky M. Contributions of AMPK and p53 dependent signaling to radiation response in the presence of metformin. Radiother Oncol. 2013;108(3):446–50.
33. Zi F-M, He J-S, Li Y, Wu C, Yang L, Yang Y, Wang L-J, He D-H, Zhao Y, Wu W-J. Metformin displays anti-myeloma activity and synergistic effect with dexamethasone in in vitro and in vivo xenograft models. Cancer Lett. 2015;356(2):443–53.
34. Saeed HK, Sutar Y, Patel P, Bhat R, Mallick S, Hatada AE, Koomoa D-LT, Lange I, Date AA. Synthesis and characterization of lipophilic salts of metformin to improve its repurposing for cancer therapy. ACS Omega. 2021;6(4):2626–37.
35. Cheng G, Zielonka J, Ouari O, Lopez M, McAllister D, Boyle K, Barrios CS, Weber JJ, Johnson BD, Hardy M. Mitochondria-targeted analogues of metformin exhibit enhanced antiproliferative and radiosensitizing effects in pancreatic cancer cellsenhanced antiproliferative effects of mito-metformin. Cancer Res. 2016;76(13):3904–15.
36. Vlotides G, Tanyeri A, Spampatti M, Zitzmann K, Chourdakis M, Spöttl G, Maurer J, Nölting S, Göke B, Auernhammer CJ. Anticancer effects of metformin on neuroendocrine tumor cells in vitro. Hormones. 2014;13(4):498–508.

37. Kalyanaraman B, Cheng G, Hardy M, Ouari O, Sikora A, Zielonka J, Dwinell MB. Modified metformin as a more potent anticancer drug: mitochondrial inhibition, redox signaling, antiproliferative effects and future EPR studies. Cell Biochem Biophys. 2017;75(3):311–7.
38. Pateliya B, Burade V, Goswami S. Enhanced antitumor activity of doxorubicin by naringenin and metformin in breast carcinoma: an experimental study. Naunyn Schmiedeberg's Arch Pharmacol. 2021;394(9):1949–61.
39. Amable G, Martínez-León E, Picco ME, Di Siervi N, Davio C, Rozengurt E, Rey O. Metformin inhibits β-catenin phosphorylation on Ser-552 through an AMPK/PI3K/Akt pathway in colorectal cancer cells. Int J Biochem Cell Biol. 2019;112:88–94.
40. Huang Y, Zhou Z, Zhang J, Hao Z, He Y, Wu Z, Song Y, Yuan K, Zheng S, Zhao Q. lncRNA MALAT1 participates in metformin inhibiting the proliferation of breast cancer cell. J Cell Mol Med. 2021;25(15):7135–45.
41. Keshandehghan A, Nikkhah S, Tahermansouri H, Heidari-Keshel S, Gardaneh M. Co-treatment with sulforaphane and nano-metformin molecules accelerates apoptosis in HER2+ breast cancer cells by inhibiting key molecules. Nutr Cancer. 2020;72(5):835–48.
42. Laskar J, Sengupta M, Choudhury Y. Treatment with the anti-diabetic drug metformin ameliorates betel-nut induced carcinogenesis in a murine model. Pharmacological Reports. 2019;71(6):1115–24.
43. Ozel AB, Dagsuyu E, Aydın PK, Bugan I, Bulan OK, Yanardag R, Yarat A. Brain boron level, DNA content, and myeloperoxidase activity of metformin-treated rats in diabetes and prostate cancer model. Biol Trace Elem Res. 2022;200(3):1164–70.
44. Saber S, Ghanim AM, El-Ahwany E, El-Kader EMA. Novel complementary antitumour effects of celastrol and metformin by targeting IκBκB, apoptosis and NLRP3 inflammasome activation in diethylnitrosamine-induced murine hepatocarcinogenesis. Cancer Chemother Pharmacol. 2020;85(2):331–43.
45. Ma L, Wei J, Wan J, Wang W, Wang L, Yuan Y, Yang Z, Liu X, Ming L. Low glucose and metformin-induced apoptosis of human ovarian cancer cells is connected to ASK1 via mitochondrial and endoplasmic reticulum stress-associated pathways. J Exp Clin Cancer Res. 2019;38(1):77.
46. Chiu S-J, Hsaio C-H, Tseng H-H, Su Y-H, Shih W-L, Lee J-W, Chuah JQ-Y. Rosiglitazone enhances the radiosensitivity of p53-mutant HT-29 human colorectal cancer cells. Biochem Biophys Res Commun. 2010;394(3):774–9.
47. H. Zhao, H. Gu, H. Zhang, J.-H. Li, W.-E. Zhao, PPARγ-dependent pathway in the growth-inhibitory effects of K562 cells by carotenoids in combination with rosiglitazone, Biochimica et Biophysica Acta (BBA)-General Subjects, 1840;(2014):545–555.
48. Huang H, Campbell SC, Bedford DF, Nelius T, Veliceasa D, Shroff EH, Henkin J, Schneider A, Bouck N, Volpert OV. Peroxisome proliferator-activated receptor γ ligands improve the antitumor efficacy of thrombospondin peptide ABT510. Mol Cancer Res. 2004;2(10):541–50.
49. Kim S, Lee J-J, Heo DS. PPARγ ligands induce growth inhibition and apoptosis through p63 and p73 in human ovarian cancer cells. Biochem Biophys Res Commun. 2011;406(3):389–95.
50. Papi A, Rocchi P, Ferreri AM, Orlandi M. RXRγ and PPARγ ligands in combination to inhibit proliferation and invasiveness in colon cancer cells. Cancer Lett. 2010;297(1):65–74.
51. Lennon AM, Ramaugé M, Dessouroux A, Pierre M. MAP kinase cascades are activated in astrocytes and preadipocytes by 15-deoxy-Δ12–14-prostaglandin J2 and the thiazolidinedione ciglitazone through peroxisome proliferator activator receptor γ-independent mechanisms involving reactive oxygenated species. J Biol Chem. 2002;277(33):29681–5.
52. Moss PE, Lyles BE, Stewart LV. The PPARγ ligand ciglitazone regulates androgen receptor activation differently in androgen-dependent versus androgen-independent human prostate cancer cells. Exp Cell Res. 2010;316(20):3478–88.
53. Shin SJ, Kim JY, Kwon SY, Mun K-C, Cho CH, Ha E. Ciglitazone enhances ovarian cancer cell death via inhibition of glucose transporter-1. Eur J Pharmacol. 2014;743:17–23.
54. Soller M, Dröse S, Brandt U, Brüne B, von Knethen A. Mechanism of thiazolidinedione-dependent cell death in Jurkat T cells. Mol Pharmacol. 2007;71(6):1535–44.

55. Niho N, Takahashi M, Shoji Y, Takeuchi Y, Matsubara S, Sugimura T, Wakabayashi K. Dose-dependent suppression of hyperlipidemia and intestinal polyp formation in Min mice by pioglitazone, a PPARγ ligand. Cancer Sci. 2003;94(11):960–4.
56. Grommes C, G.E. Landreth, M. Sastre, M. Beck, D.L. Feinstein, A.H. Jacobs, U. Schlegel, M.T. Heneka, Inhibition of in vivo glioma growth and invasion by peroxisome proliferator-activated receptor γ agonist treatment, Molecular pharmacology, 70(2006):1524–1533.
57. Ban JO, Kwak DH, Oh JH, Park E-J, Cho M-C, Song HS, Song MJ, Han SB, Moon DC, Kang KW. Suppression of NF-κB and GSK-3β is involved in colon cancer cell growth inhibition by the PPAR agonist troglitazone. Chem Biol Interact. 2010;188(1):75–85.
58. Salcher S, Spoden G, Huber JM, Golderer G, Lindner H, Ausserlechner MJ, et al. Repaglinide silences the FOXO3/Lumican axis and represses the associated metastatic potential of neuronal cancer cells. Cells. 2019;9(1):1.
59. El Sharkawi FZ, El Shemy HA, Khaled HM. Possible anticancer activity of rosuvastatine, doxazosin, repaglinide and oxcarbazepin. Asian Pac J Cancer Prev. 2014;15(1):199–203.
60. El-Fattah A, Eslam E, Saber S, Youssef ME, Eissa H, El-Ahwany E, Amin NA, Alqarni M, Batiha GE-S, Obaidullah AJ. AKT-AMPKα-mTOR-dependent HIF-1α activation is a new therapeutic target for cancer treatment: a novel approach to repositioning the antidiabetic drug sitagliptin for the management of hepatocellular carcinoma. Front Pharmacol. 2022;12:4018.
61. Varela-Calviño R, Rodríguez-Quiroga M, Dias Carvalho P, Martins F, Serra-Roma A, Vázquez-Iglesias L, Paez de la Cadena M, Velho S, Cordero OJ. The mechanism of sitagliptin inhibition of colorectal cancer cell lines' metastatic functionalities. IUBMB Life. 2021;73(5):761–73.
62. Amritha C, Punnagai Kumaravelu D. Evaluation of anti cancer effects of DPP-4 inhibitors in colon cancer-an in vitro study. J Clin Diagn Res. 2015;9(12):FC14.
63. Jang J-H, Baerts L, Waumans Y, De Meester I, Yamada Y, Limani P, Gil-Bazo I, Weder W, Jungraithmayr W. Suppression of lung metastases by the CD26/DPP4 inhibitor Vildagliptin in mice. Clin Exp Metastasis. 2015;32(7):677–87.
64. Iwaya C, Nomiyama T, Komatsu S, Kawanami T, Tsutsumi Y, Hamaguchi Y, Horikawa T, Yoshinaga Y, Yamashita S, Tanaka T. Exendin-4, a glucagon-like peptide-1 receptor agonist, attenuates breast cancer growth by inhibiting NF-κ B activation. Endocrinology. 2017;158(12):4218–32.
65. Zhang Y, Xu F, Liang H, Cai M, Wen X, Li X, Weng J. Exenatide inhibits the growth of endometrial cancer Ishikawa xenografts in nude mice. Oncol Rep. 2016;35(3):1340–8.
66. Zhao H, Wang L, Wei R, Xiu D, Tao M, Ke J, Liu Y, Yang J, Hong T. Activation of glucagon-like peptide-1 receptor inhibits tumourigenicity and metastasis of human pancreatic cancer cells via PI3K/Akt pathway. Diabetes Obes Metab. 2014;16(9):850–60.
67. Nomiyama T, Kawanami T, Irie S, Hamaguchi Y, Terawaki Y, Murase K, Tsutsumi Y, Nagaishi R, Tanabe M, Morinaga H, Tanaka T, Mizoguchi M, Nabeshima K, Tanaka M, Yanase T. Exendin-4, a GLP-1 receptor agonist, attenuates prostate cancer growth. Diabetes. 2014;63(11):3891–905.
68. Nie Z-J, Zhang Y-G, Chang Y-H, Li Q-Y, Zhang Y-L. Exendin-4 inhibits glioma cell migration, invasion and epithelial-to-mesenchymal transition through GLP-1R/sirt3 pathway. Biomed Pharmacother. 2018;106:1364–9.
69. Badi R, Khaleel E, El-Bidawy M, Satti H, Mostafa D. Exendin-4 induces cytotoxic autophagy in two ovarian cancer cell lines through inhibition of Mtorc1 mediated by activation of AMPK and suppression of Akt. Folia Biol. 2020;66(5/6):186–203.
70. Tawfik MK, Mohamed MI. Exenatide suppresses 1,2-dimethylhydrazine-induced colon cancer in diabetic mice: effect on tumor angiogenesis and cell proliferation. Biomed Pharmacother. 2016;82:106–16.
71. He W, Yu S, Wang L, He M, Cao X, Li Y, Xiao H. Exendin-4 inhibits growth and augments apoptosis of ovarian cancer cells. Mol Cell Endocrinol. 2016;436:240–9.
72. Tanaka Y, Iwaya C, Kawanami T, Hamaguchi Y, Horikawa T, Shigeoka T, Yanase T, Kawanami D, Nomiyama T. Combined treatment with glucagon-like peptide-1 receptor agonist exendin-4 and metformin attenuates breast cancer growth. Diabetol Int. 2022;13(3):480–92.

73. Lu R, Yang J, Wei R, Ke J, Tian Q, Yu F, Liu J, Zhang J, Hong T. Synergistic anti-tumor effects of liraglutide with metformin on pancreatic cancer cells. PLOS ONE. 2018;13(6):e0198938. https://doi.org/10.1371/journal.pone.0198938.
74. Kanda R, Hiraike H, Wada-Hiraike O, Ichinose T, Nagasaka K, Sasajima Y, Ryo E, Fujii T, Osuga Y, Ayabe T. Expression of the glucagon-like peptide-1 receptor and its role in regulating autophagy in endometrial cancer. BMC Cancer. 2018;18(1):1–11.
75. Zhao H, Wei R, Wang L, Tian Q, Tao M, Ke J, Liu Y, Hou W, Zhang L, Yang J, Hong T. Activation of glucagon-like peptide-1 receptor inhibits growth and promotes apoptosis of human pancreatic cancer cells in a cAMP-dependent manner. Am J Physiol Endocrinol Metab. 2014;306(12):E1431–41.
76. Zhu X-X, Feng Z-H, Liu L-Z, Zhang Y. Liraglutide suppresses the proliferation of endometrial cancer cells through the adenosine 5′-monophosphate (AMP)-activated protein kinase signaling pathway. Chin Med J. 2021;134(05):576–8.
77. Yamamoto L, Yamashita S, Nomiyama T, Kawanami T, Hamaguchi Y, Shigeoka T, Horikawa T, Tanaka Y, Yanase T, Kawanami D, Iwasaki A. Sodium-glucose cotransporter 2 inhibitor canagliflozin attenuates lung cancer cell proliferation in vitro. Diabetol Int. 2021;12(4):389–98. https://doi.org/10.1007/s13340-021-00494-6.
78. Kaji K, Nishimura N, Seki K, Sato S, Saikawa S, Nakanishi K, Furukawa M, Kawaratani H, Kitade M, Moriya K. Sodium glucose cotransporter 2 inhibitor canagliflozin attenuates liver cancer cell growth and angiogenic activity by inhibiting glucose uptake. Int J Cancer. 2018;142(8):1712–22.
79. Zhou J, Zhu J, Yu S-J, Ma H-L, Chen J, Ding X-F, Chen G, Liang Y, Zhang Q. Sodium-glucose co-transporter-2 (SGLT-2) inhibition reduces glucose uptake to induce breast cancer cell growth arrest through AMPK/mTOR pathway. Biomed Pharmacother. 2020;132:110821.
80. Papadopoli D, Uchenunu O, Palia R, Chekkal N, Hulea L, Topisirovic I, Pollak M, St-Pierre J. Perturbations of cancer cell metabolism by the antidiabetic drug canagliflozin. Neoplasia. 2021;23(4):391–9.
81. Shiba K, Tsuchiya K, Komiya C, Miyachi Y, Mori K, Shimazu N, Yamaguchi S, Ogasawara N, Katoh M, Itoh M. Canagliflozin, an SGLT2 inhibitor, attenuates the development of hepatocellular carcinoma in a mouse model of human NASH. Sci Rep. 2018;8(1):1–12.
82. Kato J, Shirakami Y, Ohnishi M, Mizutani T, Kubota M, Sakai H, Ibuka T, Tanaka T, Shimizu M. Suppressive effects of the sodium-glucose cotransporter 2 inhibitor tofogliflozin on colorectal tumorigenesis in diabetic and obese mice. Oncol Rep. 2019;42(6):2797–805.
83. Seo E-J, Sugimoto Y, Greten HJ, Efferth T. Repurposing of bromocriptine for cancer therapy. Front Pharmacol. 2018;9:1030.
84. Gao H, Wang F, Lan X, Li C, Feng J, Bai J, Cao L, Gui S, Hong L, Zhang Y. Lower PRDM2 expression is associated with dopamine-agonist resistance and tumor recurrence in prolactinomas. BMC Cancer. 2015;15(1):1–10.
85. Yan B, Peng Z, Xing X, Du C. Glibenclamide induces apoptosis by activating reactive oxygen species dependent JNK pathway in hepatocellular carcinoma cells. Biosci Rep. 2017;37(5):BSR20170685.
86. Bailey CJ. Metformin: historical overview. Diabetologia. 2017;60(9):1566–76.
87. Bailey CJ. Biguanides and NIDDM. Diabetes Care. 1992;15(6):755–72.
88. He X, Yang Y, Yao M-W, Ren T-t, Guo W, Li L, Xu X. Full title: high glucose protects mesenchymal stem cells from metformin-induced apoptosis through the AMPK-mediated mTOR pathway. Sci Rep. 2019;9(1):1–9.
89. Wu K, Tian R, Huang J, Yang Y, Dai J, Jiang R, Zhang L. Metformin alleviated endotoxemia-induced acute lung injury via restoring AMPK-dependent suppression of mTOR. Chem Biol Interact. 2018;291:1–6.
90. Liu X, Chhipa RR, Pooya S, Wortman M, Yachyshin S, Chow LM, Kumar A, Zhou X, Sun Y, Quinn B. Discrete mechanisms of mTOR and cell cycle regulation by AMPK agonists independent of AMPK. Proc Natl Acad Sci. 2014;111(4):E435–44.

91. Pierotti M, Berrino F, Gariboldi M, Melani C, Mogavero A, Negri T, Pasanisi P, Pilotti S. Targeting metabolism for cancer treatment and prevention: metformin, an old drug with multi-faceted effects. Oncogene. 2013;32(12):1475–87.
92. Xie Y, Wang JL, Ji M, Yuan ZF, Peng Z, Zhang Y, Wen JG, Shi HR. Regulation of insulin-like growth factor signaling by metformin in endometrial cancer cells. Oncol Lett. 2014;8(5):1993–9.
93. Chowdhury S, Yung E, Pintilie M, Muaddi H, Chaib S, Yeung M, Fusciello M, Sykes J, Pitcher B, Hagenkort A. MATE2 expression is associated with cancer cell response to metformin. PLoS One. 2016;11(12):e0165214.
94. Greenfield JR, Chisholm DJ. Thiazolidinediones-mechanisms of action. Aust Prescr. 2004;27:67–70.
95. Blanquicett C, Roman J, Hart CM. Thiazolidinediones as anticancer agents. Cancer Ther. 2008;6(A):25.
96. Duronio V. The life of a cell: apoptosis regulation by the PI3K/PKB pathway. Biochem J. 2008;415(3):333–44.
97. Elrod HA, Sun S-Y. PPARγ and apoptosis in cancer. PPAR Res. 2008;2008:704165.
98. Yoshizumi T, Ohta T, Ninomiya I, Terada I, Fushida S, Fujimura T, Nishimura G-I, Shimizu K, Yi S, Miwa K. Thiazolidinedione, a peroxisome proliferator-activated receptor-γ ligand, inhibits growth and metastasis of HT-29 human colon cancer cells through differentiation-promoting effects. Int J Oncol. 2004;25(3):631–9.
99. Elnemr A, Ohta T, Iwata K, Ninomia I, Fushida S, Nishimura G, Kitagawa H, Kayahara M, Yamamoto M, Terada T. PPARgamma ligand (thiazolidinedione) induces growth arrest and differentiation markers of human pancreatic cancer cells. Int J Oncol. 2000;17(6):1157–221.
100. Palakurthi SS, Aktas H, Grubissich LM, Mortensen RM, Halperin JA. Anticancer effects of thiazolidinediones are independent of peroxisome proliferator-activated receptor γ and mediated by inhibition of translation initiation. Cancer Res. 2001;61(16):6213–8.
101. Wexler D (2022) Sulfonylureas and meglitinides in the treatment of type 2 diabetes mellitus. UpToDate
102. Landgraf R. Meglitinide analogues in the treatment of type 2 diabetes mellitus. Drugs Aging. 2000;17(5):411–25.
103. Xiao ZX, Chen RQ, Hu DX, Xie XQ, Yu SB, Chen XQ. Identification of repaglinide as a therapeutic drug for glioblastoma multiforme. Biochem Biophys Res Commun. 2017;488(1):33–9.
104. Gallwitz B. Clinical use of DPP-4 inhibitors. Front Endocrinol (Lausanne). 2019;10:389.
105. Barnett AH, Dixon A, Bellary S, Hanif M, O'hare J, Raymond N, Kumar S. Type 2 diabetes and cardiovascular risk in the UK South Asian community. Diabetologia. 2006;49(10):2234–46.
106. Crepaldi G, Carruba M, Comaschi M, Del Prato S, Frajese G, Paolisso G. Dipeptidyl peptidase 4 (DPP-4) inhibitors and their role in type 2 diabetes management. J Endocrinol Investig. 2007;30(7):610–4.
107. Busek P, Duke-Cohan JS, Sedo A. Does DPP-IV inhibition offer new avenues for therapeutic intervention in malignant disease? Cancers. 2022;14(9):2072.
108. Hartmann B, Thulesen J, Kissow H, Thulesen S, Orskov C, Ropke C, Poulsen S, Holst J. Dipeptidyl peptidase IV inhibition enhances the intestinotrophic effect of glucagon-like peptide-2 in rats and mice. Endocrinology. 2000;141(11):4013–20.
109. Yorifuji N, Inoue T, Iguchi M, Fujiwara K, Kakimoto K, Nouda S, Okada T, Kawakami K, Abe Y, Takeuchi T. The dipeptidyl peptidase-4 inhibitor sitagliptin suppresses mouse colon tumorigenesis in type 2 diabetic mice. Oncol Rep. 2016;35(2):676–82.
110. Jain AB, Ali A, Gorgojo Martínez JJ, Hramiak I, Kavia K, Madsbad S, Potier L, Prohaska BD, Strong JL, Vilsbøll T. Switching between GLP-1 receptor agonists in clinical practice: Expert consensus and practical guidance. Int J Clin Pract. 2021;75(2) 10.1111/ijcp.v75.2 https://doi.org/10.1111/ijcp.13731.
111. Padda IS, Mahtani AU, Parmar M. Sodium-glucose transport protein 2 (SGLT2) inhibitors. In: StatPearls. Treasure Island: StatPearls Publishing; 2022.
112. Warburg O. On the origin of cancer cells. Science. 1956;123(3191):309–14.

113. Gatenby RA, Gillies RJ. Why do cancers have high aerobic glycolysis? Nat Rev Cancer. 2004;4(11):891–9.
114. Xu D, Zhou Y, Xie X, He L, Ding J, Pang S, Shen B, Zhou C. Inhibitory effects of canagliflozin on pancreatic cancer are mediated via the downregulation of glucose transporter-1 and lactate dehydrogenase A. Int J Oncol. 2020;57(5):1223-1233.
115. DeFronzo RA. Bromocriptine: a sympatholytic, D2-dopamine agonist for the treatment of type 2 diabetes. Diabetes Care. 2011;34(4):789–94.
116. Kline CLB, Ralff MD, Lulla AR, Wagner JM, Abbosh PH, Dicker DT, Allen JE, El-Deiry WS. Role of dopamine receptors in the anticancer activity of ONC201. Neoplasia. 2018;20(1):80–91.
117. Shivaprasad C, Kalra S. Bromocriptine in type 2 diabetes mellitus. Indian J Endocrinol Metab. 2011;15(Suppl 1):S17–24.
118. Sarkar C, Chakroborty D, Chowdhury UR, Dasgupta PS, Basu S. Dopamine increases the efficacy of anticancer drugs in breast and colon cancer preclinical models. Clin Cancer Res. 2008;14(8):2502–10.
119. Ciccarelli E, Camanni F. Diagnosis and drug therapy of prolactinoma. Drugs. 1996;51(6):954–65.
120. Yang Y, Mamouni K, Li X, Chen Y, Kavuri S, Du Y, Fu H, Kucuk O, Wu D. Repositioning dopamine D2 receptor agonist bromocriptine to enhance docetaxel chemotherapy and treat bone metastatic prostate cancer repurposing bromocriptine for prostate cancer therapy. Mol Cancer Ther. 2018;17(9):1859–70.
121. Traina AN, Kane MP. Primer on pramlintide, an amylin analog. Diabetes Educ. 2011;37(3):426–31.
122. Joy SV. Clinical pearls and strategies to optimize patient outcomes. Diabetes Educ. 2008;34(3_Suppl):54S–9S.
123. Broz DK, Mello SS, Bieging KT, Jiang D, Dusek RL, Brady CA, Sidow A, Attardi LD. Global genomic profiling reveals an extensive p53-regulated autophagy program contributing to key p53 responses. Genes Dev. 2013;27(9):1016–31.
124. Al-Keilani MS, Alsmadi DH, Darweesh RS, Alzoubi KH. Pramlintide, an antidiabetic, is antineoplastic in colorectal cancer and synergizes with conventional chemotherapy. Clin Pharmacol Adv Appl. 2018;10:23.
125. Yang Y, Peng Z, Flores ER, Kleinerman ES. Pramlintide: a novel therapeutic approach for osteosarcoma through metabolic reprogramming. Cancers. 2022;14(17):4310.
126. Chae YK, Arya A, Malecek M-K, Shin DS, Carneiro B, Chandra S, Kaplan J, Kalyan A, Altman JK, Platanias L. Repurposing metformin for cancer treatment: current clinical studies. Oncotarget. 2016;7(26):40767.
127. Shukla SK, Chan A, Parvathaneni V, Gupta V. Metformin-loaded chitosomes for treatment of malignant pleural mesothelioma—a rare thoracic cancer. Int J Biol Macromol. 2020;160:128–41.
128. Mogheri F, Jokar E, Afshin R, Akbari AA, Dadashpour M, Firouzi-amandi A, Serati-Nouri H, Zarghami N. Co-delivery of metformin and silibinin in dual-drug loaded nanoparticles synergistically improves chemotherapy in human non-small cell lung cancer A549 cells. J Drug Deliv Sci Technol. 2021;66:102752.
129. Amirsaadat S, Jafari-Gharabaghlou D, Alijani S, Mousazadeh H, Dadashpour M, Zarghami N. Metformin and Silibinin co-loaded PLGA-PEG nanoparticles for effective combination therapy against human breast cancer cells. J Drug Deliv Sci Technol. 2021;61:102107.
130. Farag MM, Abd El Malak NS, Yehia SA, Ahmed MA. Sonocomplexation as an effective tool to enhance the antitumorigenic effect of metformin: preparation, in vitro characterization, molecular dynamic simulation & MiaPaCa-2 cell line hypoxia evaluation. J Drug Deliv Sci Technol. 2020;59:101968.
131. Xiong Y, Zhao Y, Miao L, Lin CM, Huang L. Co-delivery of polymeric metformin and cisplatin by self-assembled core-membrane nanoparticles to treat non-small cell lung cancer. J Control Release. 2016;244:63–73.

132. Saber MM, Al-mahallawi AM, Stork B. Metformin dampens cisplatin cytotoxicity on leukemia cells after incorporation into cubosomal nanoformulation. Biomed Pharmacother. 2021;143:112140.
133. Nurani M, Akbari V, Taheri A. Preparation and characterization of metformin surface modified cellulose nanofiber gel and evaluation of its anti-metastatic potentials. Carbohydr Polym. 2017;165:322–33.
134. Kumar CS, Raja MD, Sundar DS, Antoniraj MG, Ruckmani K. Hyaluronic acid co-functionalized gold nanoparticle complex for the targeted delivery of metformin in the treatment of liver cancer (HepG2 cells). Carbohydr Polym. 2015;128:63–74.
135. Yang Q, Zhang T, Wang C, Jiao J, Li J, Deng Y. Coencapsulation of epirubicin and metformin in PEGylated liposomes inhibits the recurrence of murine sarcoma S180 existing CD133+ cancer stem-like cells. Eur J Pharm Biopharm. 2014;88(3):737–45.
136. Lee J-Y, Shin DH, Kim J-S. Anticancer effect of metformin in herceptin-conjugated liposome for breast cancer. Pharmaceutics. 2020;12(1):11.
137. Li Y, Luo J, Lin M-T, Zhi P, Guo W-W, Han M, You J, Gao J-Q. Co-delivery of metformin enhances the antimultidrug resistant tumor effect of doxorubicin by improving hypoxic tumor microenvironment. Mol Pharm. 2019;16(7):2966–79.
138. Saber MM, Al-mahallawi AM, Nassar NN, Stork B, Shouman SA. Targeting colorectal cancer cell metabolism through development of cisplatin and metformin nano-cubosomes. BMC Cancer. 2018;18(1):822.
139. Banala VT, Sharma S, Barnwal P, Urandur S, Shukla RP, Ahmad N, Mittapelly N, Pandey G, Dwivedi M, Kalleti N. Synchronized ratiometric codelivery of metformin and topotecan through engineered nanocarrier facilitates in vivo synergistic precision levels at tumor site. Adv Healthc Mater. 2018;7(19):1800300.
140. Liu Y, Sun J, Huang Y, Chen Y, Li J, Liang L, et al. Metformin-conjugated micellar system with intratumoral pH responsive de-shielding for co-delivery of doxorubicin and nucleic acid. Biochem Pharmacol. 2021;189:114453.
141. Javidfar S, Pilehvar-Soltanahmadi Y, Farajzadeh R, Lotfi-Attari J, Shafiei-Irannejad V, Hashemi M, et al. The inhibitory effects of nano-encapsulated metformin on growth and hTERT expression in breast cancer cells. Journal of drug delivery science and technology. 2018;43:19–26.
142. Kamazani FM, Siadat SD, Pornour M, Sheikhpour M. A success targeted nano delivery to lung cancer cells with multi-walled carbon nanotubes conjugated to bromocriptine. Sci Rep. 2021;11(1):1–15.

SPRINGER NATURE

GPSR Compliance

The European Union's (EU) General Product Safety Regulation (GPSR) is a set of rules that requires consumer products to be safe and our obligations to ensure this.

If you have any concerns about our products, you can contact us on ProductSafety@springernature.com

In case Publisher is established outside the EU, the EU authorized representative is:

Springer Nature Customer Service Center GmbH
Europaplatz 3
69115 Heidelberg, Germany

The manufacturer's authorised representative in the EU is Springer Nature Customer Service Centre GmbH, Europaplatz 3, 69115 Heidelberg, Germany. If you have any concerns regarding our products, please contact ProductSafety@springernature.com

Printed and bound by CPI Group (UK) Ltd, Croydon, CR0 4YY
01/10/2026
02225859-0004